Air Pollution and Public Health: Challenges, Interventions and Sustainable Solutions

Air Pollution and Public Health: Challenges, Interventions and Sustainable Solutions

Editors

Nour Shafik El-Gendy, *Ph.D.*
Professor
Petroleum & Environmental Biotechnology
Egyptian Petroleum Research Institute, Cairo, Egypt
Professor
October University for Modern Sciences & Arts (MSA), Giza, Egypt
Professor
Faculty of Nanotechnology for Postgraduate Studies
Cairo University, Giza, Egypt
E-mail: nourepri@yahoo.com

Vartika Mathur, *Ph.D.*
Assistant Professor
Department of Zoology, Sri Venkateswara College
New Delhi, India
E-mail: vmathur@svc.ac.in

A Publication of the Centre for Science & Technology
of the Non-Aligned and Other Developing Countries
(NAM S&T Centre)

ALLIED PUBLISHERS PVT. LTD.

New Delhi • Mumbai • Kolkata • Chennai • Bangalore • Hyderabad

ALLIED PUBLISHERS PRIVATE LIMITED

1/13-14 Asaf Ali Road, **New Delhi**–110002
Ph.: 011-23239001 • E-mail: delhi.books@alliedpublishers.com

17 Chittaranjan Avenue, **Kolkata**–700072
Ph.: 033-22129618 • E-mail: cal.books@alliedpublishers.com

15 J.N. Heredia Marg, Ballard Estate, **Mumbai**–400001
Ph.: 022-42126969 • E-mail: mumbai.books@alliedpublishers.com

No. 25/10, Commander-in-Chief Road, Ethiraj Lane (Next to Post Office)
Egmore, **Chennai**–600008
Ph.: 044-28223938 • E-mail: chennai.books@alliedpublishers.com

P.B. No. 9932, No. 15, 3rd Floor (Next to Vijaya Bank),
5th Cross, Gandhinagar, Karnataka, **Bangalore**–560009
Ph.: 080-41530285 / 22386239
E-mail: bngl.journals@alliedpublishers.com /apsabng@airtelmail.in

Sri Jayalakshmi Nilayam, No. 3-4-510, 3rd Floor (Above More Super Market)
Barkatpura, **Hyderabad**–500027
Ph.: 040-27551811, 040-27551812 • E-mail: hyd.books@alliedpublishers.com

Website: www.alliedpublishers.com

A Publication of the Centre for Science & Technology of the Non-Aligned and Other Developing Countries (NAM S&T Centre), New Delhi, India

ISBN: 978-93-90951-00-0

Contributing Editor: **Mr. Abhay Nambiar**, Research Associate, NAM S&T Centre

Published by Sunil Sachdev and printed by Ravi Sachdev at Allied Publishers Pvt. Ltd.,
1/13-14 Asaf Ali Road, New Delhi–110002

अखिल भारतीय आयुर्विज्ञान संस्थान
अंसारी नगर, नई दिल्ली-110029, भारत

शरीरमाद्यं खलु धर्मसाधनम्

All India Institute of Medical Sciences
Ansari Nagar, New Delhi-110029, India

दूरभाष/Phones :(का./Off.): +91-11-26588000, 26594800, 26594805
फैक्स सं./Fax No.: +91-11-26588663, 26588641
Phone (नि./ Res.) +91-11-26594500
ई–मेल/E-mail : director.aiims@gmail.com

आचार्य रणदीप गुलेरिया, पद्मश्री
निदेशक

PROF. RANDEEP GULERIA, Padma Shri
MD, DM (Pulmonary Medicine), FAMS, FIMSA,
DIRECTOR

21st April, 2020

Foreword

It is a privilege for me to write a Foreword for this book, not only because the book is being published by an Inter-governmental Organisation of repute engaged in *"South-South Cooperation in Science and Technology"*, viz., the *Centre for Science & Technology of the Non-Aligned and Other Developing Countries (NAM S&T Centre),* but also as a practicing pulmonologist, I have closely observed and have been associated with the management of adverse effects of air pollution for several decades.

It is now well recognized that air pollution is one of the top five causes of mortality linked to respiratory and heart diseases worldwide. The effect of long term exposure to air pollution is now being recognized and it may lead to weakening of the immune system and other negative health outcomes including strokes, progression of chronic respiratory conditions such as Asthma, Chronic Obstructive Pulmonary Disease (COPD) and development of lung cancer, allergic disorders as well as cataracts, coronary artery diseases etc. Moreover, in women, another severe negative consequence of air pollution may be the development of breast cancer. In the developing world indoor air pollution continues to be a major health hazard. It has been estimated by the WHO that indoor air pollution leads to the death of more than 2000 children, under the age of five, daily. The combination of exposure to both outdoor and indoor air pollution in many parts of the world is the leading cause for many chronic diseases in the younger age group and this is a cause of concern.

In order to deliberate on various issues relating to the insights on air pollution from the developing world and air pollution as a burden on *'One Health'*, the NAM S&T Centre, jointly with Sri Venkateswara College, University of Delhi, India organised an International Workshop on 'Air Pollution and Public Health: Challenges and Interventions' in New Delhi, India during 5–7 February 2020. This book is being published by the NAM S&T Centre as a follow up of the workshop in order to disseminate the information to the people at large on the current status and research and development work on the subject being carried out in the participating developing countries.

This book includes scientific papers which were presented by the participants during the workshop, and also contributed by other experts in relevant fields, on a diverse range of topics such as, emerging causes and hazards of air pollution, current scenario of the effects of air pollution on public health, various traditional and scientific approaches to mitigate air pollution and its harmful effects on health, international cooperation on mitigating the hazards of air pollution on public health, etc.

The book is aimed to help develop a shared vision and understanding on air pollution and its concomitant challenges, possible methods for human intervention and success stories on curbing air pollution from across the developing world. I believe that the data, annotated examples, and study results cited by various authors in this book would help scientists, researchers and professionals to prepare action plans, and undertake collaborative projects on problems relating to air pollution as a burden to *'One Health'*. In short, this book would offer indispensable information for anyone working in the field of air pollution mitigation and related health issues and help them explore and investigate new options which may improve human health and ensure sustainable development of the *Global South*. The book also includes the *New Delhi Resolution on Air Pollution and Public Health* which was unanimously adopted at the end of the Workshop that gives a set of valuable recommendations addressed to the policy makers and the scientific communities in the developing countries.

I hope that this book will become an important resource material for scientists, doctors, medical and pharmaceutical professionals, technologists, researchers and academicians working in the relevant fields in various countries to plan and implement air pollution mitigation policies and programmes.

Last but not the least, I would like to extend my appreciation to the organizers for holding such a wonderful international workshop on an important issue of public health concern and for inviting me to take part in various sessions of the workshop. I wish them great success in all their future endeavours.

Thank you!

(Prof. Randeep Guleria)

Introduction

Air pollution poses an increasing and serious threat to public health. Despite the on-going global efforts to combat rising air pollution and improve air quality, about 90% of the world's population live in places where air quality level is below the World Health Organisation (WHO) guidelines. Millions of people worldwide succumb to diseases linked to air pollution. The WHO reports that more than four million deaths annually can be attributed to ambient (outdoor) air pollution and another three and a half million deaths to household (indoor) air pollution.

In developing countries, effects of rising level of air pollution on public health are much higher. This is largely due to population explosion coupled with incessant industrialisation along with poor planning and unorganised development of urban centres. Furthermore, in developing countries household air pollution is a significant contributor to health hazards when compared with developed countries. This is primarily attributed to higher dependence of biomass, such as, wood, coal and other solid fuels, to meet day-to-day energy requirements. Women and children in particular have higher exposure to household air pollution. Therefore, it is much desired that the national authorities in developing countries put in place a robust monitoring system which would help in effectively evaluating levels of exposure to toxic air and subsequently implementing measures to reduce its impact on the population.

In order to deliberate on the effects of air pollution on public health and measures required to be taken to mitigate the same in the developing countries, the *Centre for Science and Technology of the Non-Aligned and Other Developing Countries (NAM S&T Centre)* in partnership with *Sri Venkateswara College, University of Delhi, India* organized an International Workshop on '*Air Pollution and Public Health: Challenges and Interventions*' in New Delhi, India during 5–7 February 2020, which brought together 78 experts and professionals from diverse background viz., scientists, researchers, medical and pharmaceutical professionals, industry representatives and policy makers from several developing countries. This book '*Air Pollution and Public Health: Challenges, Interventions and Sustainable Solutions*' is being published by the NAM S&T Centre as a follow up of the New Delhi International Workshop, which includes 20 scientific and research papers contributed by the experts/participants from 15 countries.

I appreciate the meticulous and sincere efforts of Dr. Nour Sh. El-Gendy from Egypt and Dr. Vartika Mathur from India for technical editing of the manuscripts.

I would also like to extend my sincerest gratitude to Prof. Randeep Guleria, DM, Director of the All India Institute of Medical Science (AIIMS), New Delhi and Dr. P. Hemalatha Reddy, Principal, Sri Venkateswara College, University of Delhi for their keen interest and valuable support for the successful organisation of the workshop. I am particularly grateful to Dr. Guleria for sparing his valuable time to write a Foreword for this publication.

I also acknowledge the interest and valuable efforts of the entire team of the NAM S&T Centre, especially Mr. Abhay Nambiar, Research Associate for compilation, proof reading and overseeing the publication process and Mr. M. Bandyopadhyay, Senior Adviser for his guidance and supervision. I am also thankful to Mr. Pankaj Buttan, Data Processing Manager, NAM S&T Centre for his valuable support in bringing out this publication.

A special thanks to Prof. Dr. Arun P. Kulshreshtha, Former Director General, NAM S&T Centre for his keen interest in this important and valuable event and for conceptualising the idea.

I am also thankful to Mr. Sharad Gupta, Publishing Consultant and Mr. Jagdish Singh from the Publication Department, Allied Publishers, New Delhi for their significant efforts in bringing out this publication.

I am sure that this book will serve as a valuable reference material for all those associated with issues related to air pollution and public health, particularly the researchers, medical professionals, policy and decision makers in the developing countries.

Amitava Bandopadhyay

(Amitava Bandopadhyay, Ph.D.)
Director General
NAM S&T Centre
New Delhi, India

Preface

Air pollution is assumed to be the main cause of climate change. Consequently, it causes many adverse effects on human health, availability of water resources, food and agricultural security, ecosystems, tourism, coastal development and biodiversity. Not only this, but worldwide, air pollution is also recognized as one of the top five causes of mortality due to respiratory and heart diseases. The situation is worse in developing countries, due to rise in industrialization, agricultural activities, large scale deforestation, significant biomass use for cooking and a general lack of awareness. Accordingly, the largest burden of diseases and mortality due to air pollution is being borne by low- and middle-income countries in the South-East Asia and Western Pacific Regions, with a total of 3.8 million deaths linked to indoor air pollution and 2.3 million deaths related to outdoor air pollution.

Clean air is a fundamental right of every individual, and therefore it is essential for all the stakeholders, including the scientists, doctors, policy makers and industries, to come together to deliberate on how to improve traditional approaches besides science, technology and innovation to mitigate air pollution and health issues associated with it.

This book stems from a series of discussions and deliberations that ensued during the International Workshop on "Air Pollution and Public Health: Challenges and Interventions" organized by the 'Centre for Science & Technology of the Non-aligned and Other Developing Countries (NAM S&T Centre), New Delhi, India and Sri Venkateswara College, University of Delhi, India, during 5–7 February 2020, in addition to some invited authors who shared their experience on the causes of air pollution and some suggested solutions to overcome such problem.

This book is divided into three sections; first, dealing with the sources of air pollution, the effect of air pollution on ecosystem and human health, and finally some suggested solutions; second, reporting some case studies that have been performed in some developing countries; and then it ends with a third section, reporting some action plans of some of the developing countries.

This book also illustrates various sustainable, multidisciplinary and interdisciplinary approaches for mitigation of air pollution, such as, (1) valorization of organic waste biomass into different valuable products, which consequently reduces the greenhouse gas emissions, (2) reducing carbon dioxide emissions through the production of synthetic natural gas, (3) applying artificial intelligence, (4) applying new eco-friendly membranes and filters and (5) performing serious action plans.

Nour Shafik El-Gendy, ***Ph.D.***
(Egypt)

Vartika Mathur, ***Ph.D.***
(India)

Contents

SECTION II: Case Studies from Developing World

SECTION III: Examples of Action Plans from Developing World

New Delhi Resolution

Section I

Air Pollution— Effects and Mitigation

Fossil Fuels and Air Pollution

James G. Speight*
CD&W Inc., 2476 Overland Road, Laramie, WY 82070-4808, USA
E-mail: JamesSp8@aol.com

ABSTRACT: Fossil fuel sources (gas, liquid, and solid) are those sources that can be used to produce fuels (gas, liquid, and solid), which are combustible or energy-generating molecular species that can be harnessed to create mechanical energy, especially heat energy. However, if no forms of control are in place, fossil fuels emit harmful air pollutants such as release oxides of carbon, nitrogen and sulfur as well as other pollutants into the atmosphere, during use. The gaseous oxides contribute to the formation of smog and acid rain. Furthermore, most of the oxides released in the developed countries are due to human activity associated with transportation, electric power generation, and agriculture which can upset the natural balance of the environment. This chapter presents a description of (i) the various aspects of fossil fuel combustion, (ii) the current controls applied to combustion systems, and (iii) the methods of monitoring the emissions.

Keywords: Fossil Fuels, Combustion Chemistry, Gas Cleaning, Combustion Mechanism, Combustion Systems, Monitoring Combustion Emissions, Environmental Legislation, The Future.

INTRODUCTION

By definition, fuel sources (gas, liquid, and solid) are those sources that can be used to produce gaseous, liquid, and solid fuels which are combustible or energy-generating molecular species that can be harnessed to create mechanical energy, especially heat energy. These fuels include natural gas and gas hydrates, coal-bed methane, crude oil, heavy oil, extra heavy oil, tar sand bitumen, coal (from which coal liquids are produced), and oil shale (from which shale oil is produced) (Table 1). All fossil fuel can be combusted, the limited exception is oil shale which is typically in excess of 85% w/w mineral matter, although to some extent and under the most appropriate conditions the organic

*Web page: https://www.drjamesspeight.com

material in oil shale (kerogen) will ignite and combust. However, if no forms of control are in place, fossil fuels emit harmful air pollutants such as release oxides of carbon, nitrogen, and sulfur as well as other pollutants into the atmosphere, during use (Table 2). The gaseous

Table 1: Informal Sub-division of Fossil Fuels by the Phase in the Natural State

Type	*Naturally-Occurring Fuel*
Gas	Unassociated gas Associated gas Gas hydrates
Liquid	High API gravity: >30° Medium API gravity: 25–30° Low API Gravity: 20–25°
Solid	Extra heavy oil* Tar sand bitumen Biomass Peat Coal: lignite, subbituminous, bituminous and anthracite Oil shale

* Liquid (mobile) in the deposit, near-solid or solid (immobile) at ambient conditions.

Table 2: Common Pollutants, Other than Carbon Dioxide, Produced During the Combustion of Fossil Fuels*

Pollutant	*Comment*
Carbon monoxide (CO)	Colorless, odorless, non-irritating but very poisonous gas. A product by incomplete combustion of carbonaceous fuels. Vehicular exhaust is a major source of carbon monoxide.
Nitrogen oxides (NOx)	Formed during high-temperature combustion.
Particulate matter	Small particles of solid or liquid suspended in a gas. Formed during combustion of fossil fuels in vehicles, power plants and various industrial processes.
Sulfur oxides (SOx)	Fossil fuel typically contains sulfur. Further oxidation of sulfur dioxide to sulfur trioxide which then forms sulfuric acid and thus acid rain.
Volatile organic compounds	An important outdoor air pollutant. Hydrocarbons are also significant greenhouse gases.

*Listed alphabetically rather than by occurrence.

oxides contribute to the formation of smog and acid rain. Furthermore, most of the oxides released in the developed countries are due to human activity associated with transportation, electric power generation, and agriculture which can upset the natural balance of the environment.

It is not the intent of this chapter to present details of these fossil fuels and their products since those topics are covered in detail elsewhere and the reader is referred to these works for detailed information (Scouten, 1990; Lee, 1991; Parkash, 2004; Gary *et al.*, 2007; Speight, 2013, 2014; Hsu and Robinson, 2107; Speight, 2017, 2019, 2020a). This chapter presents an overview of the combustion of fossil fuels and the types of emissions that can be expected from the combustion process as well as the potential effects of the emissions on the atmosphere.

TYPES OF FOSSIL FUELS

Fossil fuels exist in gaseous, liquid, or solid form. Natural gas, distillates (such as kerosene and gas oil), and coal, perhaps the most widely used examples of these three forms, each is a complex mixture of reacting and inert compounds. Each type of fossil fuel requires a different system to convert the fossil fuel to power.

Of the spectrum of fuels currently in widespread use, the simplest in composition is natural gas, which consists primarily of methane but includes a number of other constituents as well. The compositions of other gaseous fuels are generally more complex, but they are, at least, readily determined. Information on the composition of liquid or solid fuels is generally much more limited than that for gaseous fuels. Rarely is the molecular composition known since liquid fuels are usually complex mixtures of a large number of hydrocarbon species. The most commonly reported composition data are derived from the ultimate analysis, which consists of measurements of the elemental composition of the fuel, generally presented as mass fractions of carbon, hydrogen, sulfur, oxygen, nitrogen, and ash, where appropriate. The heating value, a measure of the heat release during complete combustion, is also reported with the ultimate analysis. For liquid fuels, the specific gravity or API gravity, viscosity (possibly at several temperatures), flash point (a measure of the temperature at which the fuel is sufficiently volatile to ignite readily), and distillation profiles (fraction vaporized as a function of temperature) may be reported. The properties of solid fuels vary even more widely than those of liquid fuels. The most common solid fuel is coal followed by coke produced during crude pol refining.

Formed by biological decomposition and geological transformation of plant debris, coals (the common solid fuels) is classified by rank, a measure of the degree to which the organic matter has been transformed from cellulose. Low-rank fuels such as peat or lignite have undergone relatively little change, whereas high-rank anthracite is nearly graphitic in structure. Low-rank fuels contain large amounts of volatile matter that are released upon heating. High-rank fuels contain much more fixed carbon, which remains after the volatiles are released. Solid fuels are characterized by the ultimate analysis and by the so-called proximate analysis, which identifies the degree of coalification of a solid fuel (Speight, 2013). Coal samples that have been air dried are subjected to a number of standardized tests to determine the amount of moisture inherent to the coal structure, the quantity of volatile matter released by the coal upon heating for several minutes, and the mass of ash or noncombustible inorganic (mineral) impurities that remains after low temperature oxidation (Speight, 2013). The difference between the initial mass of coal and the sum of masses of moisture, volatile matter, and ash is called fixed carbon. The conditions of these standardized tests differ markedly from typical combustion environments, and the values reported in the proximate analysis do not necessarily represent yields actually encountered in practical combustors.

FOSSIL FUEL COMBUSTION

The need to burn increasing amounts of fossil fuels and derived fuels mainly arises from the need to replace fossil fuels as the feedstock for combustion in furnaces, such as those in fossil fuel-fired power plants. In addition to satisfying normal economic constraints, existing and new combustion processes must be able to burn low-grade fuels and satisfy local environmental requirements. Some requirements cause conflict, e.g. the combustion of low-grade fuels can be relatively costly due to reduced plant capacity and efficiency, and the costs incurred in controlling pollution levels.

Not all fossil fuels have the same composition. For example, natural gas is composed of approximately 90% v/v methane (CH_4) while liquid fuels are composed of medium-to-high boiling constituents, and solid fuels contain complex hydrocarbon constituents that often contain non-hydrocarbon constituents that contain nitrogen and/or sulfur.

Since different hydrocarbon fuels have different ratios of hydrogen to carbon, they produce different ratios of water to carbon dioxide. In general, the longer and more complex the molecule, the greater the ratio of carbon to hydrogen. For this reason, combustion of equal amounts of different hydrocarbon derivatives will yield different quantities of carbon dioxide, depending on the ratio of carbon to hydrogen in molecules of each.

Combustion is used in a range of applications which vary from domestic fires to large industrial furnaces and utility boilers. The oxidant is usually air and the fossil fuel may be in any degree of dispersion, such as microscope droplets if the fuel is liquid in the normal state or small particles if the fossil fuel is solid in the normal state. Moreover, fossil fuel combustion provides the majority of consumable energy to the world and despite the continuing search for alternate sources of energy (whether they are other fossil fuels or non-fossil fuels such as biomass) there is little doubt that fossil fuel combustion will remain important source of energy throughout the 21st Century.

A major concern in the present day combustion of fossil fuels and fuels produced therefrom is the performance of the process in an environmentally acceptable manner through the use of a variety of environmentally acceptable technologies such as the use of a low-sulfur fossil fuel or through the use of post-combustion cleanup of the off-gases (Speight, 2013, 2014, 2020a). Thus, there is a marked trend in the current marketplace to initiate and develop more efficient methods of fossil fuel combustion. In fact, the ideal combustion system would be a system that is able to accept fossil fuel without a pre-combustion treatment, and/or without the need for post-combustion treatment, and/or without emitting objectionable amounts of sulfur and nitrogen oxides and particulates.

In order to improve existing combustors and to develop new combustion techniques, it is necessary to gain an improved understanding of the complex processes that occur in and around particles during combustion. For example, better insights are needed into ignition stability, the attainment of rapid burnout of particles, the nature of the various homogeneous and heterogeneous processes involved in generating gaseous pollutants and the routes by which inorganic constituents of fossil fuels are converted into ash.

The environmental issues in fossil fuel combustion can be categorized into gaseous emissions such as oxides of nitrogen and oxides of sulfur, fine particulate matter, and greenhouse gas emissions. The greenhouse gas emission is more related to the efficient use of fossil fuel rather than the properties of fossil fuel. The formation of these emissions during fossil fuel combustion is very much dependent upon the nature of the fossil fuel and the combustion system.

Chemistry and Physics

Combustion occurs, chemically, by initiation and propagation of a self-supporting exothermic (heat-producing) reaction. The physical processes involved in combustion are principally those which involve the transport of matter and the transport of energy. The conduction of heat, the diffusion of chemical species, and the bulk flow of the gas all follow from the release of chemical energy in an exothermic reaction. Thus, combustion phenomena arise from the interaction of chemical and physical processes. For simple chemical purposes, the fossil fuel is usually represented by carbon which can react with oxygen in two ways, producing either carbon monoxide or carbon dioxide.

$$2C_{fuel} + O_2 \rightarrow 2CO$$

$$C_{fuel} + O_2 \rightarrow CO_2$$

In direct combustion, fossil fuel is burned (i.e., the carbon and hydrogen in the fossil fuel are oxidized into carbon dioxide and water) to convert the chemical energy of the fossil fuel into thermal energy after which the sensible heat in the products of combustion can be converted into steam that can be converted into horsepower (such as in a gas turbine). In fact, the combustion process actually represents a mean of achieving the complete oxidation of fossil fuel in which the elements occurring in fossil fuel (carbon, hydrogen, nitrogen, and sulfur) are converted to their respective oxides:

$$C_{fuel} + O_2 \rightarrow CO_2$$

$$H_{fuel} + O_2 \rightarrow H_2O$$

$$C_{fuel} + H_2O \rightarrow CO + H_2$$
$$N_{fuel} + O_2 \rightarrow NOx$$
$$S_{fuel} + O_2 \rightarrow SOx$$

However, it is the formation of oxides of nitrogen and sulfur, in addition to the carbon dioxide, which cause serious environmental problems and require removal from any product gas streams.

In more general terms, the combustion of fossil fuels, which contain hydrogen and other elements as well as carbon, involves a wide variety of reactions between the many reactants, intermediates, and products. The reactions occur simultaneously and consecutively (in both forward and reverse directions) and may approach a condition of equilibrium. Furthermore, there is a change in the physical and chemical structure of the fuel particle as it burns.

The release of the sulfur and nitrogen from the fossil fuel is not as simple as represented here and the equations are simplifications of what are, presumably, much more complex processes. The sulfur dioxide that escapes into the atmosphere is either deposited locally or is converted to sulfurous acid and/or sulfuric acid by reaction with moisture in the atmosphere.

$$SO_2 + H_2O \rightarrow H_2SO_3$$
$$2SO_2 + O_2 \rightarrow 2SO_3$$
$$SO_3 + H_2O \rightarrow H_2SO_4$$

Thus:

$$2SO_2 + O_2 + 2H_2O \rightarrow 2H_2SO_4$$

Nitrogen oxides (Morrison, 1980; Crelling *et al.*, 1993) also contribute to the formation and occurrence of acid rain, in similar manner similar to the production of acids from the sulfur oxides, yielding nitrous and nitric acids.

$$NO + H_2O \rightarrow H_2NO_3$$
$$2NO + O_2 \rightarrow 2NO_2$$
$$NO_2 + H_2O \rightarrow HNO_3$$

Thus:

$$4NO_2 + O_2 + 2H_2O \rightarrow 4HNO_3$$

In addition to causing objectionable stack emissions, mineral ash and volatile inorganic material generated by thermal alteration of mineral matter in the fossil fuel will adversely affect heat transfer processes by fouling the heat-absorbing and radiating surfaces and will also influence the performance of the combustion system by causing corrosion, and operating procedures must therefore provide for effective countering of all these hazards.

Mechanism

The combustion of most solid fuels involves two major steps: (i) the thermal decomposition (pyrolysis, devolatilization) that occurs during the initial heating, accompanied by drastic physical and chemical changes which usually involve the particle becoming plastic then re-hardening, and (ii) the subsequent combustion of the porous solid residue (char) from the first step. The burning rate of the solid depends in part on the size of the char particle and the nature of its pore structure. These physical properties, together with important chemical properties, are affected by changes during the first step. The first step is rapid, the second is slow.

The initial step is the transfer of reactant (i.e. oxygen) through the layer of gas adjacent to the surface of the particle. The reactant is then adsorbed and reacts with the solid after which the gaseous products diffuse away from the surface. If the solid is porous, much of the available surface can only be reached by passage of the oxidant along the relatively narrow pores and this may be a rate controlling step. Rate control may also be exercised by: (a) adsorption and chemical reaction, which are considered as chemical reaction control; and (b) pore diffusion, by which the products diffuse away from the surface. This latter phenomenon is seldom a rate-controlling step. After the initial stages of combustion, volatile material is evolved (which is also combustible) and a nonvolatile carbonaceous residue (coke and/or char) remains, which can comprise up to 90% w/w of the original mass of the fossil fuel.

In summary, it is more appropriate to consider the combustion of fossil fuel (which contains carbon, hydrogen, nitrogen, oxygen, and sulfur) as involving a variety of reactions between (i) the reactants; (ii) the intermediate, or transient, species; and (iii) the products. The reactions can occur both simultaneously and consecutively (in both forward and reverse directions) and may even approach steady state (equilibrium) conditions. Where, there is a change in the physical and chemical structure of the fuel particle during the process.

COMPLETE AND INCOMPLETE COMBUSTION

Complete combustion of hydrocarbon fuels occurs when there is an excess of oxygen to allow the fuel to react completely to produce carbon dioxide and water. The combustion process is not necessarily favorable to the maximum degree of oxidation, and it can be temperature-dependent. For example, sulfur trioxide (SO_3) is not produced quantitatively by the combustion of sulfur. Nitrogen oxides (NOx) appear in significant amounts above about 1,540°C (2,800°F), and more is produced at higher temperatures.

In most industrial applications, air is the source of oxygen (O_2). However, in the air, each mole of oxygen is mixed with approximately 3.7 mol of nitrogen of nitrogen which does not take part in combustion, but at high temperatures some nitrogen will be converted to nitrogen oxides (NO_x, mostly NO, with much smaller amounts of NO_2). On the other hand, when there is insufficient oxygen to combust the fuel completely, some of the fuel carbon is converted to carbon monoxide (CO). Therefore, a representation (in equation form) of the hydrocarbon combustion requires an additional calculation for the distribution of oxygen between the carbon and hydrogen in the fuel. Models that do not take this into account will cause serious errors in the process.

Incomplete combustion will occur when there is insufficient oxygen to allow the fuel to react completely to produce carbon dioxide and water. As is the case with complete combustion, water is produced by incomplete combustion but carbon (soot) and carbon monoxide are produced instead of carbon dioxide. For most fuels, such as petroleum-derived fuels, and coal, pyrolysis occurs before combustion. In the lack of sufficient oxygen (incomplete combustion), the products of pyrolysis remain unburned and the emissions contain noxious particulate matter and gases.

Carbon monoxide is one of the products from incomplete combustion. Carbon monoxide is released in the normal incomplete combustion reaction, forming soot (particulate matter). Since carbon monoxide is considered as a poisonous gas, complete combustion is preferable, as carbon monoxide can also lead to respiratory troubles and death. Breathing carbon monoxide causes headache, dizziness, vomiting, and nausea. If carbon monoxide levels are high enough, humans become unconscious or die. Exposure to moderate and high levels of carbon monoxide over long periods is positively correlated with risk of heart disease. People who survive severe carbon monoxide poisoning may suffer long-term health problems. The carbon monoxide absorbed in the lungs which then binds with hemoglobin in the red blood cells of humans which reduces the capacity of the red blood cells to carry oxygen throughout the body.

Also, the oxides produced during fossil fuel combustion (COx, NOx, and SOx) combine with water and with oxygen in the atmosphere to create carbonic acid (H_2CO_3), nitrous acid (HNO_2), nitric acid (HNO_3), sulfurous acid (H_2SO_3), and sulfuric acid (H_2SO_4) which return to the surface of the Earth as acid deposition (commonly referred to as *acid rain*). Acid deposition harms aquatic organisms and kills trees. Due to its formation of certain nutrients that are less available to plants such as calcium and phosphorus, it reduces the productivity of the ecosystem and farms. An additional problem associated with the nitrogen oxides is that they, along with hydrocarbon pollutants, contribute to the formation of tropospheric ozone, a major component of smog. The troposphere the lowest region of the atmosphere, extending from the surface of the Earth to a height of approximately 3.7 to 6.2 miles—the lower boundary of the stratusphere (the layer of the atmosphere of the Earth above the troposphere, extending to approximately 32 miles above the surface of the Earth).

Smoldering is the slow, low-temperature, flameless form of combustion of fuels that is sustained by the heat evolved when oxygen directly attacks the surface of a condensed-phase fuel; it is typically incomplete combustion reaction. The process can go unnoticed for many days before smoke starts to be seen. The process was common in the days when coal mining waste (mine tipple) was collected in heaps on the surface near to the mine. The difference between smoldering and flame-combustion combustion is that smoldering occurs on the surface of the solid fuel rather than in the gas phase. It is a surface phenomenon but can propagate to the interior of a porous solid fuel if the solid is permeable to flow. The characteristic temperature and heat released during smoldering are low compared to those in the flame-combustion (i.e., approximately 600°C – approximately 1,110°F – compared to approximately 1500°C–2,730°F). The smoldering propagates in a slow manner (sometime referred to as creeping manner) which is approximately ten times slower than flames spread over a solid fuel. Fires occurring many feet below the surface are a type of smoldering event. For example, subsurface fires in coal mines, when active they can smolder for long periods of time (months, years, or even decades), emitting enormous quantities of combustion gases into the atmosphere, causing deterioration of air quality and subsequent health problems. Smoldering fires are fed by the oxygen in the small but continuous flow of air through natural pipe networks, fractured strata, cracks, openings, or abandoned coal mine shafts which permit the air to circulate into the subsurface. The reduced heat losses and high thermal inertia of the underground together with high fuel availability promote long-term smoldering combustion and allow for creeping but extensive propagation. These fires are difficult to detect and frustrate most efforts to extinguish them. Nevertheless, in spite of the so-called weak combustion characteristics of smoldering, the process is a significant fire hazard—toxic gases (such as carbon monoxide) are emitted at a higher yield than flame-fires and leaves behind a significant amount of solid residue (char, coke). The emitted gases are flammable and could later be ignited in the gas phase, triggering the transition to flaming combustion. Finally, spontaneous combustion is a type of combustion which occurs by self-heating (an increase in temperature due to exothermic internal reactions), followed by thermal runaway (self-heating which rapidly accelerates to high temperatures), and finally, ignition—sometimes explosive ignition.

COMBUSTION SYSTEMS

Combustion systems convert essentially all of the fuel into carbon dioxide, water, and heat but there are other products of the combustion process. The combustion system consists of a combustor, burners, igniter, and flame monitors. The *combustor* must bring the gas to a controlled uniform temperature with minimum impurities and minimum loss of pressure. The major problems of combustor design, in addition to clean combustion and proper mixing of gases, are flame stabilization; mitigation of pressure loss; and maintenance of a steady and tightly controlled outlet temperature. The gas turbine power plants are arranged with single or multiple combustors. The single combustor design is easier to control but is larger and less compact than the multiple types.

The combustor is placed between the turbine and the compressor within turbine casing. It is a direct-fired heater in which fuel is burnt to supply heat energy to the gas turbine. The air flow through the combustor has three functions: to oxidize the fuel, to cool the metal parts, and to adjust the extremely hot combustion to the desired turbine inlet. The temperature of the gas in the combustors and entering the turbine can reach up to 750°C (1350°F) or even more. Usually, low NO_X burners are used to reduce the concentration of NO_X in the exhaust gas to less than 25 ppm at full load. Water or steam can be injected into the combustors to reduce the concentration of NO_X in the exhaust gas.

The primary zone of the combustor accommodates *burners*. The fuel (typically in the gaseous state) is admitted directly into each burner and mixes with air. When liquid fuel is used it is atomized in the nozzle using high-pressure air. The atomized fuel/air mixture is then sprayed into the combustor. During oil firing, water is mixed into the oil to ensure low NO_X emission level.

Igniters are used for initial ignition of the fuel-air mixture during start-up of the gas turbine. The main fuel then flows into the combustor through burners and is ignited there by the ignition flame. The flame then spreads from burner to burner without further sparking once these burners are supplied with fuel. After ignition is accomplished, the spark plug is switched off. *Flame monitors* are installed in the combustor to indicate the presence or absence of flame. The flame monitor is sensitive to the presence of radiation, typically emitted by a hydrocarbon flame. A gas-fired power plant is a type of fossil fuel power station in which chemical energy stored in natural gas is converted successively into thermal energy, mechanical energy, and electrical energy.

In a simple cycle gas-turbine, also known as open cycle gas-turbine (OCGT), hot gas drives a gas turbine to generate electricity. The combined cycle gas-turbine (CCGT) power plant consists of simple cycle gas-turbines followed by a heat recovery steam generator and a steam turbine. The most common configuration is two gas-turbines supporting one steam turbine. They are more efficient than simple cycle plants and can achieve efficiencies slightly over 60% and dispatch times of around half an hour. Heavy fuel oil was once a significant source of energy for electric power generation. After oil price increases of the 1970s, oil was displaced by coal and later natural gas. Distillate oil is still important as the fuel source for diesel engine power plants used especially in isolated communities not interconnected to a grid. Liquid fuels may also be used by gas turbine power plants, especially for peaking or emergency service. Of the three fossil fuel sources, oil has the advantages of easier transportation and handling than solid coal, and easier on-site storage than natural gas. In a power ant in which crude oil or a crude oil derivative is used for power generation, the liquid fuel is generally sprayed into a combustor as relatively fine droplets. Volatile fuels, such as kerosene, vaporize completely prior to combustion. Heavy fuel oils may partially vaporize, leaving behind a carbonaceous solid or liquid residue that then undergoes surface oxidation. The nature of the combustion process and pollutant emissions depends strongly on the behavior of the condensed-phase fuel during combustion.

The porous medium burner for liquid fuels is more advantageous than the conventional open spray flame burner for several reasons; these include enhanced evaporation of droplet spray owing to regenerative combustion characteristics, low emission of pollutants, high combustion intensity with moderate turn-down ratio and compactness (Mujeebu *et al.*, 2009).

Spray combustion is a powerful method of burning relatively involatile liquid fuels; indeed, it remains the major way of burning heavy (viscous) feedstocks although such feedstocks they can be burned in fluidized bed combustors (Glassman and Yetter, 2008; Lackner, 2011). The basic process involved is the disintegration or atomization of the liquid fuel to produce a spray of small droplets to increase the surface area so that the rates of heat and mass transfer during combustion are greatly enhanced. The spray combustion of liquid fuels, particularly involving air blast atomizers, is widely used in the iron and steel industries for applications such as billet reheating furnaces, direct-fired heat treatment furnaces, hardening and tempering furnaces, forge furnaces, soaking pits, blast furnaces, and a wide range of other metallurgical applications, many of which require carefully controlled atmospheres of combustion products. Solid fuels are burned in a variety of systems, some of which are similar to those fired by liquid fuels. In large industrial furnaces, particularly boilers for electric power generation, coal is pulverized to a fine powder (typically, 50 micron mass mean diameter and 95% smaller by mass than about 200 micron) which is sprayed into the combustion chamber and burned in suspension. The combustion in the pulverized coal system has many similarities to the combustion of heavy fuel oils. Smaller systems generally utilize fixed- or fluidized-bed combustors that bum larger particles. Air is fed into a fluidized bed at a sufficiently high velocity to levitate the particles, producing a dense suspension that appears fluid-like. Heat transfer in the bed must be high enough and heat release rates low enough to keep the bed relatively cool and prevent the ash particles from fusing together to form large ash agglomerates known as clinkers. Noncombustible solids are often used to dilute the fuel and keep the temperature low; most commonly, limestone is used in order to retain the sulfur in the bed at the same time. In contrast to the rapid mixing in a fluidized bed, only a fraction of the air comes in contact with the fuel in a fixed bed, or stoker, combustion system, with the remainder being introduced above the bed of burning fuel. Large amounts of excess air are required to achieve reasonable combustion efficiency, and even with the large airflows, hydrocarbon and carbon monoxide emissions can be quite high, due to poor mixing above the bed. The increased air requirements lower the thermal efficiency of stoker units, so pulverized coal or fluidized-bed combustion is favored for large systems. Most large systems currently in use bum pulverized coal.

The basic elements of a solid fuel combustor are (i) fuel feeder, (ii) combustor fuel bed, (iii) under fire air mechanism, (iv) over fire air mechanism, (v) combustor enclosure, and (vi) ash discharge mechanism. There are many types of fuel bed design and examples are: (i) stationary bed, (ii) moving grate, (iii) vibrating grate, (iv) traveling grate, and (v) rotating drum. Also, a bubbling fluidized bed, circulating fluidized bed and rotary kiln may be used. Fuel feeders are used to transfer solid fuels into the combustor and the feeders may be auger (screw feeder), chain, conveyor, conveyer belt, or ram feeder. In the first stage of the process, any moisture contained in the fuel is removed after which (the second stage) the organic material is vaporized and partially combusted in the dried fuel after which combustion occurs.

A major factor in fossil fuel combustion is the reduction of the emissions, and the combustor is the primary contributor to a gas turbine's emissions of which there are five major types of emissions (i) carbon dioxide (CO_2), (ii) carbon monoxide (CO), (iii) sulfur oxides (SOx), (iv) nitrogen oxides (NO_x), and (v) unburned hydrocarbon derivatives.

Carbon dioxide is a product of the combustion process, and it is primarily mitigated by reducing fuel usage. Carbon monoxide is an intermediate product of combustion, and it is eliminated by oxidation. Sulfur oxides are produced in the combustion zone and arise from the presence of sulfur in the feedstock. Like sulfur oxides, nitrogen oxides are also produced in the combustion zone. Unburned hydrocarbon derivatives represent a portion of the fuel that was not completely combusted. Much of the unburned hydrocarbon derivatives and carbon monoxide are formed within the combustion chamber.

TYPES OF POLLUTANTS

Fossil fuel combustion releases various types of pollutants which, for convenience of identification are classed as primary pollutants and secondary pollutants. Briefly, fossil fuel production and use produces chemical waste. If this chemical waste is not processed in a timely manner, it can become a pollutant.

A pollutant is a substance present in a particular location (ecosystem) when it is not indigenous to the location or is present in a greater than natural concentration. The substance is often the product of human activity. The pollutant, by virtue of its name, has a detrimental effect on the environment. Pollutants can also be subdivided into two classes: primary and secondary.

$$\text{Source} \rightarrow \text{Primary pollutant} \rightarrow \text{Secondary pollutant}$$

A **primary pollutant** is a pollutant that is emitted directly from the source. In terms of atmospheric pollutants, examples are carbon oxides, sulfur dioxide, and nitrogen oxides from fuel combustion operations:

$$2[C]_{feedstock} + O_2 \rightarrow 2CO$$

$$[C]_{feedstock} + O_2 \rightarrow CO_2$$

$$2[N]_{feedstock} + O_2 \rightarrow 2NO$$

$$[N]_{feedstock} + O_2 \rightarrow NO_2$$

$$[S]_{feedstock} + O_2 \rightarrow SO_2$$

$$2SO_2 + O_2 \rightarrow 2SO_3$$

Hydrogen sulfide and ammonia are produced from processing sulfur-containing and nitrogen containing feedstocks:

$$[S]_{feedstock} + H_2 \rightarrow H_2S + \text{hydrocarbons}$$

$$2[N]_{feedstock} + 3H_2 \rightarrow 2NH_3 + \text{hydrocarbons}$$

The question of classifying nitrogen dioxide and sulfur trioxide as primary pollutants often arises, as does the origin of the nitrogen. In the former case, these higher oxides can be formed in the upper levels of the combustion reactor.

A secondary pollutant is a pollutant that is produced by the interaction of a primary pollutant with another chemical. A secondary pollutant may also be produced by dissociation of a primary pollutant, or other effects within a particular ecosystem. Again, using the atmosphere as an example, the formation of the constituents of acid rain is an example of the formation of secondary pollutants:

$$SO_2 + H_2O \rightarrow H_2SO_3 \text{ (sulfurous acid)}$$

$$SO_3 + H_2O \rightarrow H_2SO_4 \text{ (sulfuric acid)}$$

$$NO + H_2O \rightarrow HNO_2 \text{ (nitrous acid)}$$

$$3NO_2 + 2H_2O \rightarrow HNO_3 \text{ (nitric acid)}$$

In many cases, these secondary pollutants can have significant environmental effects, such as the formation of acid rain and smog.

Any pollutant, either primary or secondary can have a serious effect on the various ecological cycles such as the industrial cycle and the water cycle. Therefore, understanding the means by which a chemical pollutant can enter these ecosystems and influence the future behavior of the ecosystem, is extremely important.

GAS CLEANING

Flue and waste gases from power plants and other industrial operations where fossil fuel is used as a feedstock invariably contain constituents and toxins that are damaging to the climate or environment—these constituents are for example; carbon dioxide (CO_2), nitrogen oxides (NOx), sulfur oxides (SOx), dust and particles; and toxins such as dioxin and mercury. The processes that have been developed for gas cleaning (Mokhatab *et al.*, 2006; Speight, 2019, 2020a) vary from a simple single-pass wash operation to complex multi-step systems with options for recycle of the gases (Mokhatab *et al.*, 2006). In some cases, process complexities arise because of the need for recovery of the materials used to remove the contaminants or even recovery of the contaminants in the original or altered form.

The purpose of preliminary cleaning of gases which arise from fossil fuel utilization is the removal of materials such as mechanically carried solid particles (either process products and/or dust) as well as liquid vapors (i.e. water, tars, and aromatics such as benzenes and/or naphthalenes); in some instances, preliminary cleaning might also include the removal of ammonia gas. Pretreatment cleaning processes are also successful methods for removing inorganic sulfur but they do not affect the organic sulfur content. Thus, even before combustion begins, some of the sulfur can be removed from the fuel. For instance, commercially available processing methods can remove a considerable amount of the sulfur, even all of the sulfur if the fuel is passed through a hydrotreating operation.

In more general terms, gas cleaning is divided into removal of particulate impurities and removal of gaseous impurities. For the purposes of this chapter, the latter operation includes the removal of hydrogen sulfide, carbon dioxide, sulfur dioxide, and the like. There is also need for subdivision of these two categories as dictated by needs and process capabilities: (a) coarse cleaning whereby substantial amounts of unwanted impurities are removed in the simplest, most convenient, manner; (b) fine cleaning for the removal of residual impurities to a degree sufficient for the majority of normal chemical plant operations, such as catalysis or preparation of normal commercial products; or cleaning to a degree sufficient to discharge an effluent gas to atmosphere through a chimney; (c) ultra-fine cleaning where the extra step (as well as the extra expense) is justified by the nature of the subsequent operations or the need to produce a particularly pure product.

Thus, various processes exist to achieve clean combustion. For example, the fuel can be pretreated to reduce the sulfur content, in order to reduce emissions of sulfur oxides (SOx). In addition, the combustion process can be optimized by using a technique known as fuel-lean combustion which reduces the peak temperatures in the combustor leading to a decline in the emissions of nitrogen oxides (NOx) emissions. Finally, exhaust-gas cleaning (flue-gas cleaning) can be applied to clean the combustion off-gas. This can be achieved directly in the combustor, such as capturing of the sulfur oxide emissions by limestone in fluidized-bed combustors or downstream, in a flue-gas scrubbing unit or in an electrostatic precipitator.

MONITORING COMBUSTION EMISSIONS

Monitoring the emissions from the combustion of fossil fuels must be through the application of various standard test methods as prescribed in many countries by the ASTM International (ASTM, 2020). These test methods will assist to define the properties of the emissions and the methods of (gas) cleaning that should be applied before the combustion gases are emitted to the atmosphere (Speight, 2015).

Sampling

The reliability of a sampling method is the degree of perfection with which the identical composition and properties of the entire whole fossil fuel or fossil fuel products are obtained (preserved completely without alteration) in the sample—such a sample is referred to as a *representative sample.* The reliability of the storage procedure is the degree to which the sample remains unchanged thereby guaranteeing the accuracy and usefulness of the analytical data. At this point, a review of the sampling methods applied to fossil fuels and their respective products is worth of inclusion.

Since the value of any product is judged by the characteristics of the sample as determined by laboratory tests, the ability to collect and preserve a sample that is representative of the site is a critically important step. Furthermore, obtaining a representative sample of a fossil fuel or product is always a challenge due to the heterogeneity of the materials. Additional difficulties are encountered due to the ranges in volatility, solubility, and adsorption potential of individual constituents. And the procedures used for sample collection and preparation must be analytically reproducible with a high degree of accuracy. The sample to be used for the test methods must be representative of the bulk material or data will be produced that are not a true representation of the material and will, to be blunt, incorrect no matter how accurate or precise the test method and not matter how much *fine tuning* is applied by the submitter of the sample. In addition, the type and cleanliness of sample containers are important if the container is contaminated or is made of material that either reacts with the product or is a catalyst, the test results may be wrong.

Method Validation

Method validation is the process of demonstrating that an analytical method is acceptable for its intended purpose. In general, methods for product specifications and regulatory submission must include studies on specificity, linearity, detection limit, and quantitation limit as well as accuracy and precision.

The process of method development and validation covers all aspects of the analytical procedure and the best way to minimize method problems is to perform validation experiments during development. However, method validation cannot mitigate out all of the potential problems, the process of method development and validation should address the most common (anticipated) problems.

The first step in the method development and validation cycle should be to set minimum requirements, which are essentially acceptance specifications for the method. A complete list of criteria should be agreed on during method development and the end users before the method is developed so that expectations are clear. Once the validation studies are complete, the method developers should be confident in the ability of the method to provide good quantitation in their own laboratories. The remaining studies should provide greater assurance that the method will work well in other laboratories, where different operators, instruments, and reagents are involved and where it will be used over much longer periods of time.

QUALITY CONTROL AND QUALITY ASSURANCE

Quality control (QC) and quality assurance (QA) programs are key components of all analytical protocols in all areas of analysis, including environmental, pharmaceutical, and forensic testing, among others

(Patnaik, 2004). These programs mandate that the laboratories follow a set of well-defined guidelines to achieve valid analytical results to a high degree of reliability and accuracy within an acceptable range. Although such programs may vary depending on the regulatory authority, certain key features of these programs are more or less the same.

Quality Control

Quality controls are single procedures that are performed in conjunction with the analysis to help assess in a quantitative manner the success of the individual analysis. Examples of quality controls are blanks, calibration, calibration verification, surrogate additions, matrix spikes, laboratory control samples, performance evaluation samples, determination of detection limits, etc. The success of the quality control is evaluated against an acceptance limit. The actual generation of the acceptance limit is a function of quality assurance; it would not be termed a quality control.

Unlike quality assurance plans that mostly address regulatory requirements involving comprehensive documentation, quality control programs are science-based, the components of which may be defined statistically. The two most important components of quality control are (i) determination of precision of analysis and (ii) determination of accuracy of measurement.

Quality Assurance

Quality assurance is an umbrella term that is correctly applied to everything that the laboratory does to assure product reliability. As the product of a laboratory is information, anything that is done to improve the reliability of the generated information falls under quality assurance.

Quality assurance includes all the quality controls, the generation of expectations (acceptance limits) from the quality controls, plus a great number of other activities such as (i) analyst training and certification, (ii) data review and evaluation, (iii) preparation of final reports of analysis, (iv) information given to clients about tests that are needed to fulfill regulatory requirements, (v) use of the appropriate tests in the laboratory, (vi) obtaining and maintaining laboratory certifications/

accreditations, (vii) conducting internal and external audits, (viii) preparing responses to the audit results, (ix) the receipt, storage, and tracking of samples, and (x) tracking the acquisition of standards and reagents (Speight, 2015).

Thus, the objective of quality assurance (usually in the form of a *quality assurance plan*) is to obtain reliable and accurate analytical results that may be stated with a high level of confidence (statistically), so that such results are legally defensible. The key features of any plan involve essentially documentation and record keeping. In short, quality assurance program involves documentation of sample collection for testing, the receipt of samples in the laboratory, and their transfer to the individuals who perform the analyses. The information is recorded on chain-of-custody forms stating the dates and times along with the names and signatures of individuals who carry out these tasks. Also, other pertinent information is recorded, such as any preservatives added to the sample to prevent degradation of test analytes, the temperature at which the sample is stored, the temperature to which the sample is brought prior to its analysis, the nature of the container (which may affect the stability of the sample), and its holding time prior to testing.

ENVIRONMENTAL LEGISLATION

An *environmental regulation* is a legal mechanism that spells determines how the policy directives of an environmental law are to be carried out. An *environmental policy* is a requirement that specifies operating procedures that must be followed. An *environmental guidance* is a document developed by a governmental agency that outlines a position on a topic or which give instructions on how a procedure must be carried out. It explains how to do something and provides governmental interpretations on a governmental act or policy.

The environmental aspects of fossil fuel use have been a major factor in the various processes, and the see-sawing movement of the fossil fuel base between natural gas, crude oil, and coal increased the need for pollutant control for large, fossil fuel-fired power plants. These power plants emit pollutants which, by atmospheric chemical transformations, may become even more harmful secondary pollutants. In fact, it has become very apparent over the last three decades that abatement of air pollution needs to be mandatory now and in the future. Four main avenues of action are open to decrease the amount of sulfur dioxide emitted from stacks of power generating plants (a) burn low-sulfur fuels; (b) desulfurize available fuels; (c) remove sulfur oxides from flue gases; or (d) generate power by nuclear reactors.

In the past, a certain amount of pollution was recognized as being almost inevitable, perhaps even fashionable! But now this is not the case. Any industry found guilty of emitting noxious materials can suffer heavy fines. And there is also the possibility of a jail term for the offending executives! Pollution of the environment will not be tolerated.

Thus, whilst industry marches on using many of the same processes that were in use in the early days of the century, more stringent methods for clean-up are necessary before any product/by-product can be released to the atmosphere. And this is where gas processing becomes an important aspect of industrial life. Furthermore, gas-cleaning processes are now required to be more efficient than ever before.

THE FUTURE

There have been many suggestions about the future of the fossil fuel industry and the reserves of fossil fuels that are available. Nevertheless, the future of fossil fuel combustion is currently unknown in any degree of detail but it is becoming apparent that fossil fuel combustion will, at some time, in the future be passed out. But, hopefully, only after suitable substitutes for the production of energy have been tried and tested. At present the future of fossil fuel combustion is looked upon as a major environmental concern, with some justification.

There is no doubt that human activities result in emissions of four principal greenhouse gases: carbon dioxide (CO_2), methane (CH_4), nitrous oxide (N_2O) and the halocarbon derivatives (a group of gases containing fluorine, chlorine, and bromine). These gases accumulate in the atmosphere, causing concentrations to increase with time-assuming that the gases do not react with other constituents of the atmosphere or decay to by-products. Significant increases in all of these gases have occurred in the industrial era and all of these increases have been attributed to human activities (Speight, 2019). Moreover, the increase in atmospheric pollution through measurement of carbon dioxide in the atmosphere seems to point to fossil fuel combustion as the major culprit. This conclusion is reached though the measurement of past atmospheric concentrations of carbon dioxide by measurement of the carbon dioxide in ice cores.

The major issue is related to the meaning of the ice-core data in terms of the current concentration of carbon dioxide in the atmosphere and the concentration of carbon dioxide in the atmosphere of past times. Furthermore, the common assumption in interpreting ice-core carbon dioxide records is that diffusion in the ice does not affect the concentration profile. However, this assumption has been tested and the solubility of carbon dioxide in the ice melt water leads to errors in the estimation of the carbon dioxide concentration in the ice core (Anklin *et al.*, 1995; Ahn *et al.*, 2008; Aizebeokhasi, 2009) thereby throwing considerable doubt not only on the origin of carbon dioxide in ice core sample but also (and especially) on the amount of carbon dioxide in various ice core samples.

There is no doubt that carbon dioxide in the atmosphere is a contributor to global climate change but the issue relating to the measurement of carbon dioxide in ice cores as an indicator of the increase of carbon dioxide in the atmosphere must be questioned. This raises doubt related to the subsequent interpretation of the analytical data relating the relative concentration of carbon dioxide in ice cores as the major culprit behind global climate change. As a result, long-term global predictions may seemingly be beyond current capabilities, especially when all of the available facts are not considered and emotion is allowed to control the science (Speight, 2020b).

REFERENCES

[1] Ahn, J., Headly, M., Wahlen, M., Brook, E.J., Mayewski, P.A. and Taylor, K.C. (2008). CO_2 Diffusion in Polar Ice: Observations from Naturally Formed CO_2 Spikes in the Siple Dome (Antarctica) Ice Core. *Journal of Glaciology*, 54(187): 685–695.

[2] Anklin, M., Barnola, J.M., Schwander, J., Stauffer B. and Raynaud, D. (1995). Processes Affecting the Carbon Dioxide Concentrations Measured in Greenland Ice. Tellus, 47B: 461–470.

[3] ASTM (2020). Annual Book of Standards, ASTM International, West Conshocken, Pennsylvania.

[4] Aizebeokhasi, A.P. (2009). Global Warming and Climate Change: Realities, Uncertainties and Measures. *International Journal of Physical Sciences*, 4(13): 868–879.

[5] Delbianco, A. and Montanari, R. (2009). Encyclopedia of Hydrocarbons, Volume III/New Developments: Energy, Transport, Sustainability. Eni S.p.A., Rome, Italy.

[6] Gary, J.G., Handwerk, G.E. and Kaiser, M.J. (2007). Petroleum Refining: Technology and Economics, 5th Edition. CRC Press, Taylor & Francis Group, Boca Raton, Florida.

[7] Glassman, I., and Yetter, R.A. (2008). Combustion 4th Edition. Elsevier, Burlington, Massachusetts.

[8] Hsu, C.S. and Robinson, P.R. (Editors) (2017). Handbook of Petroleum Technology. Springer International Publishing AG, Cham, Switzerland.

[9] Lackner, M. (2011). Combustion. In: Ullmann's Encyclopedia of Industrial Chemistry, Wiley-VCH Verlag GmbH & Co. KGaA, Weinheim.

[10] Lee, S. (1991). Oil Shale Technology. CRC Press, Taylor & Francis Group, Boca Raton, Florida.

[11] Mokhatab, S., Poe, W.A. and Speight, J.G. Handbook of Natural Gas Transmission and Processing. Elsevier, Amsterdam, Netherlands.

[12] Mujeebu, M.A., Abdullah, M.Z., Abu Bakar, M.Z., Mohamad, A.A. and Abdullah, A.K. (2009). A Review of Investigations on Liquid Fuel Combustion in Porous Inert Media. *Progress in Energy and Combustion Science*, 35(2): 216–230.

[13] Parkash, S. (2003). Refining Processes Handbook. Gulf Professional Publishing, Elsevier, Amsterdam, Netherlands.

[14] Scouten, C. (1990). In Fuel Science and Technology Handbook. J.G. Speight (Editor). Marcel Dekker Inc., New York.

[15] Simpson, D.A., Lea, J.F. and Cox, J.C. (2003). Coal Bed Methane Production. Paper SPE 80900. Proceedings. Production and Operations Symposium. Society of Petroleum Engineers, Richardson, Texas. March.

[16] Speight, J.G. (2012). Shale Oil Production Processes. Gulf Professional Publishing Company, Elsevier, Oxford, United Kingdom.

[17] Speight, J.G. (2013). The Chemistry and Technology of Coal 3rd Edition. CRC Press, Taylor & Francis Group, Boca Raton, Florida.

[18] Speight, J.G. (2014). The Chemistry and Technology of Petroleum 5th Edition. CRC Press-Taylor and Francis Group, Boca Raton, Florida.

[19] Speight, J.G. (2015). Handbook of Petroleum Product Analysis 2nd Edition. John Wiley & Sons Inc., Hoboken, New Jersey.

[20] Speight, J.G. (2017). Handbook of Petroleum Refining. CRC Press, Taylor & Francis Group, Boca Raton, Florida.

[21] Speight, J.G. (2019). Natural Gas: A Basic Handbook 2nd Edition. GPC Books, Gulf Publishing Company, Houston, Texas.

[22] Speight, J.G. (2020a). Synthetic Fuels Handbook: Properties, Processes, and Performance 2nd Edition. McGraw-Hill, New York.

[23] Speight, J.G. (2020b). Global Climate Change Demystified. Scrivener Publishing, Beverly, Massachusetts.

[24] United States Congress (1976). Public Law FEA-76-4. Congress of the United States of America, Washington, DC.

Bio-valorization of Organic Waste Biomass into Different Valuable Products as Out-of-Box Approaches for Mitigation of Air Pollution

Nour Sh. El-Gendy* and Hussein N. Nassar

Egyptian Petroleum Research Institute, Nasr City, Cairo, Egypt, PO 11727

October University for Modern Sciences and Arts, 6th of October City, Giza, Egypt, PO 12566

Nanobiotechnology Program, Faculty of Nanotechnology for Postgraduate Studies, Cairo University, Sheikh Zayed Branch Campus, Sheikh Zayed City, Giza, Egypt, PO 12588

*Head Manager of Petroleum Biotechnology Lab., Egyptian Petroleum Research Institute, 1 Ahmed El-Zomor St., Nasr City, Cairo, Egypt, PO 11727. Center of Excellence, October University for Modern Sciences and Arts (MSA), 26 July Mehwar Rd. Intersection with Wahat Rd., 6th of October City, Giza, Egypt, PO 12566.

*E-mail: nourepri@yahoo.com; nshelgendy@msa.eun.eg

ABSTRACT: Air pollution is assumed to be the main cause of climate change. Consequently, it causes many adverse effects on human health, availability of water resources, food and agricultural security, ecosystems, tourism, coastal development and biodiversity. Thus, to overcome such problem and to catch the wave of social, health, economic and environmental sustainability, thinking out of the box is mandatory. For example, as a solution, one of the most up to date and worldwide research fields is the valorization of agro-industrial wastes, waste biomass, wastewater ... etc., which can be considered as sustainable solution for environmental pollution, waste management, energy, climate change, health issues ... etc. problems. This chapter summarizes some of the successful efforts and case studies about (1) sustainable and full integrated processes for bioremediation of organic xenobiotic polluted environment (2) sustainable bio-upgrading of petroleum and its fractions through-

out the application of biodesulfurization, biodenitrogenation... etc. (3) sustainable biosorption and wastewater treatment with the achievement of zero-waste (4) sustainable recycling and valorization of different kinds of agro-industrial wastes to produce valuable products; e.g. nano-materials, biosurfactants, biocides, catalysts to be applied in biodiesel production, corrosion inhibitors, green-catalysts with high photo-catalytic degradation properties, wastewater treatment ... etc. (5) sustainable production of different valuable industrial products from algae, which have many applications in industries of food, cosmetics, pharmaceuticals, food supplements, animal feed ... etc. (6) sustainable production of different kinds of biofuels as complementary and/or alternative to petro-fuels, for example; biodiesel from waste cooking oil and micro-algae, bioethanol from macro-algae, different lingo-cellulosic wastes, sugarcane and sugar beet molasses and solid biofuel.

Keywords: Valorization; Agro-industrial Wastes; Sustainable Bio-remediation; Sustainable Biodesulfurization and Biodenitrogenation; Green Biocides; Green Photocatalysts; Sustainable Green Synthesis of Nanoparticles; Sustainable Biofuel Production.

INTRODUCTION

In both developed and developing countries, climate change negatively impact human health, biodiversity, reverberating through the economy, from threatening water availability to sea-level rise and extreme weather impacts to coastal regions and tourism, thus, threatening development and economic stability. The Intergovernmental Panel on Climate Change (IPCC) has an enormous influence over world thinking on climate change. The IPCC notes that greenhouse gas (GHG) emissions growth in the decade leading to 2010 has been larger, averaging an increase of 2.2% per year, than in the previous three decades from 1970 to 2000, where annual increase was at 1.3%. With these emissions, temperatures have continued to rise, with each of the past three decades being successively warmer than the preceding decades since 1850. Impacts on physical systems, such as coastal erosion, melting glaciers, receding rivers and lakes, and increased drought incidences, have been accompanied by changes in extreme events; for instance, the increasing reported frequency of heat waves in large parts of Europe, Asia, and Australia, or a 0.50°C increase in near-surface temperatures over most of Africa and a reduction in precipitation (El-Gendy and Speight, 2015).

Present situation in the Middle East and the speculation in the stock exchange, among other factors; have caused great fluctuations in oil price, which consequently negatively impacted on the world economy. That comes in parallel with the ever increasing demand for energy and fast depletion of petroleum resources, specially the high quality low sulfur and nitrogen content crude oil. In addition, the intensive utilization of fossil fuels has led to the increase in the generation of polluting gases, which adds to the problem of global climate change and adversely affected the human health. Thus to overcome the problem of climate change, the worldwide environmental legislation has put limits for the sulfur and nitrogen levels in the transportation fuels to decrease both the tailpipe and evaporative combined emissions of non-methane organic gas, NOx and SOx emissions (Sadare *et al.*, 2017). Consequently, refineries must have economically feasible techniques to remove sulfur and nitrogen from crude oil and refinery streams to the extent needed to mitigate these unwanted effects. This can be done via biodesulfurization and biodenitrogenation of non-renewable fuels and/or via the usage of renewable alternative biofuels.

Biotechnology is now accepted as an attractive ecofriendly sustainable means of improving the efficiency of any industrial processes, and resolving serious environmental problems. Bioprocesses have several advantages compared with chemical processes, including: (i) Microbial enzyme reactions are often more selective; (ii) Biotransformation processes are often more energy-efficient; (iii) Microbial enzymes are active under mild conditions; and (iv) Microbial enzymes are environment friendly biocatalysts. Although many bioprocesses have been described, only a few of these have been used as part of an industrial process. Biotechnology has been successfully applied at the industrial scale in the medical, fine chemicals, agricultural, and food sectors. However, many opportunities remain in this area. For example, petroleum biotechnology is based on biotransformation processes (Speight and El-Gendy, 2017). Current applied research on petroleum biotechnology encompasses oil spill remediation, fermenter- and wetland-based hydrocarbon treatment, bio-filtration of volatile hydrocarbons, microbial enhanced oil recovery, oil and fuel bio-refining, bio-production of fine chemicals, and microbial community based site assessment.

Millions tons of agro-industrial wastes are annually produced. These wastes are not economically reused, create air, soil and water pollution.

Thus, the waste management represents another important challenge in the agro-food based industries and demands an integrated approach in the context of recycling, up-cycling, reuse and recovery of such wastes. Furthermore, the worldwide depletion in water resources is another problem to be solved. This would be done, via minimizing the industrial water utilization and the economic wastewater treatment and reuse. Upon the valorization of agro-industrial wastes into nano-particles with high adsorption capacity and photocatalytic degradation properties for xenobiotic organic pollutants, a triple positive impact would be achieved. That is, sustainable solid waste management, wastewater treatment and production of valuable products with different applications.

Globally there is an increased interest in alternative fuels, especially liquid transportation fuels. The production of biofuels in large volumes is now a reality, although there are some concerns about the use of land, water, and crops to produce fuels. Here come the biofuels from agro-industrial wastes, lignocellulosic wastes, waste oils and micro- and macro-algae (Abo-State *et al.*, 2013; Ismail *et al.*, 2018; Aboelazayem *et al.*, 2018; Soliman *et al.*, 2018).

Bioethanol from lignocellulosic waste biomass is one of the most employed liquid biofuels due to the easy adaptability of this fuel to existing engines and because this is a cleaner fuel with higher octane rating than gasoline. Moreover, it is biodegradable and reduces CO_2 emission by more than 78% compared to conventional gasoline. However, the bioconversion of lignocellulosic waste biomass to fermentable sugars which is further fermented to liquid fuels is difficult because of its physio-chemical structural and lignin content which hinder the enzymatic digestibility of cellulose and hemicellulose. To overcome such problem and maximize the yield of fermentable sugars an efficient ecofriendly and economic pretreatment method of such waste biomass is essentially required. Another obstacle to be solved is maximizing the yield of liquid fuels (e.g. bioethanol) via the fermentation of both pentoses and hexoses. Nevertheless, to reach to the point of zero waste and decrease the overall-cost of the process, the valorization of the spent lignocellulosic biomass wastes after extraction of sugars and the production of different valuable products from each process steps are the worldwide focus of research. In another word; full conversion of biomass into liquid biofuels and other valuable products in a market competitive and environmentally sustainable way is now-a-days worldwide mandates (Kheiralla *et al.*, 2018a,b).

Biodiesel is an attractive alternative fuel for diesel engines because it is nontoxic, renewable, and biodegradable and has higher combustion efficiency. Its combustion emission profile is favorable because of the low emissions of CO, CO_2, SOx and particulate matter. However, more than 95% of biodiesel production feedstock comes from edible oils using chemical catalysts for the transesterification process. This adds to the cost of biodiesel, increases the food versus fuel problem and also produces a lot of lignocellulosic waste biomass. However, the valorization of non-edible waste cooking oil into biodiesel via bio-catalysts produced from organic waste biomass is a recommendable manner to meet current and future demand for renewable and sustainable fuel (El-Gendy and Deriase, 2018).

Thus, to catch the wave of social, health, economic and environmental sustainability and the integrated 17 goals of sustainable development, this chapter summarizes some recommendations for sustainable solutions and approaches for mitigation of air pollution and waste management problem. Whereas, organic waste biomass can be valorized into liquid biofuels, into corrosion inhibitors, green biocides, green catalysts, nanoparticles with different applications (e.g. bio-upgrading i.e. biodesulfurization and biodenitrogenation of petroleum and its fractions), activated carbon and other products that can be used in wastewater treatment and disinfection.

SUSTAINABLE BIO-UPGRADING OF PETROLEUM AND ITS FRACTIONS

The worldwide depletion of light and sweet crude oils has forced oil companies to the exploitation of deeper reservoirs and other un-conventional oil reserves, including heavy oil, extra heavy oil, and oil sands and bitumen, which comprise 70% of the world's total oil resources (El-Gendy and Speight, 2015). Those reserves are rich in polynuclear aromatic sulfur and nitrogen heterocyclic (PASHs and PANHs) compounds. The sulfur content in heavy crude oil varies from 0.1% to 15% (w/w) (Shang *et al.*, 2013). Generally, sulfur is typically the third most abundant element in petroleum (Chauhan *et al.*, 2015; Peng and Zhou, 2016). It can vary from 0.04% to 14% (wt.%), according to the type of crude oil, its geographical source and origin (Voloshchuk *et al.*, 2009; ICCT, 2011). For example, the Middle East crude oil is richer in sulfur content than those of others, e.g. Indonesia (Fedorak and Kropp, 1998). However, generally, the sulfur content of petroleum falls in the range 1–4% wt. Petroleum having <1% wt. sulfur are referred to as low-sulfur petroleum and those with >1% wt. as high-sulfur petroleum (El-Gendy and Speight, 2015). The major locations where sweet crude oil is found include; the Appalachian Basin in Eastern North America, Western Texas, the Bakken Formation of North Dakota and Saskatchewan, the North Sea of Europe, North Africa, Australia, and the Far East including Indonesia. While, sour crude oil is more common in the Gulf of Mexico, Mexico, South America, and Canada. Moreover, crude oil produced by OPEC Member Nations also tends to be relatively sour, with an average sulfur content of 1.77%. Table 1 summarizes the sulfur content in crude oils from different countries all over the world.

Table 1: Petroleum-Sulfur Content in Different Countries All Over the World (El-Gendy and Nassar, 2018)

Location	*Sulfur Content (wt.%)*
Argentina	0.06 – 0.42
Australia	0 – 0.1
Canada	0.12 – 4.29
Cuba	7.03
Denmark	0.2 – 0.25
Egypt	0.04 – 4.19
Indonesia	0.01 – 0.66
Iran	0.25 – 3.23
Iraq	2.26 – 3.3
Italy	1.98 – 6.36
Kuwait	0.01 – 3.48
Libya	0.01 – 1.79
Mexico	0.9 – 3.48
Nigeria	0.04 – 0.26
Norway	0.03 – 0.67
Russia	0.08 – 1.93
Saudi Arabia	0.04 – 2.92
United Kingdom	0.05 – 1.24
USA	0.05 – 5
Venezuela	0.44 – 4.99

Most of the sulfur present in crude oil is organically bound sulfur, where, more than 200 organosulfur compounds have been identified in crude oils and can be summarized in four groups; as cyclic and non-cyclic compounds, such as mercaptans (R-S-H), sulfides (R-S-R'), disulfides (R-S-S-R'), sulfoxides (R-SO-R') where R and R' are aliphatic or aromatic groups and thiophenes (i.e. thiophene, benzo-thiophene, dibenzothiophene and their derivatives) with hydrogen sulfide and elemental sulfur (Bahuguna *et al.*, 2011). Thiophenes compromise from 50% to 95% of the sulfur compounds in crude oils and its fractions (Mohebali and Ball, 2016). Moreover, one, two and three sulfur atom containing compounds reach up to 74, 11 and 11% in heavy crude oils, respectively. Mercaptans content in crude varies from 0.1 to 15% mass from total content of sulfur compounds (Ryabov, 2009). The sulfur content increases with the boiling point during distillation and sulfur concentration tends to increase progressively with increasing carbon number (Figure 1). Thus, crude fractions in the fuel oil and asphalt boiling range have higher sulfur content than those in the jet and diesel boiling range, which in turn characterized by higher sulfur content than those in the gasoline boiling range. However, the middle distillates may actually contain more sulfur than those of the higher boiling fractions as a result of decomposition of the higher molecular weight compounds during distillation (El-Gendy and Speight, 2015).

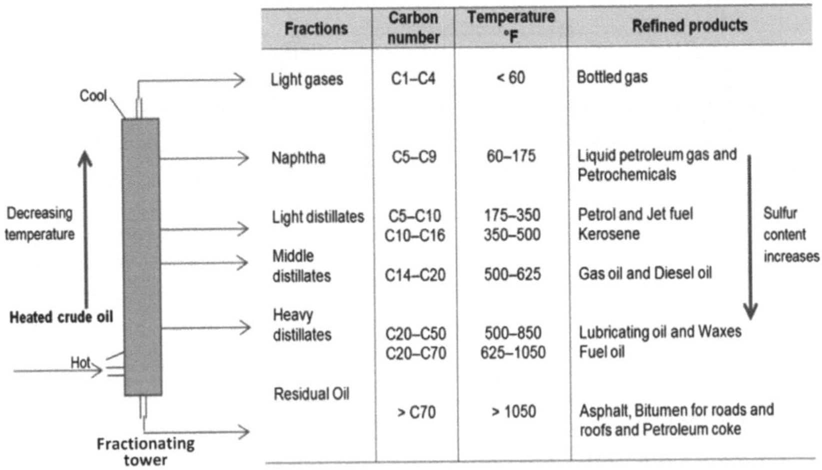

Figure 1: Distribution of Sulfur Content in Different Petroleum Distillates

The total nitrogen content of crude oils averages from 0.1 to 0.9% (Speight, 2014) of which the non-basic compounds (e.g. carbazole) comprise approximately 70–75% and the basic ones (e.g. quinoline, Qn) comprise 25–30% (Laredo *et al.*, 2002). In Egyptian crude oils, CAR and its derivatives concentrations range from 0.3 to 1% (Bakr, 2009). Oil produced from shale deposits has higher concentrations of N-compounds (0.5–2.1%) and the waste materials derived from the processing of shale oil are heavily contaminated by polyaromatic nitrogen heterocyclic compounds (PANHs) (Tissot and Welte, 1984). The nitrogen content in light cycle oil (LCO) and coker gas oil (CGO) are higher than that in straight run gas oil (SRGO) (Adam *et al.*, 2009).

Organo-sulfur and nitrogen compounds in crude oil increase its viscosity making it non-amenable to the refinery process. They have actual or potential corrosive nature, which would cause the corrosion of pipelines, pumping and refining equipment. Moreover, fire would occur upon the sudden rupturing and oil and gas leakage, due to such corrosion. The annual cost of corrosion worldwide is reported to be approximately US$ 3.3 trillion (Lateef *et al.*, 2019). Sulfur and nitrogen compounds are undesirable in refining processes as they tend to deactivate i.e. poison the catalysts used in downstream processing and upgrading of hydrocarbons (Bhadra and Jhung, 2019). Presence of sulfur in motor fuels have critical harming effect on the catalytic converters in the motorized engines, increasing the combustion related emissions i.e. the particulate matters, CO, CO_2, SOx, NOx. That consequently, increase global warming, air and water pollution (Srivastava, 2012). Moreover, the inhalation of sulfurous gases might cause lung cancer and other cardiac-related diseases such as bronchitis and asthma (Lateef *et al.*, 2019). The presence of sulfur in lubricating oils lowers the resistance to oxidation and increases the solid deposition on engine parts. Furthermore, sulfur compounds have unfavorable influence on antiknock and oxidation characteristics. High concentration of sulfur in fuels dramatically decreases the efficiency and lifetime of emission gas treatment systems in cars (Mužic and Sertić-Bionda, 2103).

Incomplete combustion of fossil fuels causes emission of aromatic sulfur and nitrogen compounds, oxidation of these compounds in the atmosphere would lead to the aerosol of sulfuric and nitric acids. For example, NO_X emission is significantly increased by 66%, corresponding to an increase in sulfur content of gasoline from 40 to 150 mg/L. It has been reported that approximately, 73% of the produced SO_2 is from anthropogenic origin and is due to the combustion of petroleum and its derivatives. The NO_X and CO_2 are thought by many to be the primary causes of "chemical smog" as well as "greenhouse gas" accumulation. It has been also reported that, sulfur is the main cause of emissions of particulate matter (PM). All of those aforementioned harmful emissions affect the stratospheric ozone, increasing the hole in the Earth's protective ozone layer (Larentis *et al.*, 2011). It has been reported that approximately 2% sulfur in diesel fuel can be directly converted to PM emissions. Whereas, the PM and SOx are known to be carcinogenic.

The visible, dark black component of smoke is carbon that has incompletely burned. The soot, resulted from the use of the lower quality of fuel (i.e. less refined one), contains large amounts of the mutagenic and carcinogenic polyaromatic hydrocarbons (PAHs). Diesel exhaust is considered to be the most carcinogenic exhausts and accounts for approximately 25% of all smoke and soot in the atmosphere. Moreover, upon the emissions of SO_2 and NO_2 in the atmosphere they react with hydrogen producing the weak sulfurous acid, strong sulfuric acid and nitric acid which are the main cause of the acid rain and haziness that reduces the average temperature of the affected area and leads to climate change. About 25% of the acidity of rain is accounted to the presence of nitric acid (HNO_3) and approximately 75% is related to the presence of sulfuric acid (H_2SO_4). Acid rains have many negative impacts on ecosystem and environment, for example; causes soil pollution, destroys green area, kills forests and damage crops, leather, cars, and buildings. It also poisons lakes and rivers leading to a devastating effect on their fauna and flora and falling in fish population. Also the presence of high levels of sulfate in water affects the human health as it causes diarrhea and dehydration. Acid rains cause degradation of many soil minerals produces metal ions that are then washed away in the runoff, causing the release of toxic ions, such as Al^{3+}, into the water streams; moreover, loss of important minerals, such as Ca^{2+}, from the soil, which would kill trees and damage crops and cause solid erosion. Acid rains have also negative impact on building and monuments.

$$2NO\,(g) + O_2\,(g) \longrightarrow 2NO_2\,(g)$$
$$3NO_2\,(g) + H_2O \longrightarrow 2HNO_3\,(aq) + NO\,(g)$$
$$2NO_2\,(g) + H_2O \longrightarrow HNO_3\,(aq) + HNO_2\,(aq)$$
$$2SO_2\,(g) + O_2\,(g) \longrightarrow 2SO_3\,(g)$$
$$SO_2\,(g) + H_2O \longrightarrow H_2SO_3\,(aq)$$
$$SO_3\,(g) + H_2O \longrightarrow H_2SO_4\,(aq)$$
$$H_2SO_4\,(aq) + CaCO_3\,(s) \longrightarrow Ca^{2+}\,(aq) + SO_4^{2-}\,(aq) + H_2O + CO_2$$

Since 1979, Canada, Japan, the United States, and the European nations have signed several agreements to reduce and monitor emissions of SO_2. Most of these agreements targeted transport fuels because they are one of the important sources of SO_2. For example, the US-EPA and other worldwide regulatory agencies, have limited the sulfur content in the transportation fuels to; 15 and 30 mg/L for diesel oil and gasoline, respectively. However, in most of the developed countries, the acceptable sulfur level in gasoline has been restricted to be <10 mg/L. Several other countries around the world are moving forward with lowering the maximum fuel sulfur content. This is particularly true in the Middle East, Russia, South Africa and some countries in Latin America. Saudi Arabia and Russia have put the limitation of 10 mg/L gasoline by 2013 and 2016, respectively. South Africa has agreed to enforce 10 mg/L gasoline by 2017. However, a much higher level (350 mg/L) is permitted for transportation fuels in Jordan and some other countries. European Fuel Quality Directive has required the on-road and off-road diesel fuel sulfur content to be set at 10 mg/L since 2011. But, in Africa, the average sulfur content is in the range of 2,000 to 3,000 mg/L for on-road diesel, and much higher for off-road one. While, South Africa planned to reach to 10 mg/L fuels by 2017. Briefly, with the exception of Africa, all regions worldwide, are projected to reach an average on-road sulfur content of < 20 mg/L by 2035 (Shang *et al.*, 2013; Martinez *et al.*, 2015; Agarwal *et al.*, 2016; Al-Degs *et al.*, 2016). The projected sulfur content in gasoline and diesel oil for 2012–2035 are shown in Table 2 Moreover, there are stringent regulations that limited the N-content in diesel fuel from > 70 mg/L in 2003 to < 0.1 mg/L in 2010.

Consequently, refineries must have the capability to remove sulfur and nitrogen from crude oil and refinery streams to the extent needed to mitigate these unwanted effects. Moreover, international cooperation is

Table 2: Expected Regional Gasoline and Diesel Sulfur Content (OPEC, 2013)

Location	*S-Content in Gasoline mg/L*					
	2012	*2015*	*2020*	*2025*	*2030*	*2035*
US and Canada	30	30	10	10	10	10
Latin America	520	255	130	45	30	20
Europe	13	10	10	10	10	10
Middle East	605	235	75	25	16	10
FSU	315	115	35	20	12	10
Africa	795	493	245	165	95	65
Asia-Pacific	205	130	65	35	20	15

Location	*S-content in On-Road Diesel mg/L*					
	2012	*2015*	*2020*	*2025*	*2030*	*2035*
US and Canada	15	15	15	10	10	10
Latin America	1085	440	185	40	35	20
Europe	13	10	10	10	10	10
Middle East	1725	415	155	70	20	10
FSU	440	175	60	15	10	10
Africa	3810	2035	930	420	175	95
Asia-Pacific	400	200	100	45	25	15

required to limit and control these harmful emissions. It is believed that refining industries will spend about US$ 37 billion on new desulfurization equipment and an additional US$ 10 billion on annual operating expenses during the next decade to meet the new sulfur regulations. There are various reported desulfurization methods to remove sulfur from fossil fuels. The energy-intensive and high operational cost hydrodesulfurization (HDS) is currently considered as the most important one. HDS is currently used for the bulk desulfurization of approximately 50 million gallons of gasoline/d (daily gasoline consumption in the U.S.). But, it requires expensive catalysts, high temperature and pressure to desulfurize the recalcitrant PASHs compounds and produces the life-threatening H_2S gas and greenhouse gas emissions (Lateef *et al.*, 2019). Desulfurization by ionic liquids (ILs), selective adsorption or reactive adsorption (RA), oxidative desulfurization (OD) and biological methods have also shown good potential to be a possible substitution for HDS technology or to be used in line with HDS in future refining systems (Srivastava, 2012; Mužic and Sertić-Bionda, 2013; Kitashov *et al.*, 2019; Lateef *et al.*, 2019; Lee and Valla, 2019; Li *et al.*, 2020).

Among the new process concepts, design approaches for ultra-deep desulfurization and denitrogenation focus on:

- *Adsorption and sulfur atom extraction:* Remove sulfur by using reduced metals to react with sulfur to form metal sulfides at elevated temperatures under hydrogen atmosphere without hydrogenation of aromatics.
- *Selective adsorption of sulfur or nitrogen compounds:* Remove sulfur or nitrogen by selective interaction with sulfur or nitrogen compounds in the presence of aromatic hydrocarbons under ambient or mild conditions without hydrogen.
- *Oxidation and extraction:* Oxidize sulfur or nitrogen compounds by liquid-phase oxidation reactions with or without ultrasonic radiation, followed by separation of the oxidized sulfur or nitrogen compounds.
- *Biotransformation via biodesulfurization (BDS) and biodenitrogenation (BDN):* is a process based on selective removal of sulfur or nitrogen by microorganisms, either with a natural capacity for it (as those of the genus *Rhodococcus* and *Pseudomonas*) or genetically modified to express the enzymes needed for BDS and/or BDN pathways.
- Moreover, to overcome the disadvantages of the oxidative, adsorptive and extractive desulfurization and/or denitrogenation for example, the decrease in the calorific value of the fuel and the generation of waste streams with concentrated sulfur and nitrogen compounds. Such waste streams could be subsequently treated using BDS and BDN to return the desulfurized and denitrogenized hydrocarbons into the fuels or to produce valuable compounds (Kilbane and Stark 2016).

Biodesulfurization

The most challenging goal for the industrial development of biodesulfurization (BDS) technology is the catalytic removal of recalcitrant organosulfur compounds, e.g. dibenzothiophene and its derivatives,

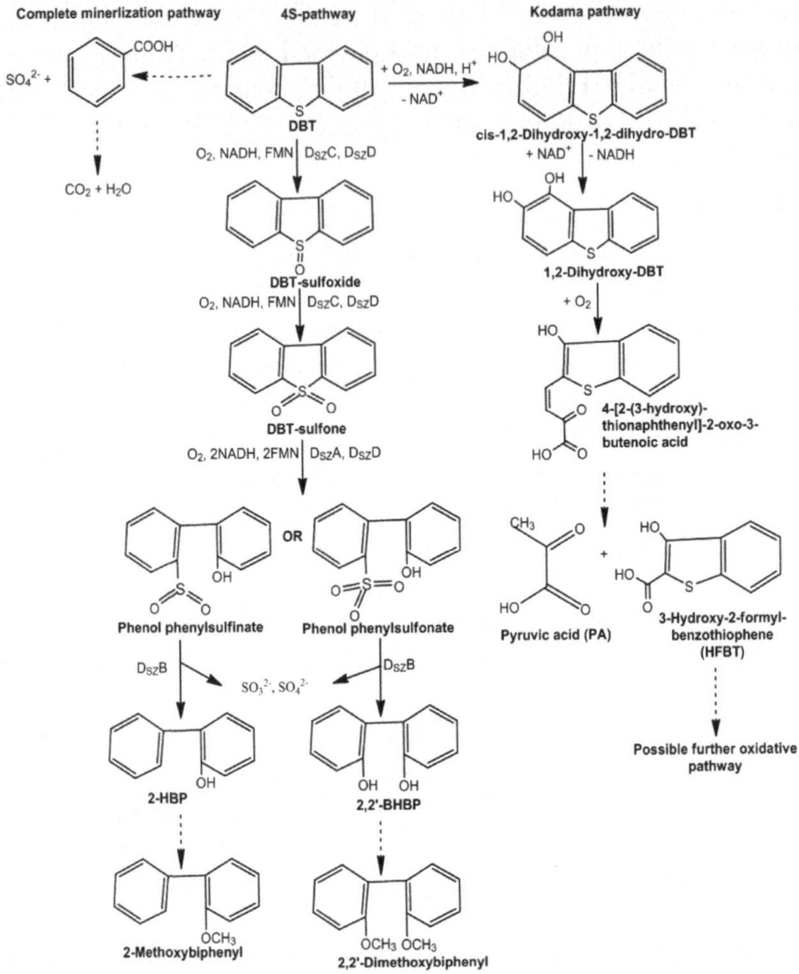

Figure 2: Aerobic Pathways for Biodesulfurization

using microbial cells or enzymes under aerobic or anaerobic conditions. Complete mineralization and Kodama pathways are not recommended for desulfurization of fuels, as carbon skeleton oxidation reduces fuel value (Figure 2).

Anaerobic biodesulfurization has been reported to be more slowly than aerobic one. However, it desulfurizes the fuels producing hydrogen sulfide same as the HDS process (Figure 3). The produced H_2S gas can be treated with existing refinery desulfurization plants (e.g. Claus process). Bio-oxidation of hydrocarbons to undesired compounds such as colored and gum forming products is minimal.

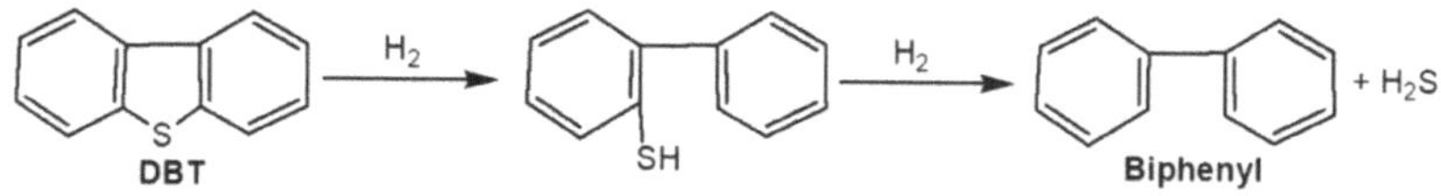

Figure 3: Anaerobic Biodesulfurization

The first report about anaerobic biodesulfurization was by year 1953 (Zobell, 1953). The main reported anaerobic biodesulfurizing micro-organisms are; *Desulfovibrio* sp., and *Thiobacillus* sp. (Eckart *et al.*, 1981; 1982; 1986; Kohler *et al.*, 1984; Pifferi *et al.*, 1990; Kim *et al.*, 1990a,b; 1995). Also, some hyperthermophiles have been reported; *Pyrococcus furiosus* (Tilstra *et al.*, 1992*), Nocardioform actinomycete* FE9 (Finnerty, 1993), *Desulfomicrobium scambium* and *Desulfovibrio longreachii* (Yamada *et al.*, 2001). *Desulfobacterium indolicum* anaerobically desulfurized kerosene with sulfur content of 48.68 mg/L to 13.76 mg/L within 72 h (Aribike *et al.*, 2008). In another study 82% of total sulfur in diesel oil has been anaerobically removed by *Desulfobacterium anilini* (Aribike *et al.*, 2009). Agarwal and Sharma (2010) reported the application of a three-step biodesulfurization–oxidative desulfurization-reactive adsorptive desulfurization (BDS-ODS-RADS) process under both aerobic and anaerobic conditions using heavy crude oil (1.88% S) and light crude oil (0.4% S). Biodesulfurization of both crudes using *Pantoea agglomerans* D23W3 resulted in 61% and 63% S removal under aerobic conditions, while, under anaerobic conditions, 63% and 69% desulfurization was achieved with heavy oil. The second step of the process, oxidative desulfurization, brought total sulfur removal of 94%, with the third step, reactive adsorptive desulfurization, brought total removal of 95%. The use of the thermophilic *Klebsiella* sp. 13T resulted in 62% S removal from light crude oil and 68% S removal from heavy crude oil during the first treatment stage. Kareem *et al.* (2012) reported biodesulfurization of diesel by *Desulfobacterium indolicum* resulted in reduction of sulfur from 166.037 mg/L to 33.412 mg/L within 72 h. *Desulfomicrobium scambium*, *Desulfovibrio longreachii* and *Desulfatiglans aniline* comb. nov., have been also reported for anaerobic biodesulfurization of kerosene (Srivastava, 2012; Kareem *et al.*, 2016).

However, due to low reaction rates, safety and cost concerns, and the lack of identification of specific enzymes responsible for anaerobic desulfurization and anaerobic microorganisms effective enough for practical petroleum anaerobic desulfurization have not been yet found, anaerobic biodesulfurization processes have not been developed up till now. Consequently, aerobic biodesulfurization has been the focus of most of the research in biodesulfurization (Mohebali and Ball, 2016).

Due to the milder and safer process conditions of aerobic biodesulfurization, significant reductions in greenhouse gas emissions and energy requirements were estimated as compared with HDS-processes. Moreover, the capital costs to set-up a biodesulfurization-process is reported to be 50% lower than that of a HDS-process. The first aerobic biodesulfurization pilot plants were built in the mid-1990s by ENCHIRA Biotechnology Corporation (ENBC) (formerly; Energy Biosystems Corporation) (Singh *et al.*, 2012). Several microorganisms have been reported since that for aerobic BDS of crude oil and its distillates (Table 3). The maximum reduction in sulfur content achieved with different microorganisms using heavy crude oil as the sulfur source are in the range of 47–68% (Setti *et al.*, 1993; EL-Gendy *et al.*, 2006; Torkamani *et al.*, 2008a, b; Agarwal and Sharma, 2010; Bhatia and Sharma, 2010, 2012; Li and Jiang, 2013; Adlakha *et al.*, 2016).

Although BDS is characterized by lower operational cost and produces valuable by-products, however, the main disadvantage of biodesulfurization is its lower reaction rates compared to HDS. In view of industrial application of BDS, cell immobilization is considered to be one of the most promising approaches (Huang *et al.*, 2012). Applying BDS as a complementary for conventional desulfurization processes is another promising technique. It has been predicted that the BDS catalyst must have S-removal capability within the range of 1–3 mmol dibenzothiophene/g dry cell weight/h in real oil feed. Thus, the efficiency of the present biocatalysts is desired to be enhanced by 500-fold to reach a commercially feasible process (Kilbane, 2006). Alves *et al.* (2015) performed a cost analysis study, comparing two BDS process designs; upstream and downstream conventional HDS. The BDS costs and emission estimations were made considering the BDS of dibenzothiophene, as model for S-compounds, while, HDS estimations were made based on crude oil HDS. The BDS downstream HDS configuration is found to be the best alternative to be applied in oil refinery, from the point of lower energy consumption, greenhouse gas emissions and operational costs, to obtain almost S-free fuels. Moreover, there are still some unsolved aspects such as stability,

Table 3: BDS of Real Oil Feed

Bacterium	*Type of Oil*	*Initial Sulfur Concentration*	*BDS Efficiency*	*O/W Phase Ratio*	*References*
Achromobacter sp.	Light gas oil	0.15% (w/w)	1.2 mg S/g DCW/h	1:3	Bordoloi *et al.* (2014)
Achromobacter sp.	HDS-diesel oil	420 mg/L	1.2 mg S/g DCW/h	1:3	Bordoloi *et al.* (2014)
*Arthrobacter sulfureus***	HDS-diesel	170 mg/L	53%	1:3	Labana *et al.* (2005)
Brevibacillus invocatus C19 (accession no. KC999852)	Diesel oil	8600 mg/L	91.31%	25%	Nassar (2015)
*Desulfobacterium aniline****	Diesel	1681 mg/L	82%	1% (v/v)	Aribike *et al.* (2009)
Escherichia coli W3110[R]	Diesel oil	250 mg/L	15%	5% (v/v)	Jae *et al.* (2003)
Gordona sp. CYKS1**	Middle distillate unit feed	1500 mg/L	70%	1:9	Rhee *et al.* (1998)
Gordona sp. CYKS1**	Light gas oil	3000 mg/L	50%	1:9	Rhee *et al.* (1998)
Gordona sp. CYKS1**	Light gas oil	3000 mg/L	35%	1:9	Chang *et al.* (2000)
Gordona sp. CYKS1**	Middle distillate unit feed	1500 mg/L	60%	1:9	Chang *et al.* (2000)
Gordona sp. CYKS1**	Diesel oil	250 mg/L	76%	1:9	Chang *et al.*(2000)
Gordonia sp. SC-10	Diesel oil	3035.3 mg/L	40.93%	1:10	Chen *et al.* (2019)
M. pheli WU-0103*	Light gas oil	1000 mg/L	52%	1:9	Ishii *et al.* (2005)

(Contd...)

(Table 3 contd…)

Bacterium	*Type of Oil*	*Initial Sulfur Concentration*	*BDS Efficiency*	*O/W Phase Ratio*	*References*
Mycobacterium goodie X7B*	Crude oil	3,600 mg/L	59%	1:9	Li *et al.* (2007a)
Mycobacterium pheli WU-F1* [&]**	B-light gas oil	350 mg/L	74%	1:1	Furuya *et al.* (2003)
Mycobacterium pheli WU-F1* [&]**	F-light gas oil	120 mg/L	65%	1:1	Furuya *et al.* (2003)
Mycobacterium pheli WU-F1* [&]**	X-light gas oil	34 mg/L	56%	1:1	Furuya *et al.*(2003)
Mycobacterium phlei SM120-1	Light gas oil	224 mg/L	66%	2% (v/v)	Srinivasaraghavan *et al.* (2006)
Mycobacterium sp. X7B*	HDS-diesel	535 mg/L	0.14 mg S/g DCW/h	1:9	Li *et al.* (2003)
Nocardia globerula R-9**	diesel oil	1807 mg/L	5.1 mmol S/kg/h	1:8	Mingfang *et al.* (2003)
Nocardia sp. strain CYKS2	Light gas oil	0.3 wt.%	0.992 mg S/L-dispersion/h	1/20	Chang *et al.* (1998)
Pantoea agglomerans D23W3	Light crude oil	3800 mg/L	71%	1:9	Bhatia and Sharma (2010)
Pantoea agglomerans D23W3	HDS-diesel	70 mg/L	26%	1:9	Bhatia and Sharma (2010)
Pantoea agglomerans D23W3	Diesel	150 mg/L	38%	1:9	Bhatia and sharma (2010)

(Contd…)

(Table 3 contd...)

Bacterium	*Type of Oil*	*Initial Sulfur Concentration*	*BDS Efficiency*	*O/W Phase Ratio*	*References*
Pantoea agglomerans D23W3	Heavy crude oil	2.61 wt.%	39%	1:9	Bhatia and sharma (2010)
Pseudomonas delafieldii R-8**	HDS-diesel	591 mg/L	91%	1:1	Guobin *et al.* (2005)
Pseudomonas delafieldii R-8	HDS-diesel	591 mg/L	47%	1:1	Guobin *et al.* (2005)
R. erythropolis ATCC 4277	Heavy gas oil	6500 mg/L	148 mg S/kg/h	40% (v/v)	Maass *et al.* (2015)
Rhodococcus erythropolis IGTS8	Diesel oil	8600 mg/L	74.42%	10%	Nassar (2015)
R. erythropolis I-19	HDS-middle distillate 1850	1850 mg/L	67%	1:9	Folsom *et al.* (1999)
R. erythropolis LSSE81[R]	Diesel oil	261.3 mg/L	73%	1:2	Xiong *et al.* (2007)
R. erythropolis NCC1	HDS-diesel oil	554 mg/L	51%	1:9	Li *et al.* (2007b)
R. erythropolis XP**	Fushun crude oil	3210 mg/L	62%	1:9	Yu *et al.* (2006a)
R. erythropolis XP**	Sudanese crude oil	1237 mg/L	47%	1:9	Yu *et al.* (2006a)
R. erythropolis XP**	Straight run gasoline	50.2 mg/L	85%	1:9	Yu *et al.* (2006a)
R. erythropolis XP**	Jilian FCC gasoline	1200 mg/L	30%	1:9	Yu *et al.* (2006a)
R. erythropolis XP**	HDS-diesel oil	259 mg/L	95%	1:9	Yu *et al.* (2006a)
R. globerulus DAQ3	Diesel oil	12600 mg/L	12%	1:4	Yang *et al.* (2007)

(Contd...)

(Table 3 contd…)

Bacterium	*Type of Oil*	*Initial Sulfur Concentration*	*BDS Efficiency*	*O/W Phase Ratio*	*References*
Rhodococcus erythropolis FSD2	HDS-diesel	666 mg/L	0.78 mg S/g DCW/ h	1:5	Zhang *et al.*, 2007a
Rhodococcus erythropolis FSD2	HDS-diesel	198 mg/L	0.26 mg S/g DCW) h	1:5	Zhang *et al.*, 2007a
Rhodococcus sp.**	HDS-diesel	170 mg/L	50%	1:3	Labana *et al.* (2005)
Rhodococcus sp. (NCIM 2891)	HDS-diesel	430 mg/L	96%	1:4	Bandyopadhyay *et al.* (2013)
Rhodococcus sp. ECRD-1	Light cycle oil	669 mg/L	92%	2% (v/v)	Grossman *et al.* (2001)
Rhodococcus sp. ECRD-1	Middle distillate fraction of Oregon Basin crude oil	20,000 mg/L	8%	1/10	Grossman *et al.* (1999)
Rhodococcus sp. IMP-S02	Diesel	500 mg/L	60%	5% (v/v)	Castorena *et al.* (2002)
Rhodococcus sp. P32C1**	HDS-diesel	303 mg/L	49%	25% (v/v)	Maghsoudi *et al.* (2001)
Sphingomonas subarctica T7B	LGO	280 mg/L	41%	1:5	Gunam *et al.* (2006)
Stachybotrys sp. WS4	Heavy crude oil	5 wt.%	76%	1/20	Torkamani *et al.* (2008a)

* Thermophilic ** Resting *** Anaerobic [R] recombinant

lifetime, inhibition effects, high costs of biocatalyst, reactor design, quantity of water in the media, separation of aqueous organic phases, low enzyme activity, low mass transfer rates and biocatalyst recovery. Also, large amounts of biomass are needed (typically 2.5 g biomass per g sulfur), and biological systems must be kept alive to function, which can be difficult under the variable input conditions found in refineries (Nehlsen *et al.*, 2005).

Nevertheless, the cost of the culture medium represents 30–40% of the total amount (El-Gendy and Nassar, 2018). Moreover, the desulfurization rate stops before the complete removal of sulfur compounds, as growing cells may be deactivated by the accumulation of 2-hydroxybiphenyl (2-HBP) (Irani *et al.*, 2011). For a cost effective BDS process, the reaction time and catalyst longevity should be of 1 and 400 h, respectively (Pacheco *et al.*, 1999). For commercialization, the desired rate of BDS was reported to should be 20 μmol DBT/min/g DCW (Nazari *et al.*, 2017), 3 mM S/g DCW/h (Singh, 2015), and in another reports 1.2–3 mM/g DCW/h (Srivastava and Kumar, 2008). Whereas the maximum rate achieved till date is 320 μM S/g DCW/h (Kilbane, 2006; Singh, 2015). El-Gendy and Nassar (2018) recommended different aspects to overcome such drawbacks. For example; (1) Isolation of new biodesulfurizing microorganism with the capabilities of utilizing wide range of organosulfur compounds; (2) Isolation of thermophilic microorganisms to be applied as a downstream process after HDS and safe time and energy consumed in cooling the feed to the BDS process; (3) Isolation of halotolerant microorganisms to decrease the requirement of fresh water; (4) Isolation of more hydrophobic microorganisms to overcome the problem of oil/water phase ratio and desulfurize more amount of feed per cycle; (5) Isolation of new biodesulfurizing microorganisms overcoming the feedback inhibition caused by 2-hydroxybiphenyl, 2, 2'-bidhydroxybiphenyl and sulfate ions; (6) Apply genetic engineering to increase the copy number of desulfurization genes and/or producing recombinant microorganisms with hydrocarbons and inhibitors tolerance and higher BDS efficiency; (7) Applying biodesulfurizing microorganisms with the capability of producing biosurfactants instead of using chemical surfactants, to overcome the problem of mass transfer, increase the contact between the cells and the organosulfur compounds and lower the operational cost; (8) Immobilization and coating cells with nanoparticles to enhance the rate of BDS, increase the lifetime of the biocatalyst and solve the problem of biocatalyst recovery; (9) Applying statistical optimization and mathematical modeling to lower the cost of process optimization; (10) Finally, producing valuable product during the BDS process.

For example, to overcome the problem of inhibition effect caused by sulfate ions, 2-HBP and its alkylated forms, it can be suggested to stop the reaction of the 4S-pathway (Figure 2) before the production of such inhibitors, at the step of producing hydroxybiphenyl sulfinate (HPBS). This compound is reported to have hydrotrope properties and to be effective as a detergent when derivatized with a long chain alkyl side chain (Johnson *et al.*, 2000). Thus it can be an integrated recommendable technique for the short-term application of bio-technology for producing 'biopetrochemicals' from oil as a spin-off of the BDS process. Moreover, that produced biosurfactants (Figure 4) can decrease the interfacial tension, and can be used in enhanced oil recovery (EOR). Another way to overcome the inhibition effect of such phenolic compounds is the isolation of bacteria capable of methoxylation of such phenols, for example; *Mycobacterium* sp. (Li *et al.*, 2003; Chen *et al.*, 2009), *Rhodococcus* sp. KT462 (Tanaka *et al.*, 2002), *Sphingomonas subarctica* T7b (Gunam *et al.*, 2006), *Rhodococcus erythropolis* FSD2 (Zhang *et al.*, 2007a), and *Rhodococcus erythropolis* strain HN2 (El-Gendy *et al.*, 2014a).

To save time, effort, energy and cost, response surface methodology based on different factorial designs of experiments has been applied to study the interactive effects of different parameters affecting the BDS process and to predict the optimum parametric levels to enhance the BDS process (Deriase *et al.*, 2012; El-Gendy *et al.*, 2014a).

Deriase *et al.* (2012) found that one of the important factors affecting biodesulfurization and/or degradation of sulfur compounds is the constituents of the media. Yeast extract for example is an excellent nitrogen source and contains all the metal ions and required micro nutrient for bacterial growth and enzymes enrichment (Sagardia *et al.*, 1975; Hirano *et al.*, 2004; Kim *et al.*, 2004; Yu *et al.*, 2006a; Deriase *et al.*, 2012; El-Gendy *et al.*, 2014a). The presence of relatively high concentrations of some volatile sulfur compounds e.g. dimethyl sulfoxide would inhibit the microbial growth and consequently the BDS process (Deriase *et al.*, 2012). Although of the reported inhibiting effect of sulfate ions on the BDS efficiency as it acts as a readily available source for the sulfur microbial requirements for cell metabolism and enzymes performance (Soleimani *et al.*, 2007). However, the presence of magnesium sulfate with certain concentrations is reported to enhance the biodesulfurization process (Deriase *et al.*, 2012).

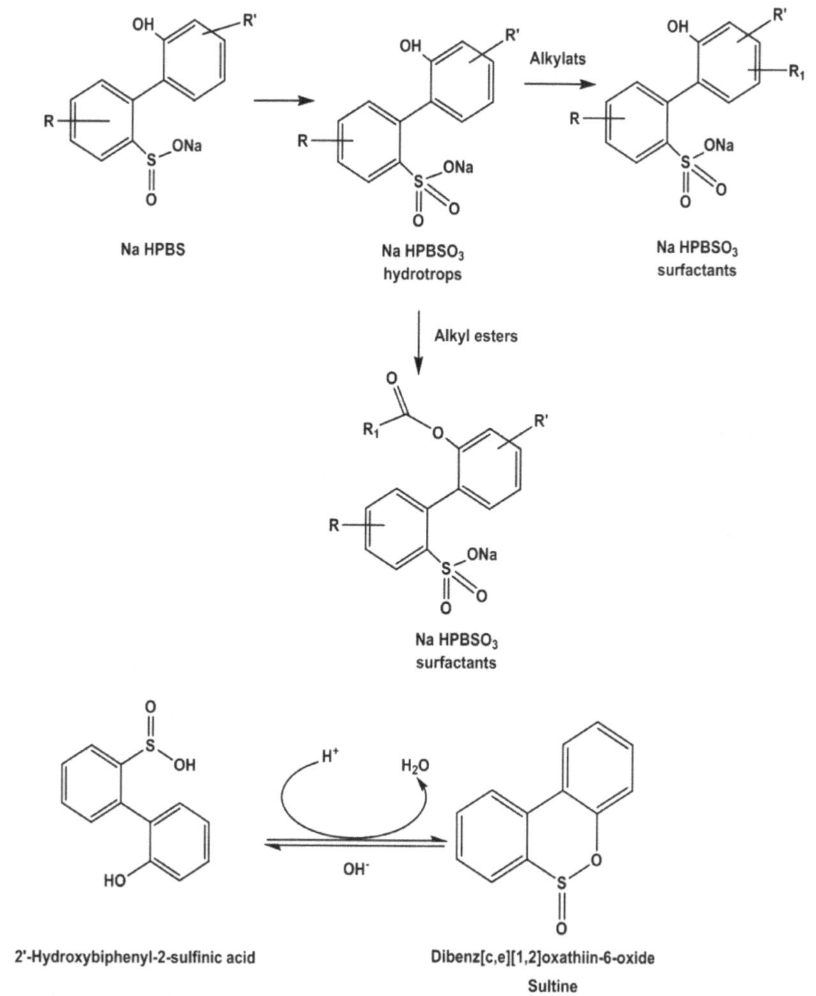

Figure 4: Conversion of the BDS Intermediate hydroxybiphenyl Sulfinate and its Alkylated Derivatives to Value-Added Surfactants

Several statistical model equations (Gaussian, polynomial and exponential functions) were investigated to assess the relationships between the variable operating conditions and both cell growth and

% BDS of *Rhodococcus erythropolis* HN2 (Nassar *et al.*, 2017a). Whereas, the statistical optimization revealed the occurrence of approximately, 80% BDS-efficiency of 1000 mg/L DBT in n-hexane, at optimum operating conditions of; pH6.69, reaction temperature 27.47°C, initial inoculum size $OD_{600\ nm}$ 0.1 (i.e. 3.73×10^4 CFU/mL) and mixing rate of 154.19 rpm.

Moreover, the presence of carbon co-substrate is important to enhance the biodesulfurization efficiency and keeping the hydrocarbon skelton of the oil feed intact as much as it could be and enlacing the calorific value of the treated fuel. El-Gendy *et al.* (2014a) proved that the concentration of such carbon co-substrate is a very critical factor in the BDS-process. Further, to gain the benefit of both waste management and cost effective biosulfurization process, the application of agro-industrial wastes and wastes from biofuels industry; recycled paper sludge hydrolysates, carob kibbles pulp liquor, sugar beet molasses, Jerusalem artichoke hydrolysates, cassava wastewater, trub, glycerol, as sources of such carbon co-substrate and other important minerals and nutrients has been reported (Alves *et al.*, 2008; Barros *et al.*, 2013; Silva *et al.*, 2013, 2014a; Alves and Paixão, 2014; Kunze, 2014; Paixão *et al.*, 2016; Porto *et al.*, 2017; Abo-State *et al.*, 2014a; El-Gendy *et al.*, 2014a).

Inhibition phenomenon is often observed when the inhibitory compound is more soluble than other substrate (Nadalig *et al.*, 2002). For example the toxicity thiophene (Th) on *Agrobacterium* MC 501 and mixed culture XACO is higher than dibenzothiophene (DBT) (Constanti *et al.*, 1996). Inhibitory effects of benzothiophene (BT) at higher concentrations were also reported by Van Hamme *et al.* (2004) and Kirkwood *et al.* (2007). Moreover, the methylation was reported to decrease aqueous solubility which might explain the lower toxicity effect of 4, 6-dimethyldibenzothiophene (4, 6-DMDBT) relative to 4-methyldibenzothiophene (4-MDBT) (Kirkwood *et al.*, 2007). Nassar *et al.* (2013) and El-Gendy *et al.* (2015) confirmed that to get a biodesulfurizing microorganism synergistically and efficiently working over a wide range of different sulfur compounds is another important factor.

Also, mathematical modeling and correlations between microbial growth and its biodesulfurization capacity is important for up-scaling of BDS process (Nassar *et al.*, 2016). Deriase *et al.* (2013) proved the importance of kinetic modeling and simulation in the estimation of the degree of toxicity of 2-HBP and 2, 2'-bihydroxybiphenyl (2, 2'-BHBP) on *Corynebacterium variabilis* Sh42 which was previously isolated for its ability to utilize different sulfur compounds. Since the studied HBPs expressed inhibitory effects on Sh42 thus according to (Nuhoglu and Yalcin, 2005), if the substrate is inhibitory, it is not possible to observe an actual maximum specific growth rate (μ_{max}). Upon applying the Haldane model, the inhibition constant (Ki), which is a measure of sensitivity by inhibitory substrates and (Ks), half-saturation constant, which is defined as the substrate concentration at which the specific growth rate (μ) equals to half the maximum specific growth rate μ_{max}, can be calculated. It has been shown that Haldane equation would go through maximal value $d\mu/ds = 0$, at substrate concentrations S* ppm. For the inhibitory substrates, S* is the concentration at which the microorganisms exhibited their maximum utilization rate, $S^* = (K_s.K_i)^{1/2}$ and the corresponding μ value is $\mu^* = \frac{\mu_{max}}{[2(K_S.K_i)\frac{1}{2}+1]}$. This reflects that the degree of inhibition is determined by [Ks /Ki] ratio and not just by Ki alone. The larger the ratio [Ks/Ki] is, the smaller the μ^* (h^{-1}) relative to μ_{max}, and thus the lower the degree of inhibition.

Deriase *et al.* (2013) concluded that, the maximum specific growth rate on 2, 2′-BHBP (0.053 h^{-1}) was greater than that on 2-HBP (0.045 h^{-1}). Considering, the fact that Ks mg/L is inversely proportional to the affinity of the microbial system for the substrate. Thus, Sh42 showed higher affinity to 2-HBP (Ks = 0.894 mg/L) than that of 2, 2′-BHBP (Ks = 1.88 mg/L). The S* for 2-HBP was smaller than that of 2, 2′-BHBP (8.21 and 11.45 mg/L, respectively). The [Ks/Ki] ratio of 2-HBP is smaller than that of 2, 2′-BHBP, recording; 0.01 and 0.027, respectively, and the corresponding μ^* recorded; 2.58×10^{-3} and 2.22×10^{-3}, respectively. The obtained data indicated that the toxicity and inhibition effects of 2-HBP on Sh42 are higher than those of 2, 2′-BHBP (Deriase *et al.*, 2013).

In another study, the logistic model equation was found to be accurately describes the change in *Rhodococcus erythropolis* HN2 biomass concentration with time using different initial DBT concentrations (Nassar *et al.* 2017b). Also an estimated kinetic model equation relating the rate of growth of microorganism and the rate of DBT consumption with yield coefficient calculated from Monod equation, allowed the prediction of the DBT-time profile along the microbial growth under the predicted optimal operating conditions. Although, a slight decrease of specific growth rate μ h^{-1} with increase of DBT initial concentrations was observed. However, the specific desulfurization rate was found to be increasing with the increase of initial DBT concentration, recording its maximum at 1000 mg/L and regardless of the initial DBT-concentration the maximum desulfurization rate occurred at 72 h incubation period. That recorded maximum specific growth and degradation rates of 0.06667 and 0.8818 h^{-1}, respectively. Thus that study, proved the strong influence of S-concentration on the rate of BDS, the well adaptation of *R. erythropolis* HN2 towards high DBT concentrations and its ability to overcome the feedback inhibition of sulfate and 2-HBP (Nassar *et al.*, 2017b). That promotes the application of *R. erythropolis* HN2 in petroleum desulfurization process.

Egyptian crude oils are characterized by being dominated with mixtures of substituted benzothiophenes and dibenzothiophenes. Specifically, Suez Gulf oils show higher abundance of dibenzothiophene derivatives relative to dibenzothiophene derivatives, as compared to Western Desert crude oils (Hegazi *et al.*, 2003). Salama *et al.* (2004) reported 45.4% total sulfur removal from Egyptian crude oil (Balayim Barry) by *Rhodococcus rhodochrous* ATCC 53968 (which is later identified as *R. erythropolis* IGTS8). The biodesulfurized crude oil was upgraded in the form of pronounced drop in sulfur content and asphaltenes fraction. In another study performed on Balayim crude oil by *Bacillus subtilis*, 38% of total sulfur removal was recorded (Ibrahim *et al.*, 2004). That was accompanied by a decrease in polynuclear aromatic, phytane, polynuclear naphthene compounds and asphaltenes. But, *B. subtilis* raised paraffins with higher carbon number (C20–30) and isoparaffins (C21–30). Both ATCC 53968 and isolated *B. subtilis* approved the bio-upgrading of Balayim crude oil throughout the concomitant decrease in asphaltene contents recording 36.5% and 22.6%, respectively. IGTS8 was also reported to remove up to 45% of dibenzothiophene moieties from Ras Badran asphaltene fraction and 63% of sulfur content of Ras Badran aromatic fraction (Moustafa *et al.*, 2006). El-Gendy *et al.* (2006) reported biodesulfurization and bio-upgrading of Balayim Mix crude oil (2.74% w/w sulfur) by the halotolerant yeast, *Candida parapsilosis* NSh45, under mild operating conditions. NSh45 reduced the sulfur content by 75%, with a decrease in the average molecular weight of asphaltene constituents by approximately 28%, and dynamic viscosity reduction by approximately 70%, compared to that of *R. erythropolis* IGTS8 which removed approximately 64% of the sulfur, with a concomitant decrease in average molecular weight of asphaltene constituents and crude dynamic viscosity of approximately 24% and 64%, respectively. The main advantage upon applying halotolerant microorganisms is avoiding the use of fresh water and minimizing the requirements from fresh water in refining processes.

BDS usually occurs in organic phases; therefore, the transference of organosulfur molecules to the cell membrane is limited. Nevertheless, it is also well documented that the first and rate-limiting step in the oxidative BDS of DBT is the transferring of DBT from the oil into the cell (Setti *et al.*, 1999). The cell membrane needs to be not only resistant but also permeable enough to capture the sulfur-containing molecules. This issue may be overcome with the use of genetically modified bacteria that are stable in oil phases or by using sorbents of organosulfur molecules (Figure 5). Moreover, in order to combine the advantages of immobilization i.e. ease of separation, biocatalyst recovery, and microbial longevity with those of free diffusion i.e. good mass transport; decorating the bacterial cells with magnetic nanoparticles (MNPs) is a very promising approach. After completion of the reaction, the bacterial cells can be separated from the products using an external magnetic field. This is a much milder and more cost-effective process than centrifugation, and allows the bacteria to

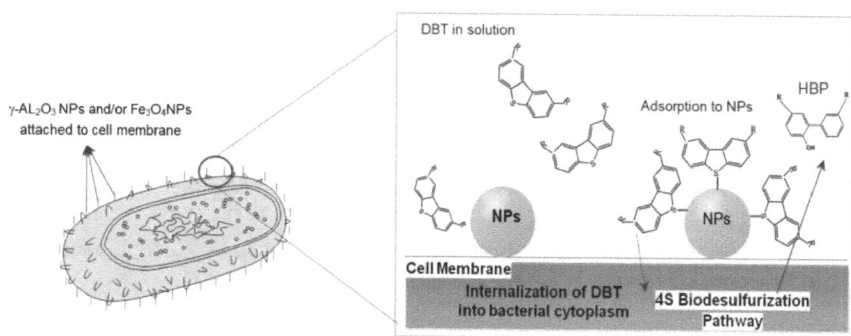

Figure 5: Enhancement of BDS Process by Using Adsorbents NPs Coated Cells

be reused many times and magnetic separation is compatible with any automated platform that can be equipped with a magnet (Ansari, 2008; Li *et al.*, 2009). In recent years, magnetic nanoparticles (MNPs) have been widely used in the field of BDS, because of their large surface-to-volume ratios, superparamagnetic properties, and low toxicities (Bardania *et al.*, 2013). The coating layer of nanoparticles does not change the hydrophilicity of the cell surface. Moreover, the coating layer has negligible effect on mass transfer because the structure of the layer is looser than that of the cell wall. Thus, the coating layer does not interfere with the mass transfer of DBT. The coated cells have good and long stability and can be reused (Shan *et al.*, 2005a; Liu *et al.*, 2010). Magnetic nanoparticles not only allow recovery of bacterial cells but also may have an important role in cell membrane permeabilization, facilitating the transport of DBT from the media to the cytoplasm of biodesulfurizing microorganisms due to the self-assembly of the nanoparticles inside the membrane, which may form pore-like structures that increases the surface conductance (Ansari *et al.*, 2009).

Some other studies, apply the integration of the entrapment immobilization technique with magnetic Fe_3O_4 nanoparticles in the BDS process, using magnetic polyvinyl alcohol (PVA) beads (Liu *et al.*, 2003; Shan *et al.*, 2003; Guobin *et al.*, 2005). Such beads were used for five successive times. Zhang *et al.* (2007b) reported the adsorption of γ-Al_2O_3 NPs onto *Pseudomonas delafieldii* R-8 increased the BDS efficiency from 14.5 mmol/kg/h to 17.8 mmol/kg/h, which was further increased to 25.7 mmol/kg/h by adsorbing the gum Arabic-modified γ-Al_2O_3NPs onto the surfaces of R-8 cells. That is due to the improvement in the dispersion and biocompatibility of γ-alumina nano-particles after modification with gum Arabic, due to the stronger affinity of gum Arabic-modified γ-Al_2O_3 NPs to the cells than the unmodified ones and its lower toxicity to the cells. Zhang *et al.* (2008) reported that, the decrease of DBT concentration is not only caused by the adsorption of Fe_3O_4 to DBT, but also partly by BDS. It is reported that adsorption between Fe_3O_4 and DBT is electrostatics so the adsorption between them is reversible and DBT can be desorbed easily. The desorbed DBT then transfer to cells for BDS. It is faster for cells to obtain DBT by this way, thus 2-HBP production rate is increased. Shan *et al.* (2005b) reported that the *Pseudomonas delafieldii* cells coated with Fe_3O_4 nanoparticles had greater desulfurizing activity and operational stability than those immobilized on Celite. Moreover, the desulfurization rate of *P. delafieldii* R-8 assembled with the nano-sorbent γ-Al_2O_3 was at least two times higher than that of the cells alone (Guobin *et al.*, 2005). Li *et al.* (2009) reported the occurrence of same BDS activity by free and Fe_3O_4 coated *Rhodococcus erythropolis* LSSE8-1 cells. But, the separated magnetic/immobilized cells were successfully reused for BDS over seven batch cycles and they retained at least 80% of its specific desulfurization activity. However, Ansari *et al.* (2009) reported approximately double increment of BDS-efficiency Fe_3O_4 nanoparticles coated *R. erythropolis* IGTS8 cells than the free cells. In another study, the activity of magnetic immobilized *Rhodococcus erythropolis* LSSE8-1-vgb cells assembled with nano γ-Al_2O_3 adsorbents expressed nearly 20% higher BDS efficiency than that of magnetic cells and decreased by less than 10% throughout three recycles (Zhang *et al.*, 2011). Dai *et al.* (2014) improved the BDS process by using the combination of magnetic nano Fe_3O_4 particles with calcium alginate-immobilized *Brevibacterium lutescens* CCZU12-1 cells. Derikvand *et al.* (2014) reported the combination of the γ-Al_2O_3 NPs with alginate immobilized cells could be very effective in BDS process. Nasab *et al.* (2015) improved the desulfurization performance of *Rhodococcus erythropolis* IGTS8 by assembling spherical mesoporous silica nano-sorbents on the surface of the bacterial cells. Moreover, in another study, the Fe_3O_4NPs decorated *Rhodococcus erythropolis* IGTS8 cells had a 15.80% higher desulfurization capacity compared to the non-decorated cells (Karimi *et al.*, 2017). Assimilation of ZnO NPs onto *P. aeroginusa* PTSOX4 and *R. erythropolis* IGTS8 expressed higher desulfurization of approximately 94% and 58% compared to free cells that recorded 68% and 48%, respectively. While Fe_3O_4NPs decorated cells expressed the same BDS efficiency as free cells (Rahpeyma *et al.*, 2017).

Nassar (2015) studied the BDS of diesel oil by using Fe_3O_4 MNPs coated cells, magnetically immobilized cells (alginate + Fe_3O_4 MNPs) and agar immobilized cells of; *Brevibacillus invocatus* C19 and *Rhodococcus erythropolis* IGST8 (Figure 6). BDS efficiency of the diesel oil (total sulfur content 8,600 mg/L) was investigated at different ratio of oil to water (O/W) (10%, 25%, 50% and 75% v/v). GC-FPD analyses revealed 10% and 25% (v/v) phase ratio were considered as the optimum O/W phase ratio of *R. erythropolis* IGTS8 and *Brevibacillus invocatus* C19, respectively. The Fe_3O_4 MNPs coated cells of *R. erythropolis* IGTS8 showed good BDS efficiency (%BDS 78.26%)

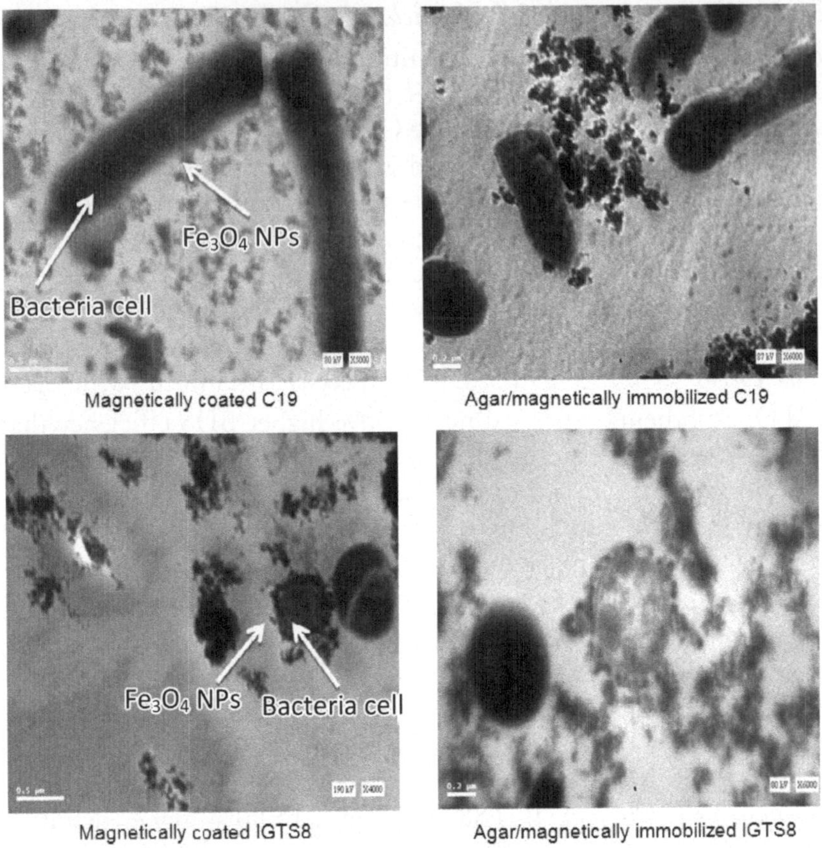

Figure 6: TEM Images, Showing the Well Adsorption of Fe_3O_4 MNPs on the Surfaces of the Biodesulfurizing Cells and Agar Immobilized Cells

in case of 10% (v/v) O/W phase ratio than that occurred by free cells (%BDS 74.42%), magnetic immobilized cells (%BDS 57.26%) and agar immobilized cells (%BDS 58.98%). The agar immobilized cells and coated cells of *Brevibacillus invocatus* C19 showed the highest BDS efficiency (%BDS 98.97% and 98.69%, respectively) at 25% (v/v) O/W phase ratio compared to free cells (%BDS 91.31%) and magnetic immobilized cells (%BDS 89.99%). Moreover, the GC-FPD proved that the BDS of DBT from diesel oil by agar immobilized C19 and coated C19 was much higher, recording; 93.71% and 92.86%, respectively and the BDS of BT was; 91.42% and 88.76%, respectively at 25% v/v optimum O/W phase ratio. While, the BDS of DBT and BT by coated IGTS8 were; 58.63% and 32.98%, respectively at optimum 10% v/v O/W phase ratio. The coated cells of

IGTS8 showed a low capability to desulfurize 4-MDBT (%BDS 25.34%) and 4,6-DMDBT (%BDS 41.56%) from diesel oil compared to coated cells of C19 (%BDS 83.04%) and agar immobilized cells of C19 (%BDS 87.16%). The effect of microbial treatment on the hydrocarbon skeleton of the total resolvable components (TRP) of the diesel oil at different ratio of oil to water was also studied using GC-FID analysis. The biodegradation capacity of total petroleum hydrocarbons (TPH) in diesel oil by agar immobilized C19 and coated C19 was found to be negligible; 10.05% and 11.73% at optimum O/W phase ratio (25% v/v), respectively. While the percent of degradation of TPH occurred by coated IGTS8 was high, recording; 34.98% at optimum O/W phase ratio (25% v/v), respectively. That study proved the importance of isolating new biodesulfurizing strains with efficient biodesulfurization capacity and higher hydrocarbons tolerance.

In an attempt of Egyptian Petroleum Research Institute to catch the wave of application of nanotechnology in the field of petroleum biotechnology, Zaki *et al.* (2013) reported the preparation of the super-paramagnetic FeO_3O_4 NPs (9 nm) with good pore size, volume and high specific surface area; 3.2 nm, 0.198 cm^3/g and 110.47 m^2/g, respectively, using a reverse water/oil micro-emulsion method. Which showed a good assembling on the Gram +ve bacterial isolates; *Brevibacillus invocatus* C19, *Micrococcus* lutes RM1 and *Bacillus clausii* BS1 (Figure 7), with remarkable adsorption capacity to different polyaromatic compounds; DBT (69 μmol/g), Pyrene (7.66 μmol/g) and carbazole (95 μmol/g) through π complexation bonding. The coated cells were characterized by higher DBT biodesulfurization, pyrene biodegradation and carbazole biodenitrogenation rates than the free cells, respectively. Moreover, these Fe_3O_4 NPs-coated cells are characterized by; higher storage and operational stabilities, and low sensitivity toward toxic by-products, and can be reused for four successive cycles without losing its efficiency and have the advantage of magnetic separation, which would resolve many operational problems in petroleum refinery (Nassar, 2015; Saed *et al.*, 2014; Zakaria *et al.*, 2015).

The application of the Fe_3O_4 NPs-coated *Brevibacillus invocatus* C19 in biodesulfurization of diesel oil (1:4 O/W) with initial sulfur concentration of 8600 ppm, showed better efficiency than that of free

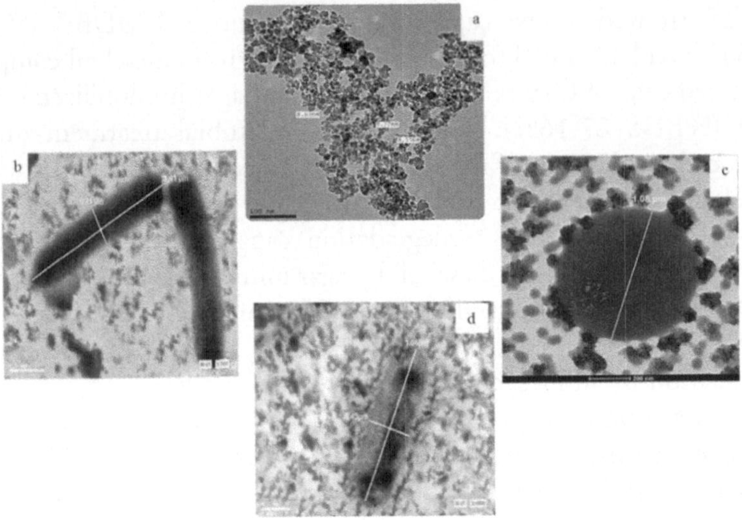

Figure 7: TEM Micro-Graphs of Magnetic Fe_3O_4 (a) and Coated Bacterial Isolates; *Brevibacillus invocatus* C19 (b), *Micrococcus* lutes RM1 (c) and *Bacillus clausii* BS1 (d)

cells, recording; approximately complete removal and 91% within 4 and 7 days of incubation at 35°C and 200 rpm, respectively. Figure 8 illustrates a preliminary bench-scale process design of high-sulfur content diesel oil BDS using magnetic NPs-coated-C19. The estimated capital cost for preparation of magnetic NPs is 3.17 US$/g cell. The summary of the estimated capital and operating costs of free and immobilized cell-systems are listed in Table 4. Taking into consideration that the free cells lost its activity after the first BDS cycle of 7 days, so for four cycles of BDS, the estimated operating and capital costs have been estimated to be 197.64 and 734.08 US$/gallon

Table 4: Estimated Operating Cost for One BDS Batch of High-Sulfur Content Diesel Oil Using Free and Magnetic NPs-coated Bacterial Cells

Fe_3O_4 NPs-coated cells decrease the S-content from 8600 to 112.66 ppm	
Operating cost US$/gallon diesel	Capital cost US$/gallon diesel
93.22	376.63
Free-cells decrease the S-content from 8600 to 747.34 ppm	
Operating cost US$/gallon diesel	Capital cost US$/gallon diesel
49.41	183.52

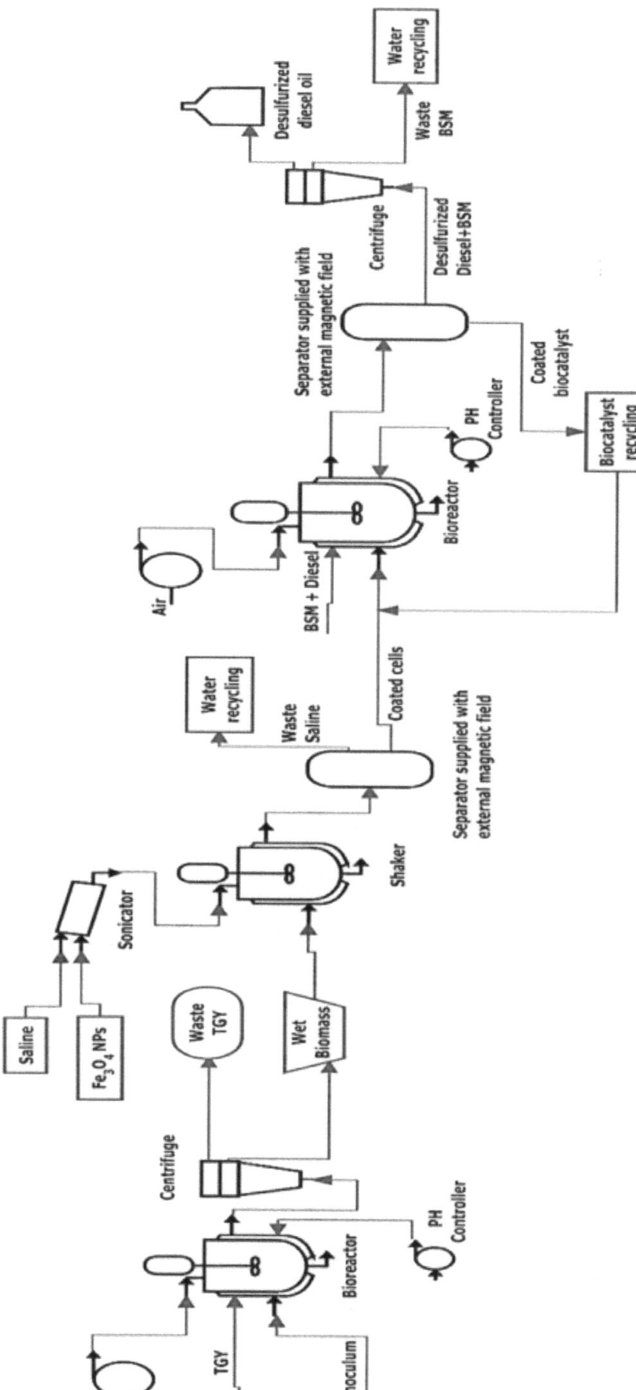

Figure 8: Preliminary Bench-Scale Process Design to Test High-Sulfur Content Diesel Oil Biodesulfurization by Magnetic NPs-Coated Bacterial Cells

diesel, respectively with a lower BDS efficiency. While for the magnetic NPs-immobilized cell system, the immobilized cells have been used in four successive cycles, each of 4 days, without losing its BDS activity and additionally with higher BDS rate (approximately complete removal of sulfur). Thus, the estimated operating and capital costs were estimated to be 163.42 and 682.87 US$/gallon diesel for the four successive cycles. The main cost in this process was found to be mainly coming from the media cost, preparation of magnetic Fe_3O_4 NPs-coated cells; 4.4 US$/g cell and the use of centrifugation for preparation of resting cells and separation of the biodesulfurized diesel oil from the aqueous phase. Thus, further work is suggested to decrease the cost of media by using low cost and readily available source of nutrients and carbon as a co-substrate. Replace the centrifugation step for harvesting the biodesulfurizing bacteria by cyclones. Decrease the cost of magnetic NPs preparation by applying the green synthesis using extract of agricultural wastes.

Biodenitrogenation

The presence of nitrogen in petroleum is of much greater significance in refinery operations than might be expected from the small amounts present. Most of the N-containing compounds in petroleum are currently removed by the conventional chemical and physical refinery processes (El-Gendy and Nassar, 2018). Hydrodenitrogenation (HDN) under high temperature and pressure is the most widely method for the removal of nitrogen polyaromatic heterocyclic compounds from petroleum, but it is expensive, hazardous and needs more severe conditions, more hydrogen and a higher energy consumption (Speight and El-Gendy, 2017). Moreover, the NH_3 produced during the HDN can also inhibit the HDS catalysts. Besides, hydrogenation of aromatic rings, is required prior the attack of the C-N bonds, for nitrogen removal. Nitrogen compounds can be responsible for the poisoning of cracking catalysts, and they also contribute to gum formation in such products as domestic fuel oil (Speight, 2014). Carbazole (CAR) as an example for the non-basic nitrogenous polyaromatic hydrocarbons (NPAHs) can directly impact the refining processes in two ways: (1) during the cracking process, CAR can be converted into basic derivatives, which can be adsorbed to the active sites of the cracking catalysts. (2) It directly inhibits the hydrodesulfurization catalysts. Thus, the removal of CAR and other nitrogen-compounds would significantly increase the extent of catalytic cracking and consequently the gasoline yield. It has been reported that, by 90% reduction in nitrogen-content, a 20% increase in gasoline yield occurs. That has a major economic improvement in low-margin; high volume refining processes (Benedik *et al.*, 1998). Basic N-compounds are more-inhibitory for catalysts than the non-basic ones. But, they can potentially be converted to basic compounds during the refining/catalytic cracking process. Thus, they are also inhibitory to catalysts. Moreover, metals like nickel and vanadium are potent inhibitors for catalysts and in petroleum; these metals are typically associated with N-compounds (Mogollon *et al.*, 1998). Thus, removal of nitrogen is preferable before the HDS process (Singh *et al.*, 2011a).

Although, specific biodesulfurization of petroleum and its distillates has been reasonably investigated, there is a little information about biodenitrogenation of oil feed without affecting its calorific value. It has been estimated that biodenitrogenation of petroleum would be beneficial for deep denitrogenation, where, the classical hydro-processing methods are costly and non-selective (Vazquez-Duhalt *et al.*, 2002). It will also eliminate the contribution of fuel nitrogen to NOx emissions. However, the economics of nitrogen-removal processes are affected by the amount of the associated hydrocarbon lost from the fuel, during the denitrogenation process.

Generally, the currently well-established CAR-biodenitrogenation pathway resembles that of dibenzothiophene-Kodama pathway (Figure 9). That is economically unfeasible due to the loss of the fuel value. However, most of the CAR degrading microorganisms, produce; 2'aminobiphenyl-2,3-diol as the first step in CAR-biodenitrogenation pathway. Thus, recovering of CAR-nitrogen as anthranilic acid (ANA) or 2'aminobiphenyl-2, 3-diol, the less inhibitory to refining catalysts, would solve part of that problem. Since, the entire carbon-content of the fuel is preserved. This can be performed by mutant or recombinant strains. Other pathway would liberate nitrogen from CAR in the form of ammonia (Rhee *et al.*, 1997). Most of the attack is aerobic, but anaerobic degradation has been also noted (Fallon *et al.*, 2010). Zakaria *et al.* (2016) have isolated Gram-positive *Bacillus Clausii* BS1 from an Egyptian coke sample. BS1 showed a higher biodenitrogenation (BDN) efficiency relevant to the well-known biodenitrogenating Gram-negative bacterium strain *Pseudomonas resinovorans*

CA10, recording 77.15 and 60.66% removal of 1000 mg/LCAR with the production of 119.79 and 102.43 mg/L anthranilic acid, and 121.19 and 90.33 mg/L catechol, as by-products, respectively. Several species of the genus *Pseudomonas* for its solvent tolerance, have been isolated that degrade carbazole and its alkyl derivatives. Moreover, other microorganisms have been reported to mineralize non-basic nitrogen-compounds, including species of *Bacillus*, *Sphingomonas*, *Xanthomonas*, *Gordonia*, *Klebsiella*, *Burkholderia*, *Arthrobacter* and *Novosphingobium* (Singh *et al.*, 2011a,b; Zakaria *et al.*, 2016). A thermophilic carbazole-degrading bacteria *Anoxybacillus rupiensis*, that can tolerate up to 80°C, with maximum activity at temperature range 55–65°C has been reported, which would be advantageous for application in real petroleum processing (Fadhil *et al.*, 2014). Larentis *et al.* (2011) reported that Tween 20 increased the dispersion of CAR in the aqueous solution and improved its bioaccessibility, thus enhanced its biodenitrogenation by *Pseudomonas stutzeri*.

From the practical point of view, a dual microbial process for both selective biodesulfurization and biodenitrogenation, with the overcome of the significant technical hurdles, such as; tolerance against solvents, high concentration of nitrogenous compounds, high oil to water ratio, would make microbial refining processes and bio-upgrading of petroleum and its fractions feasible on a large scale (Kilbane, 2006). Duarte *et al.* (2001) in PETROBRAS, the Brazilian oil company, have isolated *Gordonia* sp. strain F.5.25.8 that can utilize dibenzothiophene through the 4S-pathway and CAR as a sole source of S and N, respectively. Where, F.5.25.8 is the first reported strain that can simultaneously metabolize dibenzothiophene and carbazole (Santos *et al.*, 2005). Santos *et al.* (2006) reported that F.5.25.8 can tolerate up to 42°C, which would add to its advantageous in industrial application of biodesulfurization/biodenitrogenation as complementary to hydrotreatment process. Moreover, Kayser and Kilbane (2004) inserted genes encoding amidases downstream the artificial carA operon, to accomplish the cleavage of the final C-N bond and produce biphenyl-2,2,3-triol (Figure 9), that is reintroduced to the fuel, keeping its fuel content. That genetically engineered bacterium decreased the CAR content in a petroleum sample by approximately 95% in 2:10 petroleum/aqueous medium within 16 h using a genetically engineered bacterium. Yu *et al.* (2006b) reported the recombinant *Rhodococcus erythropolis* SN8 expressed good biodesulfurization and biodenitrogenation activities towards a wide range of recalcitrant alkyl CARs and dibenzothiophene in crude oil, in just a one-step bioprocess. Maass *et al.* (2015) reported biodesulfurization/biodenitrogenation of heavy

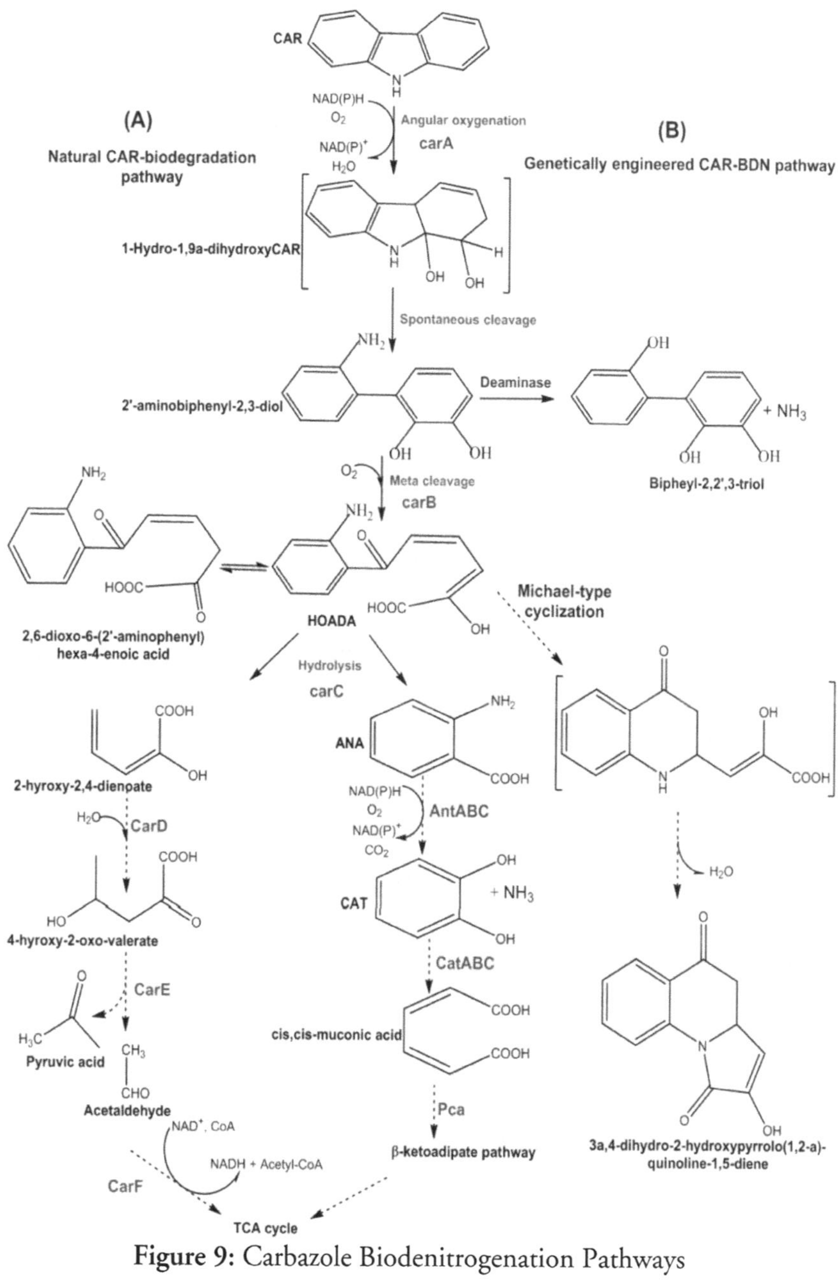

Figure 9: Carbazole Biodenitrogenation Pathways

gas oil (HGO), which is an intermediate fraction obtained from vacuum distillation used in the production of diesel and some lubricants, by *R. erythropolis* ATCC 4277 in a batch reactor. That reached maximum desulfurization and denitrogenation rate of 148 mg S/kg HGO/h and 162 mg N/kg HGO/h at 40% (v/v) HGO/water, respectively. Response surface methodology based on central composite design of experiments was applied to enhance the biodenitrogenation efficiency of *Bacillus clausii* BS1 via the addition of yeast extract and the surfactant Tween 80 (Zakaria *et al.*, 2015a). Whereas, the BDN efficiency increased from ≈ 88% without yeast extract or Tween 80 to ≈ 95% in presence of optimum concentration of 0.868 g/L yeast extract and 0.861% (v:v) Tween 80, which would represent a major economic improvement in low-margin, high-volume refining processes. Tween 80 is known to reduce the concentration of toxic metabolites around the bacterial cells, increasing the biomass concentration and activity (Feng *et al.*, 2006). The CAR-BDN by *Novosphingobium* sp. strain NIY3 (Ishihara *et al.*, 2008) and *Klebsiella* sp. strain LSSE-H2 (Li *et al.*, 2008) was enhanced in the presence of 0.2 g/L and 0.05 g/L yeast extract, respectively. The entrapped *Sphingomonas* sp. XLDN2-5 in a mixture of Fe_3O_4 nanoparticles and gellan gum expressed higher carbazole-biodegradation (3479 μg carbazole/g wet weight cell/h) than the non-magnetically immobilized (1761 μg carbazole/g wet weight cell/h and free (3092 μg carbazole/g wet weight cell/h XLDN2-5 and can be used for eight successive cycles (Wang *et al.*, 2007). Where, the specific degradation rate increased from 3479 to 4638 μg carbazole/g wet weight cell/h in the 8th cycle, due to the good growth of cells in the magnetic gellan gel beads. But, the recorded decrease in carbazole-biodegradation rate within the non-magnetically immobilized matrix is attributed to the mass transfer limitation and steric hindrance. While, the increase in the magnetically immobilized matrix is due to the presence of magnetic nanoparticles which loosen the binding of the sheets of the gellan gum matrix and the existence of many pores between the sheets of gellan gum matrix. Coating *Bacillus clausii* BS1with magnetic Fe_3O_4 NPs enhanced the rate of CAR-BDN (Zakaria *et al.* 2015b). Whereas, complete removal of 1000 mg/L CAR has been achieved by coated cells and 94.25% removal occurred with free cells with $t_{1/2}$ values of 31.36 and 64.78 h, respectively. Coated BS1 cells are characterized by higher, storage and operational stabilities, and low sensitivity toward toxic by-products, and can be reused for four successive cycles without losing its biodenitrogenation efficiency and have the advantage of magnetic separation, which would resolve many operational problems in petroleum refinery (Figure 10). Li *et al.* (2013) and Kafayati *et al.* (2013) reported that the magnetite iron oxide nanoparticles (MNPs) have negligible toxicity on the living bacterial cells. The *Sphingomonas* sp. XLDN2-5 cells/Fe_3O_4 biocomposite and free cells reported to exhibit the same CAR-biodegradation efficiency (Li *et al.*, 2013). That was attributed to the biocompatibility of MNPs, i.e. the coating layer itself, as it does not change the hydrophilicity of the cell surface. Moreover, this coating layer has a negligible effect on mass transfer, as its structure is looser than that of the cell wall. Thus, microbial cell/Fe_3O_4 biocomposite produces a system that is not limited by diffusional limitations. The activity of microbial cell/Fe_3O_4 biocomposite increased gradually during the recycling process. Where complete removal of 3500 μg CAR occurred within 9 h for sixth successive cycles, but the same amount was completely removed in only 2 h within the 7th to the tenth cycles.

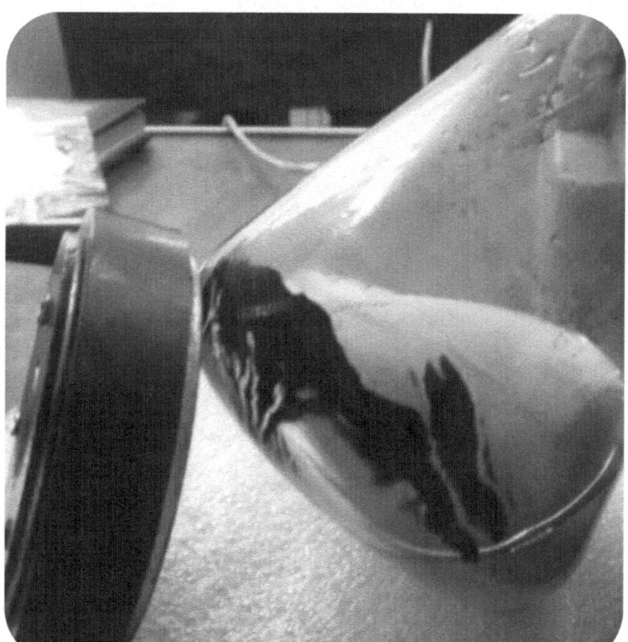

Figure 10: Recovery of Magnetically Coated Microorganisms

SUSTAINABLE BIOREMEDIATION OF ORGANIC XENOBIOTIC POLLUTED WATER

Oil spill can act as a source for global warming and adds to the problem of climate change, as it is considered as an immediate fire hazard, releasing greenhouse gasses and it is also an indirect source of air pollution via the volatile hydrocarbons (e.g. mercaptans) and harmful hydrogen sulfide gases. Moreover, once oil spill occurs, it disorders the food chain and ecosystem (Ouada *et al.*, 2018). Oil pollution directly and indirectly affects human, animals and birds, thus, it severely disturbs biodiversity and has adverse effect on economy and tourism (Ali *et al.*, 2006; El-Gendy and Moustafa, 2007; Raafat *et al.*, 2007; El-Gendy *et al.* 2014b). Oil spills or leakages are the most important sources for environmental contamination with polyaromatic sulfur hetrocyclic (PASHs) compounds. They exhibit some mutagenic and carcinogenic activity. Observations and analyses associated with oil spills and chronic pollution have indicated that PASHs are resistant to microbial degradation, accumulated to a greater degree, and departed more slowly than their polyaromatic hydrocarbons PAHs analogs (Speight and El-Gendy, 2017). For example, Organic sulfur compounds found in crude oil usually serve as marker of oil pollution in fish and shell fish. The sulfur atom imparts some fundamentally different chemical properties compared to hydrocarbon oil constituents. In that, photochemical oxidation leads to sulfur-oxidized compounds (e.g. dibezothiophene sulfoxide DBTO) which are of particular interest due to their toxic properties, or surface active properties (e.g. aromatic sulfonic acids). Despite their wides-pread occurrence and their use as environmental source markers only very little is so far known about the environmental fate of PASHs. Compared to the detailed knowledge of the microbial degradation of the toxic, mutagenic, neurotoxic and carcinogenic PAHs, PASHs have not been investigated intensively. PASHs are mainly found in asphaltenic crude oils, as asphaltene molecules has a high content of O, N and S heteroatoms as well as metals (V, Ni and Fe) (Speight, 2014). Asphaltenes are also considered to be the product of complex heteroatomic aromatic macrocyclic structures polymerized through sulfide linkages (El-Gendy *et al.*, 2006; Ali *et al.*, 2012). Thus, it is important to design a sustainable and feasible bioremediation process with special emphasis on microorganisms capable of degrading the constituents of the asphaltene pollutant. Moreover, it is essential to make this process cost effective as much as it could be. From here comes the idea of applying agro-industrial wastes to achieve a successful bioremediation process. Such wastes can act as a source of required nutrients and minerals (Obayori *et al.*, 2010; Thavasi *et al.*, 2011; Younis *et al.*, 2013; El-Mahdi *et al.*, 2014). It can also act as a self-buffering system, thus maintaining the pH in the biotreatment reactor (Ali *et al.*, 2014; El-Gendy *et al.*, 2014b). Moreover, it can be a natural emulsifier and surfactant, increasing the pollutants availability to the microbial cells and consequently the rate of biodegradation (Silvia *et al.*, 2014b; El-Mahdi *et al.*, 2015a).

El-Gendy (2006) isolated two bacterial isolates *Aureobacterium* sp. and *Enterobacter* sp. that degraded 49% and 36% of 1000 mg/L DBT, via the production of 2-hydroxybiphenyl and 2,2'-bihydroxybiphenyl, respectively, that further degraded to benzoic acid. Raafat *et al.* (2007) isolated halotolerant *Staphylococcus xylosus* which expressed good bioremediation capabilities on two polluted seawater samples with total petroleum hydrocarbon contents of 250 and 65 mg/L, recording 90% and 72%, respectively. El-Gendy and Abo-State (2008) isolated *Staphylococcus gallinarum* NK1 which is capable of producing bio-surfactants upon the biodegradation of different polyaromatic sulfur heterocyclic compounds and polyaromatic compounds. In a further study *Staphylococcus gallinarum* NK1 performed good bioremediation capabilities of polluted seawater sediment samples (El-Gendy *et al.*, 2009). *Corynebacterium variabilis* sp. Sh42 has been isolated for its capability to degrade a mixture of different polyaromatic compounds; antheracene (Ant) and phenantherene (Phe) as model compounds for tri-aromatic ring, pyrene (Pyr) as a model compound for four-aromatic ring compounds, dibenzothiophene (DBT), 4-methyldibenzothiophene (4-MDBT) and 4,6-dimethyldibenzothiophene (4,6-DMDBT) as representative models for PASHs compounds, 2-hydroxybiphenyl (2-HBP) and 2, 2'-bihydroxybiphenyl (2, 2'-BHBP) as models for recalcitrant, toxic and carcinogenic phenolic compounds (El-Gendy *et al.*, 2010). Moreover, Nassar (2010) isolated *Bacillus sphaericus* HN1 which has sufficient broad substrate specificity to degrade major organic sulfur compounds found in diesel oils. When the oil phase ratio was increased from 1/9 to 1/4 (O/W) phase ratio, the sulfur content was decreased from 9,594 mg/L to 3,393 mg/L (%BDS 64.63%) and to 1,743 mg/L (%BDS 81.83%), respectively. The efficiency of total sulfur removal was decreased to 78%, 20.94% and

14.93% at higher o/w phase ratio 2/3, 1/1 and 3/2, respectively. The biodegradation capacity of DBT from diesel oil was; 62.75%, 78.71%, 74.86%, 29.46% and 0.53% in 1/9, 1/4, 2/3, 1/1 and 3/2 (O/W) ratio cultures, respectively and the biodegradation of BT was; 60.96%, 41.34%, 35.19%, 32.49% and 24.75% in 1/9, 1/4, 2/3, 1/1 and 3/2 (O/W) ratio cultures, respectively. *Bacillus sphaericus* HN1 showed an excellent ability to remove high alkylated DBTs (4-MDBT and 4,6-DMDBT) from diesel oil. The biodegradation capacity of 4-MDBT was; 66.22%, 78.66%, 75.89%, 14.87% and 11.07% in 1/9, 1/4, 2/3, 1/1 and 3/2 (O/W) ratio cultures, respectively, while the biodegradation of 4,6-DMDBT recorded; 63.07%, 80.85%, 77.50%, 32.11% and 32.47% in 1/9, 1/4, 2/3, 1/1 and 3/2 (O/W) ratio cultures, respectively. In this study, the percent of degradation of total petroleum hydrocarbons (TPH) was; 63.05%, 86.87%, 57%, 53.90% and 39.24% in 1/9, 1/4, 2/3, 1/1 and 3/2 (O/W) ratio cultures, respectively, indicating the maximum TPH degradation occurred in 1/4 (O/W) ratio cultures. Thus, *Bacillus sphaericus* HN1 can be applied for bioremediation of oil spill was high sulfur content.

In an attempt for biodegradation of asphaltene, Ali *et al.* (2012) isolated three halotolerant bacterial species from an Egyptian oil-polluted water sample for their ability to utilize asphaltene (Asph) fraction as sole carbon and energy source. These bacteria degrade 83–96% of 2500 mg/L asphaltene within 21 d at 30°C and pH7. They were identified as *Bacillus* sp. Asph1, *Pseudomonas aeruginosa* Asph2, and *Micrococcus* sp. Asph3. To make the process of bioremediation more economically sustainable, Ali *et al.* (2014) designed an integrated process for bioremediation of oil polluted water and production of biosurfactants. Whereas, in a batch system applying halotolerant *Pseudomonas aeruginosa* Asph2, bioremediation of oil polluted seawater with total petroleum hydrocarbon content of 5 g/L using the readily available, and commercial nutrient, corn steep liquor has been achieved. *P. aeruginosa* Asph2 expressed good biodegradation capabilities for different petroleum hydrocarbon components, recording approximately 58, 64, 56, 55, and 53% for total petroleum hydrocarbon, saturates, aromatics, asphaltenes, and resins, respectively, within 21 days of incubation at 30°C, pH 7, and 150 rpm. *P. aeruginosa* Asph2 proved good uptake of crude oil with high production of rhamnolipid biosurfactants, yielded approximately 1.3 g/L biosurfactant with the consumption of approximately 2.9 g/L crude oil. In a further study applying that halotolerant and biosurfactant producer Asph2 and corn steep liquor as a carbon co-source, Asph2 showed a great efficiency in the biodegradation of the recalcitrant biomarkers pristane, phytane, terpanes, steranes and hopanes (El-Gendy *et al.*, 2014b). Thus bioaugmentation and biostimulation together significantly improved the efficiency of total petroleum hydrocarbon TPH and the recalcitrant asphaltene degradation by 23 and 17% compared to biostimulation alone. Younis *et al.* (2013) reported the superiority of the autochthonous bioaugmentation over biostimulation of a petroleum hydrocarbons-polluted water sample in a batch reactor using the halotolerant *Corynebacterium variabilis* sp. Sh42 and corn steep liquor as a readily available and cheap source of nutrients. Younis *et al.* (2013) proved also the successfulness of applying well-adapted bacterial consortium for the biotreatment of petrogenic polluted seawater, and attributed this to two different reasons. (1) The synergistic interactions among the members of the microbial association, which is complex and favor petrogenic hydrocarbons degrading mechanisms, where, one microbial species *Corynebacterium variabilis* sp. Sh42 would remove the toxic metabolites of another species *Rodococcus erythropolis* IGTS8 that would have begun the biotransformation process, or when the two species work in succession with the first partially degraded compounds and the second finished the job. (2) The broad enzymatic capacities and co-metabolic relationships of the augmented consortium with the indigenous microbial population where there is a more favorable attack and metabolize not only aliphatic or aromatic compounds but also nitrogen, sulfur and oxygen (NSO) compounds. In further study, Younis *et al.* (2020) proved the importance of applying crude enzymes extracted from halotolerant *Corynebacterium variabilis* Sh42 in batch biodegradation of a mixture of xenobiotic organic pollutants. Since, the specific biocatalytic degradation rates of the studied xenobiotics in the batch enzymatic bioreactors were higher than those obtained in batch bacterial cells bioreactors. Moreover, Yano and Koga model equation seemed to be the best adequate expression for 2-hydroxybiphenyl and benzoic acid biodegradation rates. But, Haldane biokinetic equation adequately expressed the specific biodegradation rate of catechol.

El-Sheshtawy *et al.* (2014) enhanced the bioremediation of oil polluted seawater by the application of αFe_2O_3 or $Zn_5(OH)_8Cl_2$ nanoparticles in microcosm inoculated by two halotolerant and biosurfactant producer *Pseudomonas xanthomarina* KMM 1447 and *Pseudomonas stutzeri* ATCC 17588. It is worth to know that applying of bacterial consortium enhances the rate of bioremediation. Saed *et al.* (2014) recommended also the application of magnetic nanoparticles Fe_3O_4 as a decorating nano-toxic material with superparamagnetic properties for *Micrococcus lutes* RM1 in the biodegradation of polyaromatic hydrocarbons. The $t_{1/2}$ values of the batch biodegradation of 1500 mg/L pyrene were, 17 and 24 days in coated and free cells reactors, respectively.

Response surface methodology based on central composite design was applied to study the effect of oil type polluting the seawater (i.e. heavy and light crude oil) and optimize the amount of solid waste date (SWD), as a low cost natural agro-industrial material to improve the crude oil biodegradation in contaminated seawater using *Pseudomonas aeruginosa* (El-Mahdi *et al.*, 2014). Whereas, 95% of 0.25 g/L heavy crude oil and approximately complete removal of 0.2 g/L light crude oil achieved using 0.8 and 0.2 g/L SWD within 28 d of incubation, respectively. In another study, Amer *et al.* (2014) confirmed the importance of applying bacterial consortium for the bioremediation of petroleum hydrocarbons polluted seawater. Whereas, a bacterial consortium of ten bacterial isolates belonging to five different genera; *Bacillus*, *Pseudomonas*, *Marinobacter*, *Providencia* and *Sphingomonas* isolated from a chronically oil polluted Egyptian Mediterranean sediments in El-Max district recorded approximately 77.8% w/w total petroleum hydrocarbons (TPH) removal from an oil polluted seawater (10,000 mg/L). The bacterial consortium showed broad versatility on biotransformation of the 16 polynuclear aromatic hydrocarbons (PAHs) listed by US Environmental Protection Agency EPA as priority pollutants with ≈ 85.79% biotransformation of carcinogenic five and six membered rings PAHs and n- and iso-alkanes degradation of approximately 97.97 and 72.92%, respectively.

Application of immobilized microorganisms is a promising technique in polluted water as it increases the microbial cell stability, promotes the good diffusion between substrates and products, protects the microbes from the rugged environmental conditions and finally facilitates the separation of biomass and water (Wang *et al.*, 2015). El-Gendy and Nassar (2015a) reported the successful bioremediation of diesel oil polluted seawater using alginate entrapped halotolerant *Pseudomonas aeruginosa.* The improved tolerance of the immobilized cells towards different toxic components of diesel oil has been proved and the biotransformation rate of different components of diesel oil: aliphatic derivatives, polyaromatic sulfur heterocyclic compounds, and biomarkers pristane, phytane, and 4,6-dimethyldibenzothiophene was enhanced by immobilization. Storage stability and reusability tests revealed that the diesel oil degradation ability of the immobilized cells was stable after storage at 4°C for 30 d and can be effectively reused for two batches of 56 d.

El-Mahdi *et al.* (2015b) reported the isolation of petroleum hydrocarbon degrader *Kocuria* sp. SAR1 from 'Tobruk Refinery' oil water pit, located along the Eastern Coast of Libya, that expressed 68% and 70% biodegradation of crude oil in batch systems seeded with 0.2% w/v of the agro-industrial wastes; solid waste dates and corn steep liquor, respectively. In another study, El-Mahdi *et al.* (2015c) reported the isolation of different petroleum hydrocarbon degrading bacteria from oil-contaminated sites at Al Hariga Oil Terminal and Nafoora Oilfield, in Libya, where the Gram-negative isolate; *P. aeruginosa* NAF1 expressed 91% and 97% degradation of 5% crude oil w/v, when the polluted seawater was amended with 0.5% w/v corn steep liquor and solid waste dates, respectively. El-Mahdi *et al.* (2016) reported the biosurfactants production by *P. aeruginosa* NAF1 during the bioremediation of oil polluted seawater in presence of agro-industrial wastes as carbon-co-substrates.

It is important to study the kinetics of biodegradation of single pollutant and mixture of pollutants for scaling up the bioremediation process. A mathematical model was predicted to describe the bioremoval kinetics of different polyaromatic sulfur heterocyclic compounds (PASHs); thiophene (Th), benzothiophene (BT), dibenzothiophene (DBT), 4-methydibenzothiophene (4-MDBT) and 4,6-dimethyldibenzothiophene (4,6-DMDBT) with different initial concentrations range S_o of (100–1000 mg/L) employing suspended cultures of *Bacillus sphaericus* HN1 with initial concentrations X_o range of (291.90–362.01 mg/L dry weight) in a series of batch experiments (Deriase and El-Gendy, 2010). The predicted model is based on Haldane bio-kinetic equation for substrate inhibition which is applied to describe the dependence of specific growth rate μ (h^{-1}) on the initial substrate concentration S_o (mg/L). The calculated biokinetic constants indicated that the toxicity and inhibition effects of these PASHs on HN1 can be ranked in the following order Th > BT > DBT. Whereas, the maximum specific growth rates μ_{max} were 0.165, 0.231, 2.461, 0.207 and 0.202 h^{-1}, the saturation constants K_s were 3.007, 18.425, 2004.25, 42.25 and 103.43 mg/L, while the inhibition constants K_i were 2110.42, 1752.42, 46.849, 2242 and 360.61 mg/L for Th, BT, DBT, 4-MDBT and 4,6-DMDBT, respectively. That was attributed to the water solubility of these PASHs, as with the increase of molecular weight and aromatic rings, water solubility decreases. Moreover, the toxicity and inhibition effects of dibenzothiophenes on HN1 ranked in the following order 4-MDBT > 4, 6-DMDBT > DBT. Nevertheles, regardless of the type of PASHs, growth decreased with the increase in initial substrate concentration. Experimental results have also made clear that microbial growth was always stopped before the complete removal of the PASHs. That was attributed to the biodegradation process itself which might have produced toxic metabolites to the cell. But from the time profile of each PASH, regardless of PASHs type and its initial sulfur concentration S_o (mg/L), it was clear that; the bioremoval characteristic was expressed at increasing levels during the bacterial growth cycle even after reaching its maximum growth. That indicated that the stationary phase cells have the ability to continue degrading PASHs.

The proposed mathematical model can be described by the following equations:

$$\frac{dS}{dt} = -qX^c \qquad \frac{dX}{dt} = \mu X^c - bX^c$$

where, b (h^{-1}) is the decay rate coefficient and found to be in the range (0.0001–0.009) and c is the power of microbial concentration and found to be in the range of (0.4–0.6).

The maximum specific removal rate (q_{max}, h^{-1}) recorded; 0.042, 0.063, 1.53, 0.088 and 0.053 h^{-1} for Th, BT, DBT, 4-MDBT and 4, 6-DMDBT, respectively. The bioremoval rates ranked in the following increasing order Th < BT < DBT and 4, 6-DMDBT < 4-MDBT < DBT. Furthermore, the highest recorded efficiency in DBT cultures indicated the lowest toxicity effect occurred by DBT and the well adaptation of HN1 biodegrading enzymes towards DBT as it was previously isolated and enriched on DBT (Deriase and El-Gendy,

2010). In another study, El-Gendy *et al.* (2015) reported that the removal of thiophenic compounds by *Bacillus sphaericus* HN1 ranked in the following decreasing order benzothiophene > dibenzothiophene > thiophene, whether in single- or tertiary-substrate system. But, the total sulfur removal decreased in the following order single-substrate batch system > binary-substrate batch system > tertiary-substrate batch system, recording 94%, 87%, and 75%, respectively. Moreover, Deriase *et al.* (2015) proved the importance of the kinetic modeling of the BDS-process to know the exact time, inoculum size and substrate concentration for achieving the maximum specific desulfurization rate in eight series of batch experiments of mono-, binary-, and tertiary-substrate systems using *Bacillus sphaericus* HN1. That was based on a full factorial design 2^3, by varying the studied three studied thiophenic compounds (thiophene, benozthiophene and dibenzothiophene) concentrations between 0 and 100 mg/L. A multiple comparison test was performed for growth of microorganism and substrate bioremoval to obtain pairwise comparison between the three studied systems. Both tests ANOVA1 and Kruskal-Wallis showed that there was a highly statistically significant difference in bacterial growth between negative control (free of PASHs) and all other treatments (p = 2.2105e-7 and 9.303e-4, respectively). Although, there was a statistical significant difference in the biodegradation of DBT in mono- and binary- substrate systems within the time interval 96–168 h ($p < 0.05$). However, there was a non-statistically significant difference in the biodegradation of both Th and BT in the three studied systems ($p > 0.05$). A multi-substrate form of the Monod kinetic model was applied to predict the substrate interactions in binary- and tertiary- substrate systems using the Monod parameters derived from the mono-substrate systems. Where, modeling the relationship between cell growth and substrate consumption rates was done by direct coupling of substrate consumption rate with the cell growth model applying constant cell yield and biomass decay rate coefficient during the time course of batch experiments.

$$\frac{dC_i}{dt} = -\mu_i \, {}^{X}/_{Y_i}$$

Taking into consideration, the biodegradation rate of a substrate (i) is proportional to both the specific growth rate (μ_i) on that substrate and the biomass concentration (X) in the system.

The change in microorganism concentration with time was expressed mathematically, as:

$$\frac{dX}{dt} = \mu_i X - bX$$

where, the change in biomass concentration was modeled as if; the substrate (i) was the only growth substrate present. The dependent variables were; C_i the concentration of the PASHs, that is, the substrate i (mg/L) and X is the biomass concentration (mg/L), while the independent variable was time, t (h). Y_i (mg biomass/mg substrate) is the stoichiometric biomass yield coefficient for substrate i and b is the endogenous decay rate coefficient (h^{-1}), which was estimated using the independent substrate free experiment (–ve control) with a known starting biomass concentration. The endogenous rate of microbial decay which was characterized by rate coefficient b (h^{-1}) was subtracted from the growth rate of the biomass to account for the energy lost for cell maintenance, and b was assumed to be constant for all the studied biodegradation systems. The Monod kinetic model relates growth rate to substrate concentration [$\mu_i = f(C_i)$] via the two bio-kinetic Monod parameters; K_{Si} and $\mu_{max,i}$;

$$\mu_i = \frac{\mu_{max,\, i} C_i}{K_{Si} + C_i}$$

where, $\mu_{max,i}$ is the maximum specific growth rate on substrate i (h^{-1}) and K_{Si} is the half saturation coefficient for substrate i (mg/L), the parameters estimation is challenging for Monod equation. The relationship between biomass formation and substrate consumption can be determined by the yield coefficient Y_i (dry weight of bio-mass/weight of utilized substrate) indicating the maximal conversion of unit substrate to cell mass.

$$Y_i = \frac{dX/dt}{dC_i/dt} = -\frac{dX}{dC_i} = -\frac{X - X_o}{C_i - C_{io}}$$

where, $X > X_o$ and $C_i < C_{io}$

In a multi-substrate system, the biomass growth is due to the utilization of all available hydrocarbons in the system to the micro-organism, accordingly, the specific growth rate will be the summation of all individual ones for each substrate.

$$\mu_T = \sum_{i=1}^{n} \mu_i'$$

where μ_T is the total specific growth rate (h^{-1}), μ_i is the specific growth rate on substrate i, in the multi-substrate system (h^{-1}), which can be calculated from the multi-substrate Monod growth relationship and n is the number of substrates in the system.

$$\mu_i' = \frac{\mu_{\max,\ i} C_i}{K_{Si} + \sum_{j=1}^{n} \frac{K_{Si}}{K_{Sj}} C_j}$$

Thus, according to the obtained data from mono-substrate system, the affinity of *Bacillus sphaericus* HN1 towards the studied PASHs could be ranked in the following decreasing order BT > DBT > Th, indicating the high toxicity of Th compared to BT and DBT. This was confirmed by the lowest $\mu_{max,i}$ 0.01434 h^{-1} obtained with Th (Deriase *et al.*, 2015).

Experimentally, the biodegradation rates for the individual substrates in the multi-substrate systems may be enhanced or reduced relevant to the mono-substrate system. The effect of one substrate on the degradation of another is given by $\frac{K_{si}}{K_{sj}}$ terms, where, K_{Si} and K_{Sj} are the half saturation coefficients for substrate i and j (mg/L) (Knightes and Peters, 2006). The experimental observations and modeling predictions, i.e. the simulation results for the biodegradation of the three studied PASHs in the three studied systems, revealed the following; relative to the biodegradation of the three studied PASHs in mono-substrate-batch cultures; BT enhanced the biodegradation of DBT in the binary substrate-batch cultures of DBT-BT, especially within the time interval of 96–168 h. While, in Th-BT binary substrate-cultures, the BT expressed a negative impact on the biodegradation of Th up to 96 h but it enhanced the Th-biodegradation within a relatively short time interval of 96–144 h then the antagonistic effect occurred again with longer incubation period. Although DBT enhanced the biodegradation of BT in the binary substrate batch cultures of BT-DBT especially within the time interval 24–168 h, it depleted the biodegradation of Th in the binary substrate batch cultures of Th-DBT. However, Th depleted the biodegradation of both BT and DBT in the binary substrate-batch cultures of BT-Th and DBT-Th, respectively, where, the negative impact of Th was very obvious within the time intervals of 12–120 h and 48–168 h, respectively. In the tertiary substrate-batch cultures Th-BT-DBT, the biodegradation efficiency of the three PASHs was depleted relevant to their biodegradation efficiencies in the mono-substrate batch cultures. Where the $\Sigma\frac{K_{si}}{K_{sj}}$ recorded; 5.785, 0.975, and 1.887 for the biodegradation of Th, BT, and DBT, respectively, which indicated that the antagonistic effect between the three studied PASHs on the rate of their biodegradation can be ranked in the following decreasing order Th > DBT > BT (Deriase *et al.*, 2015).

Simulation of the applied models to describe and estimate the biodegradation kinetics of the three studied PASHs in the different investigated systems was performed and the validity of the models was confirmed. The simulated effectiveness factor of the kinetic models was depicted by using a mean relative error (MRE), to show the goodness of fit between the obtained experimental and the simulated data of PASHs depletion with time.

$$MRE = \frac{1}{m}\sum_{i=1}^{m}\frac{|C_{i.cal} - C_{i.exp}|}{C_{i.exp}}$$

where, $C_{i,cal}$ and $C_{i,exp}$ are the simulated substrate concentration determined from the model and its observed experimental value at time t, respectively, while, m is the number of experimental points. The results of MRE, ranged between 0.096 and 0.282, with percentage MRE ranged from 9.6% to 28.2%, which confirmed the adequacy of the mono- and multi-substrate Monod kinetic models for estimating the biodegradation of the studied PASHs. That was also very obvious in the agreement between the measured and simulated profiles for the substrates depletion with time. Thus, the approaches of modeling and simulation for the experimental results obtained in that study performed by BT (Deriase *et al.*, 2015) are considered to be essential in order to facilitate the understanding of the interactive effect of PASHs contaminants and their biodegradation processes for better achievement of biotreatment of contaminated sites and industrial effluents.

GREEN SYNTHESIS OF NANOPARTICLES USING AGRO-INDUSTRIAL WASTES FOR TREATMENT OF POLLUTED WATER

Biological contaminants, for example the pathogenic microorganisms are also considered as serious water pollutants (Roig *et al.*, 2013). Wastewater from petroleum refineries contains toxic, carcinogenic and recalcitrant phenolic compounds (Younis *et al.*, 2020). Thus, when talking about water treatment it is important to mention the application of green synthesized nanoparticles as photo-degraders for organic xenobiotic pollutants and biocides for pathogenic microorganisms (Singh *et al.*, 2018). The bottom-up green synthesis of nanoparticles NPs overcome the drawbacks of physico-chemical methods. It is not only inexpensive but also non-toxic, less wastage of inputs, less complicated and less time consuming, safer, ecofriendly and does not produce hazardous toxic wastes (Malik *et al.*, 2014; Kumar and Kathireswari, 2016; Omran *et al.*, 2018a; El-Gendy and Omran, 2019).

Spheroidal fluorapatite (FAp, $Ca_5(PO_4)_3F$) nanoparticles with average size of 55 nm have been reported to be easily prepared by a simple calcination process of waste buffalo bones (El-Gendy *et al.*, 2016a). Moreover, quasi-spherical ZnO NPs with average particles size of 21.5 nm were prepared by wet chemical method using zinc nitrate and sodium hydroxide as precursors. Then, a monodispersed nanosphere photo-catalytic bio-composite $ZnO:Ca_5(PO_4)3F$ (ZnO-FAp) with average size of 40 nm, was prepared by mechano-mixing technique. The BET surface area were; 12.63, 8.63 and 5.87 cm^2/g and the total pore volume recorded ≈ 0.074, 0.02 and 0.016 cm3/g with average pore diameter of ≈ 0.94, 0.12 and 0.11 nm for FAp, ZnO/FAp and ZnO, respectively. One of the advantages of FAp is; it avoided the agglomeration of ZnO, as it made a shell layer around the ZnO NPs, retaining a stable dispersion of ZnO-FAp nanocomposite compared to the individual cases (Figure 11). The adsorption capacity of the prepared nano-materials towards chlorophenols has been tested. That ranked as follow ZnO/FAp > ZnO ≥ FAp, for 3-chlorophenol. But ranked as follow ZnO > FAp ≥ ZnO/FAp, for 2, 3-dichlcrophenol (2, 3-DCP). That was attributed to the pollutant itself, since the chlorophenols CPs are in the molecular form at pH < pKa and exist as anions (negative charge) at pH > pKa (pKa of 3-CP and 2, 3-DCP is

8.85 and 7.44, respectively). While, the experiments were carried out at pH6 (below the pHpzc; 8, 9 and 8.4 for FAp, ZnO and ZnO/FAp, respectively), implying that the surface of the catalysts was positively charged. That might explain the higher adsorption capacity of 2, 3-DCP compared to 3-CP. Moreover, the photocatalytic activity of ZnO was enhanced by FAp and ranked in the following decreasing order FAp > ZnO/FAp > ZnO. The higher specific surface area and porosity of FAp might explain its higher photo-catalytic activity. The photo-catalytic activity of FAp might have been caused by the generation of active superoxide anion radicals ($O_2^{2-\bullet}$), which would have been occurred due to the change in the electronic state of the surface PO_4^{3-} group under UV irradiation (El-Gendy *et al.*, 2016a).

Figure 11: SEM and TEM Micrographs of ZnO/FAp Nano-Bio-Composite

Moreover, El-Gendy *et al.* (2016a) mentioned that the enhancement of the photocatalytic activity of ZnO by FAp might have been due to the enhancement of the surface area, pore volume and diameter, surface hydroxyl groups and molecular oxygen and might have been also due to the improved charge separation and extended energy range of photo-excitation, throughout the transition of energy levels between $Ca_5(PO_4)_3F$ orbital and ZnO orbital (Figure 12).

El-Gendy *et al.* (2016a) have also illustrated the photodegradation pathway of the studied CPs reaching acetate (Figure 13). Whereas, in

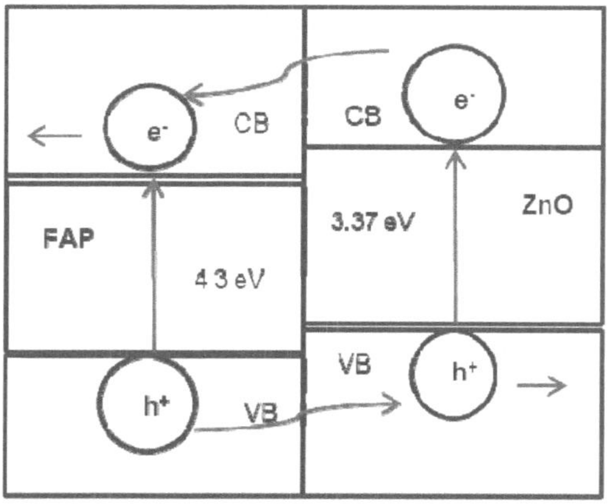

Figure 12: Proposed Energetic Model of ZnO and FAp Nano-Bio-Composite

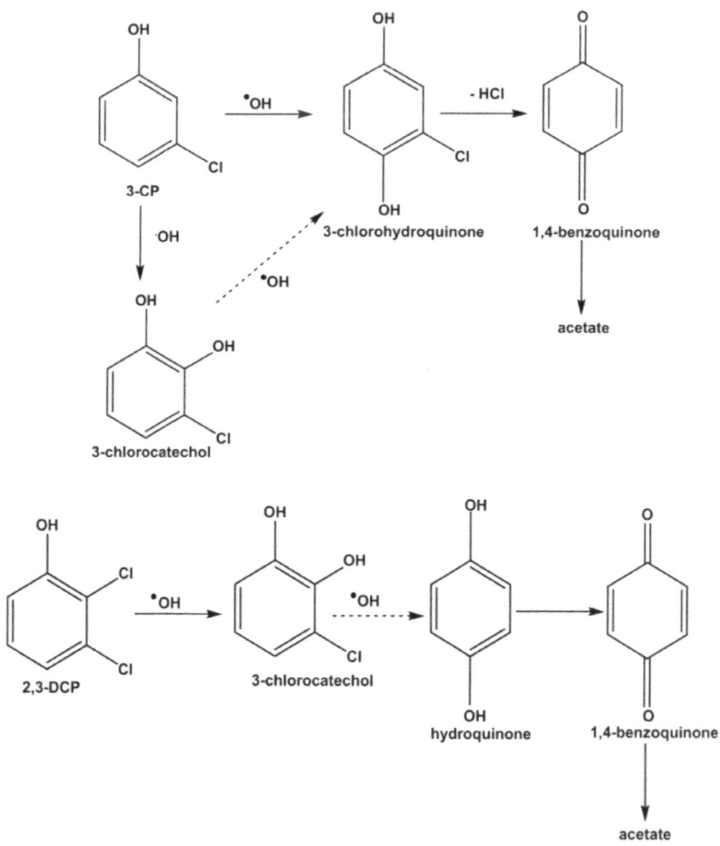

Figure 13: Photo-Catalytic Degradation Mechanism of 3-CP and 2, 3-DCP by ZnO/FAp Nano-Bio-Composite

the occurred photo-catalytic reaction, photo-generating electrons and holes were captured by O_2 and H_2O adsorbed by the photo-catalyst forming the super-active $^{\bullet}OH$ and $O_2^{2-\bullet}$ oxidants. Thus, the increased amount of the hydroxyl groups on the ZnO/FAp might have not only increased the trapping sites for photo-generated holes, but might have also increased the trapping sites for photo-generated electrons by adsorbing more molecular oxygen, which might have resulted in more hydroxyl radicles which would have participated in the photo-catalytic reaction (Figure 14).

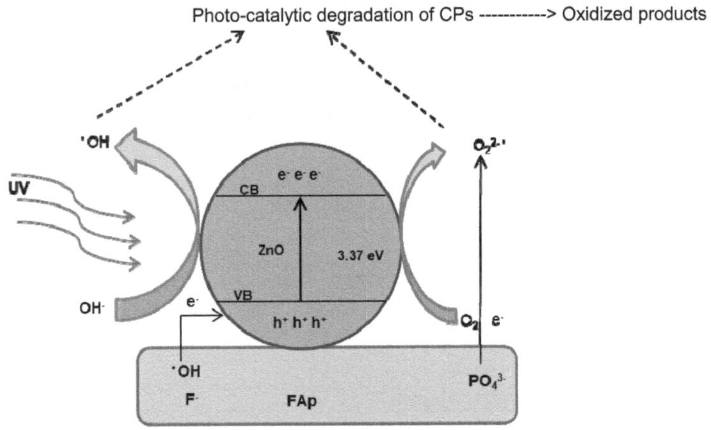

Figure 14: Photo-Catalytic Mechanism for ZnO/FAP Nano-Bio-Composite

The phenolic compounds and other main constituents of hot water extract of mandarin peels (e.g. 2, 3-dihydro-3, 5-dihyroxy-6-methyl-4H-pyran-4 one (DDMP), 5-hydroxymethylfurfural, 5-methyl-2-furancarboxaldehyde, 2-methoxy-4-vinylphenol, d-limonene, n-hexadecanoic acid, 3, 5-dihdroxy-2-methy-4H-pyran-4-one and 2, 4-dihydroxy-2, 5-dimethyl-3(2H) furan-3-one), have been reported to be involved in the bioformation and stabilization of α-Fe_2O_3 NPs (Figure 15). The prepared α-Fe_2O_3 NPs were porous, quasi-spherical, rough shaped particles with average particles size of 20–63 nm, weak ferromagnetic properties and band gap (Eg) of 2.38 eV (Ali *et al.*, 2017).

That green synthesized α-Fe_2O_3 NPs expressed high adsorption capacity towards different pollutants, recording approximately 28, 20 and 16% removal of 50 mg/L Basic Blue 41, Acid Blue 158, and 2, 6-dichlorophenol. Moreover, such iron oxide NPs expressed efficient

Figure 15: HRTEM Micrographs the Green Synthesized α-Fe_2O_3

photodegradation activity towards those pollutants under visible light, recoding ≈ 77.3, 82.5, and 81.5%, respectively. The primary photocatalytic oxidation mechanism (Figure 16) is believed to proceed by the formation of the powerful oxidizing hydroxyl radicals and/or the adsorbed O_2 would have captured the photo-generated electrons, producing $O_2^{2-\bullet}$ that can directly degrade the pollutant. Moreover, The $O^{2\bullet-}$ can also react with the photo-generated holes and then form active $OH^{\bullet}$ radicals and peroxides which also photodegrade, the pollutants (Ali *et al.*, 2017).

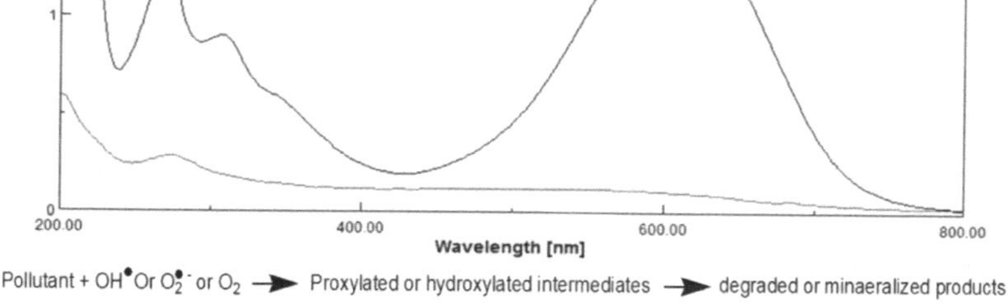

Figure 16: Pollutant Degradation with Visible Irradiation Using Green Synthesized α-Fe_2O_3 NPs

Omran *et al.* (2018b) reported that the main components of orange peels hot water extract (e.g. 4-vinyl guaiacol, eugenol, aromatics, terpenes, sugar derivatives and saturated fatty acids) are involved in the bioreduction of $AgNO_3$ into spherical shaped AgNPs with average particles size of 3–12 nm (Figure 17a). The operational production cost of such green synthesized AgNPs was estimated to be 81 US$/10 g AgNPs. According to the US Research Nanomaterials, Inc., the global cost of the chemically synthesized AgNPs is 250 US$/10 g of AgNPs. Thus, 67.6% savings can be achieved upon application of green synthesis of AgNs using waste peels. Moreover, to reach to the point of zero-waste, the spent waste orange peels after the preparation of AgNPs was valorized into activated carbon (Figure 17b) with specific surface area of 2.9 m^2/g, average pore volume and pore size of 0.02 cm^3/g and 1.9 nm, respectively, which would have different applications.

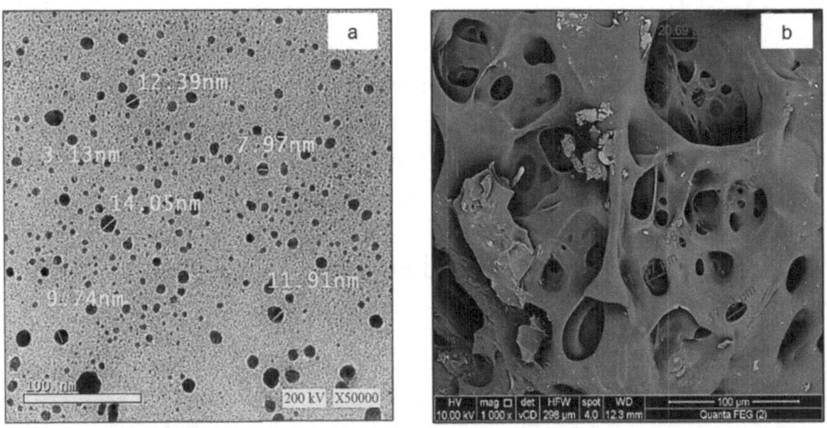

Figure 17: TEM Micrograph of AgNPs (a) and FESEM Micrograph of Activated Carbon (b) Prepared from Orange Peels

Nanoparticles with large surface area form pores in cell membranes and penetrate it. That would lead to leakage of ions and metabolites; denaturation of cytoplasmic proteins; inactivation of enzymes and respiratory chain enzymes; formation of intracellular reactive oxygen species (ROS); interactions with ribosome; interactions with nucleic acids and inhibition of signal transduction (Gurunathan *et al.*, 2014; Kedziora *et al.*, 2018).

From that concept, Omran *et al.* (2018a) proved the potent antimicrobial activities of the biosynthesized monodispersed spherical shaped

6–21 nm AgNPs against the pathogenic micro-organisms; Gram-positive *Bacillus subtilis* ATCC 6633 and *Staphylococcus aureus* ATCC 35556, Gram-negative *Pseudomonas aeruginosa* ATCC 10145 and *Escherichia coli* ATCC 23282 and a yeast *Candida albicans* IMRU3669. The minimum inhibitory concentration recorded 300, 400, 600, 500 and 300 µg/mL, respectively. But, the minimum lethal concentration was 600 µg/mL for all except *Candida albicans* and *Pseudomonas aeruginosa* which recorded 400 and 600 µg/mL, respectively. It is worth to know that the biologically synthesized AgNPs are reported to be the least toxic form of silver to cells of *Allium cepa* (Panda *et al.*, 2011). The biosynthesized AgNPs is also reported to be less toxic for fresh water fish *Oreochromis niloticus* than the chemically synthesized AgNPs (Girilal *et al.*, 2015) and did not express negative effects on the survival or growth of zebrafish (*Danio rerio*) larvae up to 1 mg/L (Cambier *et al.*, 2018). Taking into consideration that such mycosynthesis of AgNPs via *Aspergillums brasiliensis* costs 172 US$/10 g AgNPs. Thus, it is recommendable to use such cost effective green synthesized AgNPs in water disinfection from pathogenic micro-organisms. The observed biocidal activity of AgNPs is related to its large surface area, its strong reactive sites and the good attraction of the negatively charged microbial cell membrane towards the Ag^{+} ions. AgNPs penetrate the bacterial cell wall, oxidize the proteins existed on the plasma membrane and thereby cause disturbance in cellular homeostasis. AgNPs cause also the cell rupture, and/or drastically disturb its proper function like respiration and permeability (Figure 18).

Moreover, Omran *et al.* (2019a) reported the efficient biocidal activities of the biosynthesized quasi-spherical shaped, monodispersed ferromagnetic 20–27 nm Co_3O_4 nanoparticles, against the Gram-positive (*B. subtilis* and *S. aureus*) and the Gram-negative (*P. aeruginosa* and *E. coli*) bacteria. The minimum inhibitory concentration and minimum lethal concentration recorded 2.5 mg/mL for all except *S. aureus* which recorded 5 mg/mL. The cost of the mycosynthesis of such Co_3O_4 NPs using via *Trichoderma longibrachiatum* was estimated to be high, recording approximately 225 US$/10 g Co_3O_4 NPs. The Magnetic Hysteresis (M-H) curve of such bioinspired Co_3O_4 proved 17.3 emu/g magnetization saturation (Ms), 0.12 emu/g and 2.9 g remnant magnetization (Mr) and coercivity (Hci), respectively, with

Mr/Ms value of 7.05. Thus, the observed magnetic properties of that bioinspired Co_3O_4 NPs (Figure 19), added to its advantages and value. That ferromagnetic Co_3O_4 NPs might have successful applications in magnetic storage devices, water disinfection and wastewater treatment. Moreover, for decreasing the overall cost of the mycosynthesis process and retaining the point of zero-waste, the valorization of the spent waste fungus biomass used for the biosynthesis of Co_3O_4 NPs into activated carbon with good specific surface area 6.32 m^2/g and pore size ranged between 4.22 and 37.24 µm was achieved.

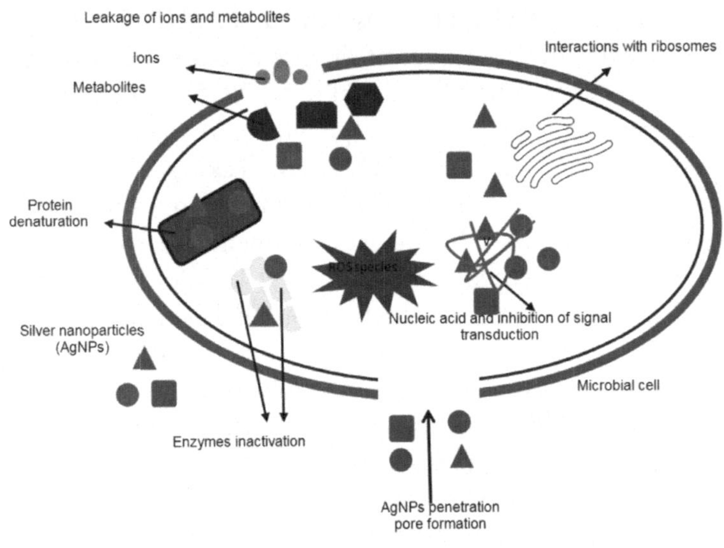

Figure 18: A Suggested Mode for Biocidal Action of AgNPs

Figure 19: Magnetic Properties of the Mycosynthesized Co_3O_4–NPs

GREEN BIOCIDES AND CORROSION INHIBITORS FROM AGRO-INDUSTRIAL WASTES

Carbon steel has been widely employed as a construction material for pipe work in the oil and gas production such as down whole tubules, flow lines, and transmission pipelines (Al-Senani, 2016; Ali and Khan, 2017). In oil fields, the corrosive hydrochloric acid solution is recommended as the cheapest way to dissolve calcium carbonate and scales inside the pipelines under most conditions (Ghulamullah *et al.*, 2015). Nevertheless, acid pickling of steel is usually carried out at elevated temperature. Temperature effect on carbon steel especially under acidic conditions is known to be very harsh (Hamdy and El-Gendy, 2013). Accordingly, corrosion inhibitors must be injected with the hydrochloric acid solution to avoid the destructive effect of acid on the pipelines. Moreover, upon the exposure of metal surfaces to seawater; they undergo a sequence of biological and inorganic changes that lead to biofouling and corrosion. Biofouling process starts by conditioning of the metal surface by organic and inorganic molecules, then microbial colonization (i.e. microbial biofouling) and finally, microbial biofouling established. Microbial influenced corrosion (MIC) occurs due to the attachment and activities of aerobic bacteria and anaerobic bacteria, mainly the sulfate-reducing bacteria (SRB) onto metal surfaces (Souza *et al.*, 2017). *Desulfovibrio* species have been revealed as the most abundant population in corrosive environments and biofilms (Zhang and Fang, 2001). Moreover, the presence of SRB is the main reason of the production of the toxic and corrosive H_2S and the occurrence of pitting corrosion and consequently the sudden breakage of the oil pipelines and consequently the oil leakage and fires in petroleum fields (Das and Deka, 2019). *Brachidontes*, *Mytilus*, *Perna*, *Dreissena*, *Modiolus* and *Corbicula* are the main macro-foulants they mainly cause clogging in pipelines (Lyons *et al.*, 1988). To control such problems, the non-ecofriendly chemical biocides, such as chloride, glutaraldehyde and quaternary ammonium salts, are usually used (Pichtel, 2016). A lot of agro-industrial wastes are annually produced and most of them are eliminated by open-burning which causes a lot of air pollution. The use of these wastes for production of sustainable, cost effective and biodegradable green corrosion inhibitors and biocides would offer an effective and promising solution for waste management, microbial induced corrosion and biofouling problems, which would lead to occurrence of sudden oil leakage. Thus, such sustainable criteria would have a double positive impact on economic and environmental problems (Saratha *et al.*, 2009; Ogbonna *et al.*, 2011; Hamdy and El-Gendy, 2013; Chaubey *et al.*, 2015; Al-Senani, 2016).

Upon these mandates, Omran *et al.* (2013) investigated the effect of the concentrated extract of waste bitter water of Egyptian lupine pickling and the hot water extracts of orange and mandarin peels as a green sustainable mitigation for biocorrosion and biofouling. The three bioextracts showed efficient biocidal effect against different strains of sulfate reducing bacteria (SRB); the non-halotolerant *Desulfovibrio sapovorans* (ATCC 33892) and the halophilic bacterium *Desulfovibrio halophilus* (ATCC 51179) and three other mixed cultures of SRB (SRB1, SRB2 and SRB3) that were previously isolated from different Egyptian oil fields with different salinities. However, the halotolerant SRB consumed higher biocidal concentrations. Moreover, such bioextracts expressed considerable biocidal effect against the macrofoulant *Brachidontes variabilis* that was sampled from Egyptian Suez Gulf water. As the recorded LC_{50} ranged between 1000 and 2000 mg/L. The most important finding, is the toxicity of such bioextracts was negligible against non-target sea organisms; *isopoda*, *amphipoda* and decapoda up to 4000 mg/L for 8 days exposure. Moreover, it was proved to be less toxic than the commercial chemical biocides most commonly used such as Bayluscide and copper sulphate. The advantages of such green biocidal action on halotolerant SRB is very important, as halophilic microorganisms have strategies that allow them not only to withstand osmotic stress, but also to function better in the presence of high salt concentrations, partly due to the synthesis of compatible solutes that allow them to balance their osmotic pressure as well as the enzymes which become active in solutions of very high ionic strength. Thus, they have higher resistance to biocides, as they are extremophile microorganisms, which can survive under stress conditions (González-Hernández, and Peña, 2002). However, the presence of quinolizidine alkaloids, phenols (4-vinyl guaiacol), furanones (2,4-dihydroxy-2,5-dimethyl-3(2H)-furan-3-one), furfurals (5-hydroxymethylfurfural) and saturated fatty acids (hexadecanoic acid), 2,3-dihydro-3,5-dihydroxy-6-methyl-4H-pyran-4-one (DDMP), and methyl methane thiosulphinate (MMTSO) in the bioextracts are the main reason for their biocidal activities. As they attack cell membranes, increase permeability, lead to cell leakage and lysis (Omran *et al.*, 2013).

In another study, El-Gendy *et al.* (2016b) investigated the effect of hot water extract of different wastes; outer brown peels of onion (A), outer peels of garlic (B), orange (C), and mandarin (D) peels as green corrosion inhibitor for C-stell in 1 M HCl. The corrosion rate decreased with the increase of bioinhibitor concentrations recording maximum efficiency at 3000 mg/L. The corrosion inhibition efficiency of such bioextracts ranked in the following decreasing order C > B > A > D. The potentiodynamic polarization measurements proved that the bioextracts act as mixed-type inhibitors where it decreased the anodic dissolution of carbon steel and the hydrogen evolution reaction. The maximum inhibition efficiency (η_{icorr} %) recorded 96, 90, 88 and 84%, respectively. Moreover, the scanning electron microscope (SEM) revealed the presence of damaged surface with obvious pitting corrosion in 1 M HCl while it disappeared in the presence of the bioextracts (Figure 20). The adsorption isotherm for all bioextracts was best fitted with Langmuir ($R^2 \geq 0.997$), with slope of almost unity. The corrosion inhibition was related to the constituents of such bioextracts. Onion peels (A) extract contains catechol and protocatechuic acid, quercetin (3, 5, 7, 3′,4-pentahydroxy flavone), quercetin-3-glucoside, and some tannin. Garlic peels (B) extract contains allyl-propyl disulfide ($C_6H_{12}S_2$), diallyl disulfide, ($C_6H_{10}S_2$), and two more two sulfur-containing compounds in addition to carbohydrates and proteins. The major components in orange and mandarin peel (C and D) extracts are eugenol and D-limonene, respectively, with 4-vinyl guaiacol, and other phenolic compounds. The presence of sulfur, oxygen, or nitrogen atoms could be used as adsorption centers for reactive sites on the metal surface preventing the supply of the aggressive ions to the surface and/or transported reaction products away from the surface. The shape of the molecule, its size, and the presence of conjugate double bonds and other groups (such as $CH_3O^{-,}$ OH^-) would determine the electron density on the adsorption centers and consequently the strength of the extract on the metallic surface. However, only hot water extract of orange and mandarin peels expressed efficient biocidal effect on SRB. But the bioextracts of outer peels of onion and garlic did not express any biocidal effect on SRB. That might be attributed to the presence of sulfide compounds in these bioextracts which would act as a substrate for such SRB (El-Gendy *et al.*, 2016b).

Figure 20: Corrosion Inhibition of C-steel in 1 M HCl Using Bioextracts of Some Agro-Industrial Wastes

In a further investigation by El-Gendy *et al.* (2018a) the thermal and surface studies, proved the high efficiency of aqueous hot extract of *Allium cepa* (onion) skin (HWEOS) as green corrosion inhibitor for petroleum pipeline under acidic condition. Although of the observed increment of corrosion rate with elevation of temperature, however, it decreased upon the application of 450 mg/L (HWEOS). The corrosion inhibition ranged between 95% and 56% within temperature range of 20 and 60°C, respectively. The main components of HWEOS as indicated by LC/MS are hydroxyl aromatic compounds; 2, 4, 6-trihydroxybenzaldehyde, quercetin, quercetin-3-glucoside, quercetin-4′-glucoside and delphinidin-3-glucosid. A presumptive corrosion inhibition mechanism was suggested based on the assumption that the adsorption would occur through the non-bonding electron pairs of the oxygen atoms as well as the π-electrons of the aromatic rings (Figure 21). Moreover, co-ordinate bond may be formed between unshared e^- pairs of un-protonated oxygen atom of the inhibitor and vacant d-orbitals of metal surface atoms (Figure 21). Taking into consideration that the adsorption of the inhibitor molecules on the metal surface can be considered as a place exchanger reaction:

$$Inh_{aq} + xH_2O_{ads} \leftrightarrow Inh_{ads} + xH_2O_{aq}$$

where x is the number of water molecules displaced by one molecule of an organic inhibitor.

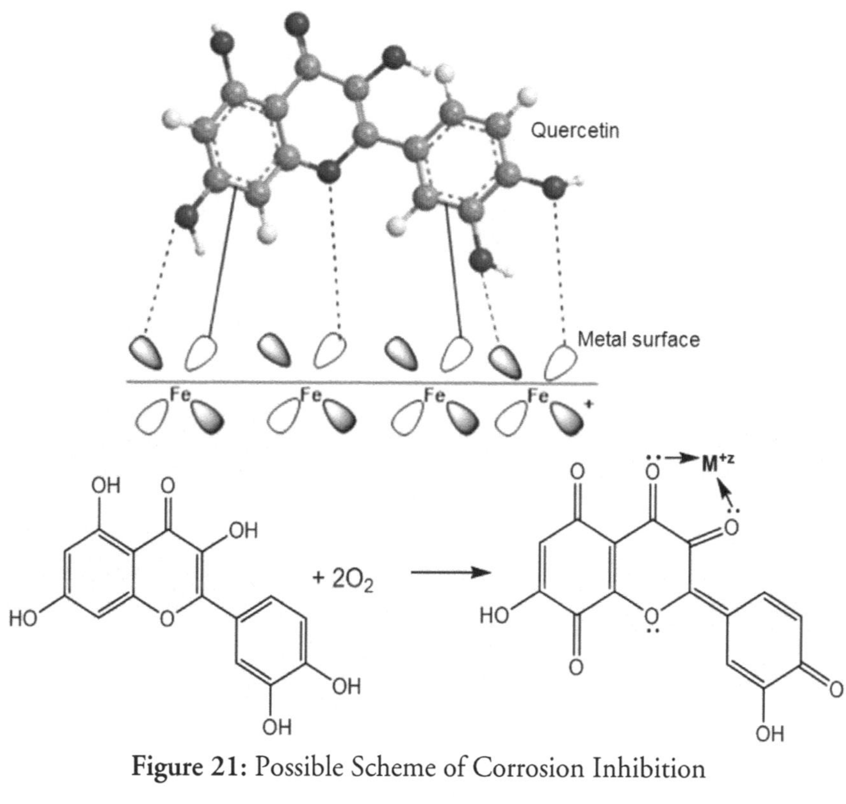

Figure 21: Possible Scheme of Corrosion Inhibition

Omran *et al.* (2019b) reported the efficient biocidal effect of biosynthesized non-agglomerated spherical, triangular and cuboid 5–11 nm AgNPs on halotolerant planktonic mixed culture of SRB isolated from an Egyptian oil field. The high resolution transmission electron microscope (HERTM) analysis of the cultures without treatment with AgNP (Figure 22 a-c) showed well shaped colonies of SRB, surrounded by some form of stable extracellular material. At different higher magnification powers, the SRB appeared as a long and thin rod- and/or vibrio-shaped flagellated cells, for motility with a regular outlined cell wall and regularly distributed cytoplasm. Cells with double membrane layered cell wall structure and/or amorphous dense coat, with electron-translucent cytoplasm free of conspicuous reserve granules, were also observed. The presence of the produced FeS NPs within the SRB cell was also obvious. But after treatment with 2000 mg/L AgNPs, HRTEM analysis (Figure 22 d-f) showed a clear evidence of an alteration in cell morphology, damaged colonies of SRB, overlaid with AgNPs. A disruption of SRB cell membranes, a lysis in cell wall, cuts in its flagella, shredding in bacterial cells and a cytoplasmic extraction with a release in cell components, were also obvious. Moreover, a loss in the stable extracellular martial and disruption in the double membrane layered cell wall structure was very obvious. Taking into consideration such mycosynthesis of AgNPs via *Trichoderma longibrachiatum* costs 94 US$/10 g AgNPs. Thus, it is recommendable to use such cost effective green synthesized AgNPs in water disinfection from SRB and/or as additives to coatings applied on petroleum pipelines.

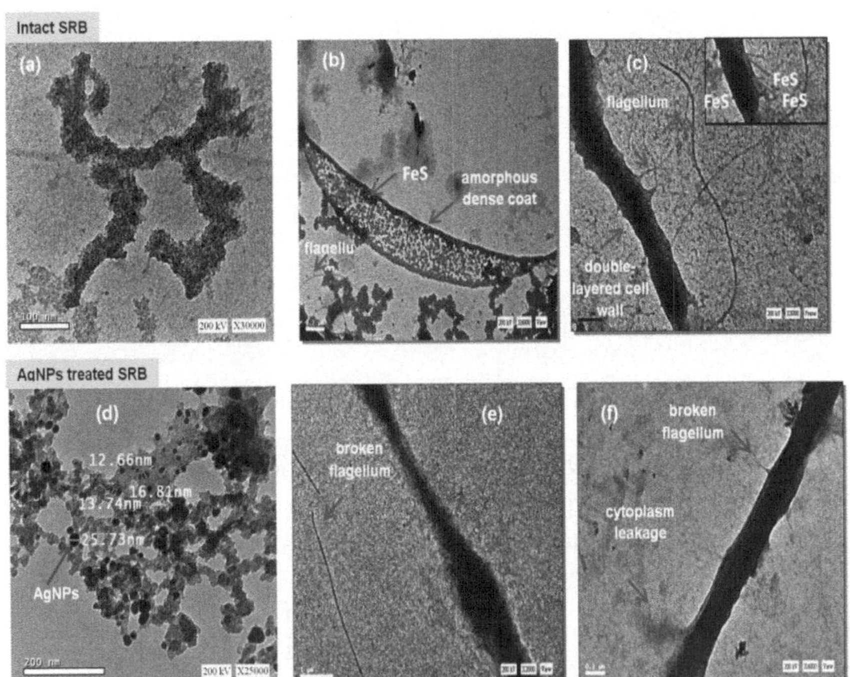

Figure 22: Different Fields Taken by HRTEM for the Untreated and Treated Planktonic SRB Mixed Culture by 2000 mg/L AgNPs

SUSTAINABLE BIODIESEL PRODUCTION AS A TOOL FOR OVERCOMING AIR POLLUTION

Biodiesel is defined as; simple alky esters of fatty acids produced from renewable sources such as; vegetable oils (edible and non-edible oils), animal oil/fats, tallow and waste frying oil (WFO) (Diamantopoulos *et al.*, 2015). Biodiesel is a clean, renewable, sustainable, biodegradable, non-toxic; reduces global hygiene and health problems. It has similar physicochemical properties like those of petro-diesel fuel (Li *et al.*, 2014). Biodiesel has approximately 95% energy content of petro-diesel and it has a positive net energy gain of approximately 3–4:1 (Jairam *et al.*, 2012). It is characterized by better lubricity and thus, enhances engine longevity, extremely low sulfur and aromatic contents fuel and lacks metals and tiny particles of solid. Biodiesel returns about 90% more energy than that used in its production (Farooq *et al.*, 2015). Thus, biodiesel is a realistic fuel alternative for diesel engines (Vashist and Ahmad, 2014). It can be used in any compression ignition engine without any required modifications. Its higher flash point makes it has a better and safer performance during its handling and storage. Moreover, its high oxygen content (approximately 10–12%) improves the combustion efficiency and reduces the greenhouse gases emissions; carbon mono- and di- oxides (CO and CO_2), soot, unburned hydrocarbons and particulate matter (PM) (Gorji and Ghanei, 2014). Biodiesel affords a sustainable reduction in SOx emissions. The CO_2 is continuously cycled and reused for growing plants instead of being released in the atmosphere, since the plants help in CO_2 fixation during the photosynthesis process. The National Renewable Energy (NREL) and US-Department of Energy (US-DOE) reported that the overall life cycle emissions of CO_2 from 100% biodiesel (i.e. B100) are approximately 78.45% lower than those of 100% petro-diesel (i.e B0) (Luque *et al.*, 2010). The blend with 20% biodiesel (i.e. B20) would reduce the net CO_2 emissions by approximately 15.66%. The substitution of B0 by B100 in buses would reduce the life cycle consumption of petro-diesel by 95%. While applying B20 would drop the life cycle consumption of petro-diesel by approximately 19%. Helwani *et al.* (2009) reported that the combustion of B100 decreases the CO, PM and unburned hydrocarbons by approximately 46.7, 66.7 and 45.2, respectively. One of the most important advantages of biofuels in general, is, politically, producing biofuels would make the nations totally or partially independent on other oil producing countries. Thus, would consequently save foreign currency. Besides, the less global warming, it has additional benefits to the society; rural revitalization and creation of new jobs opportunities (Sulaiman and Amin, 2016). But biodiesel may cause a slight increase in the NOx, which can be positively influenced by delaying the injection timing in engines (Liaquata *et al.*, 2012). It has also, some technical problems, including; the low-temperature properties, lower oxidation and storage stability (Kumar *et al.*, 2015). Antioxidant additives can be added to improve the oxidation and storage stability (Sorate and Bhale, 2015). While, oxygenates and cetane improving additives can be added to reduce the NOx emissions and increase the cetane number, respectively (Tan *et al.*, 2015).

Most of the biodiesel produced worldwide comes from edible oils (Khan *et al.*, 2014). That causes competition with the edible oil market, which consequently increase the cost of edible oils and biodiesel. In some countries, that causes deforestation. Since, forests have been felled for oil crops plantation, which adds to the problem of climate change. The cost of the feedstock accounts about 60–95% of the overall cost of biodiesel production (Niju *et al.*, 2014). Nowadays, production of second-generation biodiesel from non-edible oil crops like castor, jatropha, jojoba, neem, polanga, rubber seed, cotton seed, karanja, linseed... etc., is of great interest, since these crops can be cultivated in waste lands that are not suitable for food crops, using low quality water for irrigation. Cultivation of these crops is of a lower cost, sustain reasonably high yield without intensive care (Meira *et al.*, 2016). However, compared to edible-oils; non-edible oils contain high free fatty acids (FFAs). That requires pretreatment process before the transesterification process, which increases the overall production cost and may lower the ester yield of biodiesel below the standards (Atabani *et al.*, 2013). Moreover, although, the non-edible oil crops solve the problem of direct competition with food oils (Aboelazayem *et al.*, 2018). But, it does not overcome the problem of requirement of large plantation land area (Gui *et al.*, 2008) and it also produces a lot of solid wastes (for example the pressed seed cake) which causes a new disposal and waste management problem (Nunes *et al.*, 2009). On the other hand, animal fats cost is higher than that of vegetable oils, and usually contain higher concentration of saturated fatty acids. Thus, normally exist in solid form at room temperature, which consequently would cause many problems in biodiesel production process. It is desirable to use waste frying oil (WFO) and/or waste cooking oil (WCO) as feedstock for second-generation biodiesel, due to its economic and environmental benefits; its low price, which is approximately two to three cheaper than fresh vegetable oils (Zhang *et al.*, 2003; Ismail *et al.*, 2017a). Most of the produced WCO are annually discarded into water streams or dumped into land without any processing. Thus, valorization of the WFO and/or WCO into biodiesel would reduce waste management problems and produce sustainable green fuel (Niju *et al.*, 2014). However, the quality of WFO and/or WCO is very important, as its physicochemical properties depend on the composition of the fresh cooking oil. Not only this, but the WFO and/or WCO would contain undesirable impurities, water and high concentration of FFAs (El-Gendy and Deriase, 2015; El-Gendy *et al.*, 2015a, b). Generally, there are three types of catalyst used in the transesterification reaction; homogenous, heterogeneous and enzymes. Even, the enzymes can be either homogenous or heterogeneous depending on its mobility (Gorji and Ghanei, 2014). The main factors affecting the biodiesel yield is the reaction time and temperature, mixing rate, type of oil feedstock, catalysts and alcohol, amount of water and FFA in the oil feedstock, catalyst concentration, and alcohol to oil ratio. Transesterification can be performed at different temperatures, and the biodiesel yield increases with the increase of the reaction temperature, but to a certain limit. If the reaction temperature increased beyond the optimum level, the biodiesel yield decreased due to the acceleration of saponification reaction of triglycerides (TG) at higher temperature (El-Gendy *et al.*, 2015c). Moreover, if the temperature increases, reaching the boiling point of the alcohol used in the transesterification reaction, a lot of alcohol bubbles would be formed that would inhibit the mass transfer on the phase's interface, which would lower the reaction rate. In case of enzymes; the conversion reaction increases over the temperature range 30–55°C, otherwise denaturation of enzymes would occur losing its activity. The reaction rate seems to be sigmoidal i.e. slow at both, the beginning (to overcome the mass transfer limitation) and at the end of the reaction (since, the glycerol phase separates out taking most of the catalyst with it, as the reaction proceeds) but the rate is fast at the intermediate stage (Slinn and Kendall, 2009). Thus, the reaction time should be optimized to obtain high biodiesel yield with a reduced production cost. The transesterification reaction is either two phases in homogenous catalyzed process or three phases in heterogeneous catalyzed process. So it proceeds, as mentioned before, in three stages; mass transfer between the reactants, the transesterification reaction then the establishment of equilibrium. Mixing rate and efficiency are very important factors to overcome the mass transfer problem and improve the reaction rate, thus, consequently increase the biodiesel yield and decrease the overall production cost as it would save time and energy consumption (El-Gendy *et al.*, 2015c, 2016c). However, beyond certain limit of stirring there would be no significant increase in the biodiesel yield, especially when applying enzymes or heterogeneous solid porous catalysts (El-Gendy *et al.*, 2015b).

Nowadays, alkali-catalyzed transesterification reaction is the most widely applied process for biodiesel production. However, the maximum amount of FFA content is below 2.5 wt.%, and pretreatment step is required for feedstock with higher FFA content. Homogenous alkaline catalysts used for transesterification reaction are; NaOH, KOH and alkoxides, such as sodium and potassium methoxides. The alkali-catalyzed transesterification proceeds approximately 4000 times faster than that of acid-catalyzed ones and can be performed in relatively modest operating conditions (Wen *et al.*, 2010). Thus, the alkali-catalyzed process is the most commercially applied.

Shalaby and El-Gendy (2012) reported a two-steps alkaline transesterification process of high acid value WCO (5.1 mg KOH/g) with methanol using KOH. That yielded 85 wt% purified biodiesel yield. El-Gendy *et al.* (2014c) reported the application of response surface methodology (RSM) based on central composite design (CCD) of experiments to optimize the biodiesel production from WFO collected from different local restaurants using methanol and KOH, where nearly a complete conversion occurred at 9:1 M:O, 0.6% KOH wt.%, 1 h at 60°C and 300 rpm. The fuel properties of the produced biodiesel and bio-petro-diesel blends were measured and compared with those of petro-diesel and the American Society for Testing and Materials standards for biodiesel and bio-petro-diesel blends (ASTM D6751 and D7467, respectively), and acceptable agreement was observed. The three main advantages of the produced biodiesel were; zero sulfur content, high flash point (170°C) and compatible viscosity (5.69 cSt). Thus, no emissions of SOx, safer upon storage and can be applied in existing engines without hardware modifications. Mathematical correlations were also established to describe the changes of basic properties of the produced fuel with the volumetric percentage of the biodiesel for the bio-petro-diesel blends. Where, the changes in viscosity, flash point, calorific value, diesel index, and sulfur content were found to be following polynomial of 3rd order, while those of density, total acid number, and cetane number followed polynomial of 4th order. El-Gendy *et al.* (2015c) reported the application of RSM based on rotatable central composite design CCRD of experiments with three levels (coded by –1, 0, and +1) and $\pm\ \alpha$ of $\pm$ 1.82116, to study the significance and interactive effects of five variables; M:O molar ratio, catalyst concentration, operating temperature, reaction time and mixing rate on KOH catalyzed transesterification reaction of WCO to maximize the biodiesel yield. Approximately 99 wt.% biodiesel yield was obtained at optimum operating conditions of; 7.54:1 M:O, 0.875 wt.% KOH concentration, 52.7°C, 1.17 h and 266 rpm. Estimation for the cost of the biodiesel produced from waste cooking oil using the selected optimum conditions for the transesterification process has been done, which predicted income of approximately US$ 1.821/gallon biodiesel. Hence, biodiesel production from WCO could be a good substituent to petro-diesel, leading to savings in foreign exchange for importing and generating employment. To catch the wave of the third-generation biodiesel, Mostafa and El-Gendy (2017) reported the transesterification of lipid extract from the sustainable microalgae *Spirulina platensis* with methanol in the presence of NaOH catalyst. That yielded 16 and 3.45 wt.% biodiesel and crude glycerol relative to the weight of obtained microalga biomass. The results of that study indicated that the microalga *S. platensis* is a valuable candidate to be used for biodiesel production, due to its high growth rate 2.23 g/L/d, sufficient lipid content, requiring just a simple and inexpensive culture medium and producing other valuable byproducts (e.g. allophycocyanin, C-phycocyanin, carotenes and chlorophyll) which would decrease the overall cost of biodiesel production. The produced pigments can be used as natural coloring and therapeutic. After the performance of all extraction steps for production of the aforementioned valuable products, the waste biomass was dried, pressed and its calorific value was measured to be 39 MJ/kg. Thus it can be used as another source of bio-energy i.e. solid biofuel. Instead of that, the biomass after lipid extraction can be also used as animal food supplement, as it has considerable amounts of total proteins, carbohydrates and nucleic acids. Moreover, the carbohydrates content suggested also the possible production of other kinds of biofuels e.g. bioethanol. It is worth mentioning that the extreme conditions of salinity and pH in which *S. platensis* can survive add to its value to be used for production of biodiesel, as it dimensioned the use of fresh water. Aboelazayem *et al.* (2018) applied RSM based on RCCD to optimize the transesterification of castor oil into biodiesel using KOH. The optimum conditions for 97.82% biodiesel yield have been achieved at M:O molar ratio of 5.4:1, potassium hydroxide (KOH) catalyst concentration of 0.73%, temperature of 64°C, time of 2.5 h and stirring rate of 320 rpm. The kinetic calculations concluded that the transesterification reaction is pseudo second order with reaction rate constant, activation energy and frequency factor of 0.16 $M^{-1}min^{-1}$, 21.95 kJ/mol and 6.02 M^{-1} min^{-1}, respectively. Most of the physicochemical properties agree with both EN14214 and ASTM D6751, with zero sulfur content, thus eliminates SOx emissions. However, both density and viscosity of the produced biodiesel exceeded the limits of biodiesel standards due to the existence of the hydroxyl group in ricinoleic acid. Finally, a case study investigating the performance and emissions on a direct injection (DI) diesel engine (Figure 23) fuelled by biodiesel/petro-diesel blends (5, 10, 15 and 20% v:v) has been performed. That concluded a significant reduction of greenhouse and toxic gases (i.e. CO, CO_2, HC and PM) with the increment of biodiesel percentage in the bio/petro-diesel blends (Figure 24).

However, the NOx increased, as biodiesel is an oxygen rich fuel which accordingly increases the oxygen level in the combustion environment where oxygen emissions have reported increasingly effect. Moreover, the exhaust temperature has increased in that study by 22% from B0

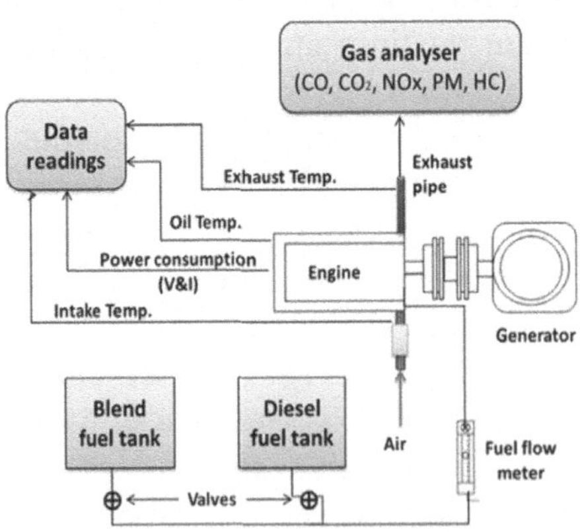

Engine specifications.

Model	Specifications
Engine Configuration	Single cylinder, vertical, compression ignition and four strokes.
Bore	85 mm
Stroke	110 mm
Compression ratio	17:1
Cooling system	Water cooled
Rated output	3.8 kW at 1500 rpm
Number of valves	2

Figure 23: Schematic Diagram of an Engine Setup

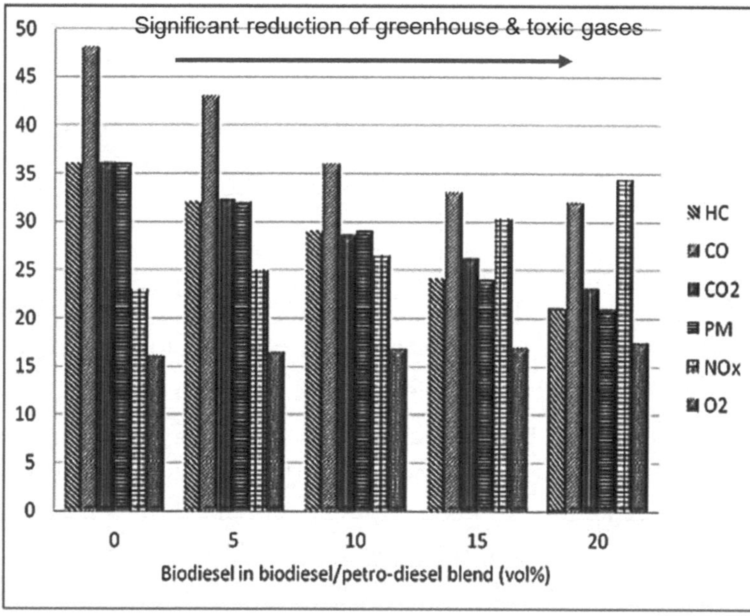

Figure 24: Exhaust Emission Analysis, where HC, CO, CO_2, PM, NOx and O_2 were Measured in; mg/L, %vol. (*103), %vol. (*10), dg/m^3, %vol.(*101), %vol., Respectively

to B20 blends. Accordingly, these reasons have introduced the suitable environment for NOx emissions to increase. It has been concluded that up to B10 blend, the physic-chemical properties agree well with the biodiesel blends standard properties (ASTM D7467). It was worth mentioning that the brake specific fuel consumption (BSFC) increased with the increment of biodiesel ratio in the blend, with a slight decrease brake thermal efficiency (BTE). That might be due to the lower cetane index, lower calorific value, smaller ignition delay and higher BSFC of the produced biodiesel relative to petro-diesel, thus, the output of combustion is reduced which consequently increase the fuel consumption and BTE.

It should be known that, the alkaline-homogenous catalysts are hygroscopic and absorb water from air during storage. The use of alkaline catalysts has also some drawbacks; the alkali reacts with the FFAs producing large amounts of unwanted soap, creating serious problems; consumes the catalyst, reduces the catalyst efficiency, thus, more catalyst will be needed and it also causes an increase in viscosity and formation of gels. It also causes difficulties in glycerol and biodiesel separation that lower the biodiesel yield and increase the overall production cost. It is generally recommended that the acidity of the oil feedstock to be less than 1 mg KOH/g oil (Mazubert *et al.*, 2013). Usually the reaction requires high alcohol/oil molar ratio (such as 6:1 or higher) instead the stoichiometric 3:1 ratio to achieve high biodiesel yield. Also, the removal of unreacted catalysts is a technical challenge and brings extra cost to the final product. A large amount of water is required for the washing step to neutralize the produced biodiesel and remove the unreacted catalyst. Thus, a huge amount of very toxic wastewater from the transesterification process is produced which produces a new waste management problem.

Lipase enzymes are highly specific, non-polluting, and chemically clean. Moreover, overcome most of the alkaline-transesterification problems; tolerates high FFAs and water contents in the oil feedstock, avoiding the soap formation. Thus, the purification of biodiesel and glycerol is much easier. But, enzymes are expensive to be used on commercial scale. The production cost of a lipase catalyst is significantly greater than that of acid or alkaline ones. High concentrations of FFAs and short chain alcohols (methanol and ethanol) would denature the enzyme and glycerol the main by-product of the trancesterification reaction has also a serious negative effect on lipase. Enzymes need more stringent control on the reaction conditions, for example; the reaction alcohol/oil molar ratio, temperature pH and the co-solvent type otherwise the deactivation of the enzyme would occur (El-Gendy *et al.*, 2015b). Enzymes require limited amount of water to be active, which should be lower than the monolayer amount of water molecules around the enzyme molecules. If the water content in the feedstock exceeds this limit; it will deactivate lipase and consequently will decrease the conversion and biodiesel yield (Shah *et al.*, 2003; Kumari *et al.*, 2009).

Ismail *et al.* (2017a) applied RSM based on face center composite—1/2 factorial fraction design (FCCD) of experiments to optimize batch transesterification process of waste cooking oil with methanol using immobilized *Candida antarctica* lipase (Novozym 435) to maximize the production of high-purity biodiesel. The optimum operating conditions were found to be: 3.63:1 M:O, 8.94 wt% Novozym 435, 45.23°C, 2.76 h, and 535.84 rpm. That produced 97 wt% biodiesel yield with % purity of ≈ 91.79%. The main drawback observed in that study is the occurrence of changes in the physical characteristics of the immobilized Novozym 435 after the first experimental run, where, it was impossible to recover the Novozym 435, for reusability. In an attempt to produce lipase for biodiesel production, Ismail *et al.* (2018) isolated mesophilic and neutrophilic *Bacillus stratosphericus* PSP8 from petroleum hydrocarbon-contaminated soil with high lipase productivity 47 U/mL. RSM based on CCF enhanced the lipase productivity by 1.6-fold. Preliminary scaling up of the validated optimized process was carried out in a batch 10 liter stirred tank bioreactor, applying the optimum predicted operating conditions; pH 6.98, 348°C, 2.2 × 10^6 cells/mL, 200 rev/min, 482 g/L tributyrine concentration, 1% sucrose and 01% yeast extract. This yielded 89 U/mL at the late log phase of bacterial growth (48 h). Logistic kinetic model effectively characterized the submerged fermentation process, and the maximum specific growth and lipase production rates were estimated to be 0.338 and 0.164 h^{-1}, respectively.

The basic heterogeneous catalysts can be classified into four groups; a single component metal oxide, zeolites, supported alkali metal and clay minerals (Kumar *et al.*, 2014). Using heterogeneous solid catalysts; such as metal hydroxides and alkali earth metal oxides; such as calcium oxide, magnesium oxide and strontium oxide overcomes most of the drawbacks of the homogenous catalysts. In addition, it is more environmentally friendly. Solid catalysts are not consumed or dissolved in the reaction mixture, can be easily separated from the products, can be regenerated and reused. No washing step for the product is required, where biodiesel and glycerol with high purity would be easily obtained. It can be applied in batch and continuous processes, which improves the economy of biodiesel production. It can catalyze the esterification and transesterification reactions, thus, the pre-esterification step in high FFAs feedstock would be eliminated. Among alkali earth metal oxides, CaO has attracted many researchers, due to its relatively high basic strength, high activity, low cost, low solubility in methanol, it is considered as environmentally benign catalyst, readily available, can be prepared from cheap sources (e.g. calcium hydroxide) and natural ones (e.g. limestone, dolomite, and natural waste shells). Moreover, CaO can be applied successfully in the transesterification of low quality oils, as it can tolerate high FFAs and moisture contents. In addition, it is characterized by long life activity and can be reused for approximately eight cycles with minimum loss of its activity.

The main drawback of CaO, that, it is easily losing its catalytic activity when exposed to air. This is due to its reaction with humidity and CO_2, producing CaOH and CaO_3, respectively. In order to overcome this problem, it is recommendable to thermally pretreat the catalyst at 700°C before its application in the transesterification reaction (Granados *et al.*, 2007; Ismail *et al.*, 2017b).

El-Gendy and Deriase (2013) studied the effect of the type of the feedstock on the biodiesel yield using chemical CaO as a basic heterogeneous catalyst. Based on 3- level D- optimal design, a statistical design of experiments (DOE) strategy has been performed to evaluate and investigate the biodiesel production process, involving as factors: methanol-to-oil molar ratio (M:O), CaO concentration (wt%), reaction time (min) and type of waste cooking oil WCO feedstock. MATLAB software was employed for experimental design and data analysis. The type of the oil feedstock found to statistically affect the biodiesel yield. The optimum values of the selected predictor variables were obtained according to response surface optimizer. The optimum conditions for the transesterification process were M:O molar ratio of 10.55:1, catalyst concentration of 6.653 wt% and reaction time 100.54 min. The perfect type of oil was waste frying corn oil WFCO. Those optimum conditions yielded 95% conversion. El-Gendy *et al.* (2014d) used the RSM based on face centered design (FCCD) of experiments to statistically evaluate and optimize the conditions for maximum production of biodiesel from waste frying sunflower oil using chemical CaO and to study the significance and interaction effects of M:O molar ratio, catalyst concentration, and reaction time on the yield. Based on an estimated quadratic model, the optimum operating conditions; 7.05:1 M:O, 8.21 CaO wt.% and 1.5 h at 60°C reaction temperature and 300 rpm mixing rate yielded 93 wt.% biodiesel. The chemical CaO catalyst has been used for five successive cycles, producing sufficient biodiesel yields. The most important advantage was; the produced biodiesel was free of sulfur thus prevent the emission of SOx. The presence of water in oil feedstock within a certain concentrations, reported to enhance the transesterification efficiency, as it would have been adsorbed on the CaO surface producing the OH^- that enhances the reaction (Ismail *et al.*, 2017b). Moreover, relatively low CaO concentration and high M:O ratio decreased the biodiesel yield due to the dilution effect. Nevertheless, the low M:O molar ratio and high CaO concentration also decreased the biodiesel yield due to the mass transfer resistance between the reactants due to the high viscous reaction mixture (Ismail *et al.*, 2017b). The importance of optimizing the exact amount of methanol during the transesterification of used domestic waste cooking oil using chemical heterogeneous basic CaO catalyst has been discussed with the elucidation of transesterification mechanism (Figure 25). Increasing the M:O molar ratio to a certain extent enhances the formation of methoxy species on CaO surface, thus shifting the equilibrium to the forward direction, increasing the biodiesel yield. Moreover, the produced glycerol would react with CaO producing less active calcium glyceroxide catalyst. However, further increase would shift the equilibrium to the reverse direction, as glycerol would dissolve in excess methanol. Optimization of the reaction temperature is also important to overcome the mass transfer resistance and vaporization of methanol. The predicted optimum conditions 4.7:1 M:O, 5.67 wt.% CaO concentration, 50.8°C, 2.28 h and mixing rate of 577.95 rpm with high quality biodiesel yield approximately 92%, and the reusability of the catalyst for five successive cycles without losing its activity recommends that cost effective process for biodiesel production (Ismail *et al.*, 2017b).

$$CaO + H_2O \longrightarrow Ca(OH)_2$$
$$Ca(OH)_2 \longleftrightarrow Ca^{2+} + 2OH^-$$
$$CaO \longleftrightarrow Ca^{2+} + O^{2-}$$
$$CH_3OH + O^{2-} \longleftrightarrow CH_3O^- + OH^-$$
$$CH_3OH + OH^- \longleftrightarrow CH_3O^- + H_2O$$

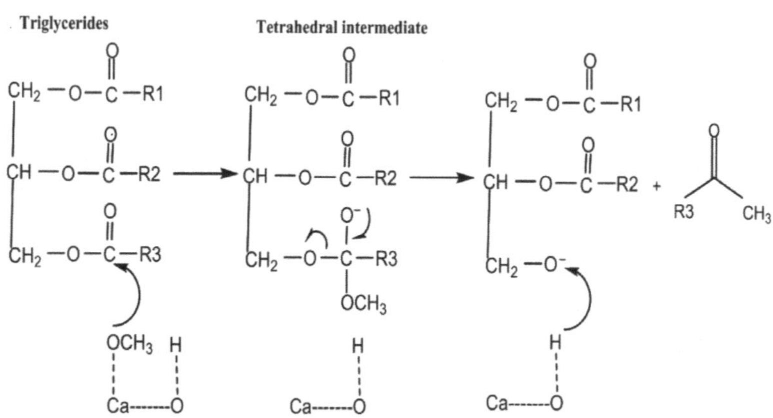

Figure 25: CaO Oxide Catalyzed Transesterification Reaction

The cost of the applied catalyst affects the overall production cost as well. Thus, applying catalysts prepared from waste materials will decrease the cost and enhance the development of a sustainable biodiesel production process (Farooq *et al.*, 2015). The preparation of inexpensive heterogeneous catalysts with high activity is very important to produce biodiesel with competitive cost and quality with petro-diesel. Moreover, waste shells and animal bones cause a lot of waste management problems and environmental pollution (such as nasty smells) as they are usually disposed in landfills without any pretreatments. The natural basic heterogeneous catalysts from such natural wastes have the advantages of sustainability, abundance, low cost, porous structure, non-toxic, high basicity and low solubility in biodiesel (El-Gendy and Hamdy, 2014).

El-Gendy and Hamdy (2014) studied the efficiency of different bio-catalysts; Novozym 435, calcined animal bones at 700°C, calcined eggshells and mollusks shells at 800°C for biodiesel production from WCO and compared their activity with that of chemical commercial CaO. Those yielded 87, 90, 94, 93 and 95 wt.% biodiesel, respectively in batch transesterification reactions at 60°C, 300 rpm, 60 min, 9:1 M:O molar ratio and 9 wt.% catalyst loading. All the tested catalysts kept their catalytic activity up to five successive transesterification processes except the Novozym which has been only used once. El-Gendy *et al.* (2014e) reported the production of biodiesel from the wastes of fish restaurants. Where, RSM based on D-optimal design of experiments (DOE) to study the significance and interactive effects of M:O molar ratio, catalyst loading, reaction time and mixing rate on the biodiesel yield, and to optimize the transesterification reaction of WCO with methanol using CaO produced by calcination of waste mollusks and crabs shells at 800°C and 700°C, respectively. The prepared green catalysts yielded 96 and 98 wt.% biodiesel yields at optimum operating conditions of 6:1 and 12:1 M:O molar ratio, 4.5 and 7.3 wt.% catalyst loading, 82 and 30 min reaction time, and 220 and 214 rpm, respectively. Moreover, the high basicity of the prepared green CaO promoted their activities over the commercial chemical CaO and Novozym 435. The later produced 83 and 35 wt.% and 78 and 40 wt.% biodiesel yield applying the two operating conditions of the prepared green catalysts, respectively. Nevertheless, the high basicity of CaO prepared from mollusks (52.18 mmol CO_2/g) was very advantageous as lower amount of methanol and catalyst are required for producing higher biodiesel yield. In another study, El-Gendy *et al.* (2014f) used calcined snails' shells at 800°C, which were collected from Egyptian shoreline, to produce biodiesel from waste frying corn oil. Where approximately 96 wt.% conversion occurred at 6:1 M:O molar ratio, 3 wt% catalyst concentration, 60 min, 200 rpm and 60°C. That process was proved to be cost effective from the points of; low preparation time, and low chemicals (alcohol), biocatalyst, and energy consumption. Moreover, the biodiesel yield was comparable with that produced using chemical CaO but higher than the produced applying Novozym 435 which yielded 87 wt.% biodiesel. El-Gendy *et al.* (2015b) reported the preparation of basic heterogeneous fluorapatite ($Ca_5(PO_4)_3F$) catalyst from waste animal bones and elucidated the possible mechanism in the transesterification reaction of WCO, where the reaction would start with the dissociation of $Ca_5(PO_4)_3F$ or hydroxyapatite $Ca_5(PO_4)_3OH$ and methanol and formation of methoxide anion from the reaction between methanol and a F^- or OH^- ion. In the first step, surface O_2^- of $Ca_5(PO_4)_3$ extracts H^+ from methanol to form the methoxide anion on the surface of $Ca_5(PO_4)_3$, which attacks the carbonyl carbon of the triglyceride to form a tetrahedral intermediate. The rearrangement of the intermediate molecule results in the formation of a mole of methylester and diglyceride. In the former step, H^+ from the surface of $Ca_5(PO_4)_3$ attacks O^- of the diglyceride anion for the formation of diglyceride molecule along with the regeneration of $Ca_5(PO_4)_3$. A new methoxide anion attacks the formed diglyceride, producing another mole of methylester and monoglyceride. Finally, another methoxide attacks the monoglyceride producing a total of three moles of methylester and a mole of glycerol. RSM based on D-optimal design of experiment was applied for optimizing the transesterification process applying such green synthesized fluorapatite and compare its efficiency with the commercial available immobilized lipase Novozym 435. The optimum combination for transesterification were found to be 7.35:1 and 6:1 M:O, 4.35 and 8.8 catalyst wt%, 91 and 96 min, and 331 and 394 rpm at 60°C, for the prepared fluorapatite and Novozym 435, respectively, with maximum biodiesel yield of ≈ 96 and 62%, respectively. That proved the superiority of fluorapatite, due to its sufficient basicity 15.31 mmol/g. Moreover, El-Gendy and Deriase (2015) proved the advantageous of using green synthesized CaO form the simple calcination of eggshells at 800°C, as there was no statistical significant effect on the biodiesel yield upon using different types of oil feedstock. Moreover, such green catalyst is recommendable for biodiesel production, from the viewpoint of short transesterification time, low energy consumption and cost effective transesterification process with high yield of biodiesel fuel (BDF). Since, nearly, complete conversion of sun flower waste cooking oil into high qualified biodiesel occurred at predicted optimum conditions of; 6:1 M:O, 3% eggshells CaO catalyst concentration, 350 rpm, 30 min at 60°C (El-Gendy *et al.*, 2015a). The process optimization based on the DOE was found to be capable and reliable to optimize biodiesel production from WCO using natural CaO prepared by calcination of waste eggshells at 800°C and chemical CaO (El-Gendy *et al.*, 2016c). Where, the activity of the produced green catalyst was comparable to that of chemical CaO, producing high yield of biodiesel ≈ 91 and 98%, respectively, at 8.57:1 M:O molar ratio, 3.99 wt.% catalyst concentration, 31 min reaction time, and 398.88 rpm mixing rate at 60°C. EL-Gendy *et al.* (2016c) attributed the recorded higher activity of chemical CaO relative to that of natural CaO prepared from eggshells to; (1) the higher basicity of chemical CaO, which recorded 42.4 and 25.3 and mmol CO_2/g, respectively, (2) its lower crystalline size (35.52 and 47.39 nm, respectively) and (3) higher; specific surface area (106.37 and 56.82 m^2/g, respectively), average pore size (4.12 and 2.89 nm, respectively), and pore volume (0.303 and 0.058 cm^3/g, respectively).

SUSTAINABLE INTEGRATION OF BIOETHANOL AND SOLID BIOFUEL PRODUCTION WITH WASTEWATER TREATMENT

Today, the transportation sector worldwide is almost entirely dependent on petroleum-based fuels. It is responsible for 60% of the world oil consumption. In addition, transportation sector accounts for more than 70% of the global carbon monoxide (CO) emissions and 19% of the global carbon dioxide (CO_2) emissions. Around the world, there were about 806 million cars and light trucks on the road in 2007. These numbers are projected to increase to 1.3 billion by 2030 and to over 2 billion vehicles by 2050 (Balat, 2011, Ray *et al.*, 2013). This growth is expected to consequently affect the stability of ecosystems and global climate as well as global oil reserves. The dramatic increase in the price of petroleum, and the finite nature of fossil fuels, in addition to the increasing concerns regarding environmental impact, especially that related to the greenhouse gas (GHG) emissions, and the health and safety considerations, are forcing the search for new energy sources and alternative ways to power the world's motor vehicles (Abo-State *et al.*, 2013; Soliman *et al.*, 2018). An alternative fuel must be technically feasible, economically competitive, environmentally acceptable, sustainable, and readily available. Biomass-based fuels, offer many advantages over petroleum-based fuels (Patel, 2012): such as, (1) biofuels can be produced from common biomass sources which contribute to sustainability, (2) they reduce CO_2 emission, although of the biomass consumption, (3) bioethanol for example, has a considerable environmentally friendly potential because it is a oxygenated fuel, (4) they are biodegradable, and (5) combustion of bioethanol results in relatively low emissions of volatile organic compounds, carbon monoxide and nitrogen oxides (Abo-State *et al.*, 2013). The main drawback of first generation bioethanol production is that it mainly depends on edible sources, for example, sugar cane in Brazil and corn in USA. Such raw materials represent 40–70% of the bioethanol production cost. This causes the food versus fuel problem, consumes a lot of water and leads also to deforestation which consequently adds to the problem of climate change (Abo-State *et al.*, 2013). The annual world production of sugarcane bagasse and rice straw as examples of lignocellulosic biomass is estimated to be ~279 million metric tons (MMT) and ~731 million tons (Ray *et al.*, 2013; Singh *et al.*, 2014). Most of it is left in landfill to decay and/or get rid of it by open-burning, which cause a lot of environmental problems (Abo-State *et al.*, 2013). The development of second-generation bioethanol using such biomass for production of bioethanol and other agro-industrial wastes such as the sugar-beet pulp and molasses has been widely studied and applied (Chandel *et al.*, 2012; Madian *et al.*, 2012; Abo-State *et al.*, 2013; Belal, 2013; El-Gendy *et al.*, 2015d; Paramjeet *et al.*, 2018), as it can increase the sustainability of feedstock and bioethanol production without competing with food production or the cultivation of farmland and from the waste-management standpoint, producing bioethanol from agricultural wastes is environmentally beneficial. However, the bioconversion of lignocellulosic biomass to fermentable sugars is difficult and expensive because of the physio-chemical structural of such biomass, as cellulose and hemicellulose are found together with lignin in an intense cross linked, rigid lingo-cellulosic complex (Abo-State *et al.*, 2013). That hinders the enzymatic digestibility of cellulose and hemicellulose. To overcome this problem, the lignocellulosic biomass should be pretreated by an efficient and cost-effective pretreatment method to improve the rate of enzymatic hydrolysis using cellulases and hemicellulase and increase the yields of fermentable sugars from cellulose or hemicellulose and consequently the bioethanol yield (Rabelo *et al.*, 2011; Han *et al.*, 2012; Kumar and Pushpa, 2012; Zheng *et al.*, 2013). Further, come the step of bioethanol fermentation applying bacteria or yeast with good efficiency to ferment both hexoses and pentose and with stand the presence of inhibitors and high bioethanol concentrations (Vane *et al.*, 2012; Abo-State *et al.*, 2014b; El-Gendy *et al.*, 2015d; Hamouda *et al.*, 2016).

Abo-State *et al.* (2013) isolated *Candida tropicalis* Y-26 and *Saccharomyces cerevisiae* Y-39 for its sufficient ability to utilize different types of sugars (pentoses, hexoses and disaccharides). Moreover, *Trichoderma viride* F-94 and *Aspergillus terreus* F-98 were also isolated for their ability to produce cellulases and hydrolyze lingo-cellulosic wastes in one solid state fermentation step without acid or alkaline pretreatment process (Madian *et al.*, 2012). Upon the application of separate biological hydrolysis and fermentation (SHF) process for bagasse by *Trichoderma viride* F-94 and *Candida tropicalis* Y-26, it yielded 226 kg of ethanol/ton bagasse. While, upon applying *Aspergillus terreus* F-98 and *Saccharomyces cerevisiae* Y-39, it yielded 185 kg of ethanol/ton bagasse (Abo-State *et al.*, 2013). Madian *et al.* (2012) reported bioethanol yield of approximately 50 and 39 gallon of ethanol/ton rice straw upon applying *Trichoderma viride* F-94 for fungal hydrolysis of rice straw in a solid state fermentation process followed by bioethanol fermentation using *Candida tropicalis* Y-26 and *Saccharomyces cerevisiae* Y-39, respectively. However, upon the application of SHF using *Aspergillus terreus* F-98 and *Candida tropicalis* Y-26 or *Saccharomyces cerevisiae* Y-39 it yielded 38 and 45 gallon of ethanol/ton rice straw, respectively (Abo-State *et al.*, 2014b). In an attempt to decrease the process optimization cost, response surface methodology based on central composite design was applied to decrease the number of experiments required for estimating the optimum conditions for maximum bioethanol production form untreated sugarcane molasses (El-Gendy *et al.*, 2013). A statistical quadratic model equation, was predicted and a maximum ethanol production of 255 g/L was obtained in a *Saccharomyces cerevisiae* Y-39 batch fermentation process at optimum operating conditions of approximately 71 h, pH 5.6, 38°C, molasses concentration 18% wt.%, and 100 rpm. In another study, El-Gendy *et al.* (201d5) applied response surface methodology based on central composite of experiments to statistically evaluate and optimize the conditions for a thermal acid pretreatment of sugar-beet pulp (SBP). The optimum applied conditions for the pretreatment of SBP for further separate fungal hydrolysis and fermentation experiments for ethanol production, were selected based on an economic consideration to be; 0.1 N HCl, 6 min, 15 wt% SBP and 120°C. That hydrolyzed approximately 57% and 40% of the hemicellulose and cellulose contents of SBP, with an increase in the ratio of cellulose/hemicellulose (C/H) from ≈ 1.38 to 1.95, and ≈ 51% total weight loss with a considerably high scarification or hydrolysis percentage of ≈ 30%. Moreover, the pectin was sharply decreased during the acid pretreatment while a large part of the protein remained insoluble, which would be favorable for further fungal hydrolysis, since pectin usually act as a physical barrier for enzymatic hydrolysis and restricts the access of hydrolytic enzymes to the digestible parts of substrate and the remained protein would provide sufficient nitrogen source to support the fungal growth. The sugars released from acid hydrolysis were glucose, galacturonic acid, arabinose, xylose and galactose. Solid state fermentation of the acid pretreated SBP has been performed using *Trichoderma viride* F94. That expressed high delignification efficiency (≈ 87%), with high cellulolytic activity, recording a decrease in the hemicellulose and cellulose of ≈ 52 and 57%, respectively. It showed also some enzymatic hydrolysis on pectin recording a pectin loss of ≈ 36%, with good utilization of the protein content of the acid pretreated SBP of ≈ 57%. The solubilized protein in the acid and fungal hydrolysates, would favor the further batch ethanol fermentation, as the yeast would use it as a nitrogen source to support its growth for ethanol production. Glucose and maltose are the main produced sugars after fungal hydrolysis. In that study, the total scarification or hydrolysis % after acid pretreatment and fungal hydrolysis reached ≈ 86%. The thermal acid pretreatment and fungal hydrolyzes were integrated with two parallel batch fermentation processes of the produced hydrolysates using *Saccharomyces cerevisiae* Y39, that yielded a total of ≈ 48 g/L bioethanol, at a conversion rate of

≈ 0.32 g bioethanol/g SBP. The proposed integrated process produced approximately 97.5 gallon of ethanol/ton SBP. Hamouda *et al.* (2015a) have applied also RSM based on CCD for optimization sugarcane molasses batch fermentation process and maximize the bio-ethanol yield using the Egyptian yeast isolate *Candida tropicalis* strain HSC-24. That yielded 0.6 g ethanol/g total sugars at the elucidated optimum conditions of; initial pH5, 20% initial molasses concentration, 35°C, 100 rpm and 66 h. In another study using another Egyptian isolate *Pichia veronae* strain HSC-22, RSM based on CCD was also applied to optimize the fermentation of untreated molasses (Hamouda *et al.*, 2015b). That yielded 0.44 g ethanol/ g total sugars at the elucidated optimum conditions of; initial pH5, 25% (w:v) initial molasses concentration, 35°C, 116 rpm, and 60 h. Moreover, El-Gendy *et al.* (2015e) performed statistical optimization of alginate immobilization process of *Candida stauntonica* strain MY1 for bioethanol production applying RSM based on CCFD. That was in an attempt to maximize the bioethanol yield, protect the fermenting microorganisms form the inhibitors produced during the fermentation process, increase its life time and reusability in successive fermentations batches and/or continuous fermentation reactors. The optimum conditions for cell immobilization were found to be 2.5 mm bead size, 2.5 g/L initial inoculum size and 5.5 g/L alginate concentration. That yielded 0.42 g ethanol/g glucose, which was about 2.3 fold higher than that produced with free cells batch fermentation operated under the same conditions of; 48 h, pH 5.5, 30°C and 100 rpm. The immobilized cells showed good stability, with long storage time 21 d and can be used for four successive batches with maximum bioethanol productivity. Kheiralla *et al.* (2018a) reported enzymatic scarification of tomato pomace yielded sufficient amount of total reducing sugars (TRS), which with a further separate batch bioethanol fermentation using *S. cervisae* yielded 0.3 g bioethanol/g tomato pomace. To decrease the overall cost of the process and make it much profitable as much it could be, 54.40 g single cell protein/L pomace hydrolysate was produced as a byproduct.

Unfortunately, bioethanol production process is still very costly in terms of the energy input. However, the possibility of valorizing the lignocellulosic materials in its entirety, via linking bioethanol production with the co-production of other valuable co-products (e.g. enzymes), thermo-chemical fuels and/or power, and valorize of the waste byproducts obtained from each step in the process into valuable products (e.g. nano metals and/or nano metal oxides), can reduce the production costs. Madian *et al.* (2012) reported the production of cellulases from rice straw via solid state fermentation process, *Trichoderma viride* F-94 and *Aspergillus terreus* F-98. Kheiralla *et al.* (2018a) reported the production of endoglucanase CMCase 173.3 IU/g protein and exoglucanase FPase 17.33 IU/g protein using *Synchytrium endobioticum*, cellobiase 19.43 IU/g protein using *Aspergillus niger* and xylanase 106.9 IU/g protein *using Penicillium chrysogenum* via submerged fermentation processes of different lignocellulosic wastes. Kheiralla *et al.* (2018b) reported the production of xylanase from an agro-industrial waste; corn cobs xylan using *Fusarium moniliforme* in a submerged fermentation process. That xylanase has been used for the scarification of some agro-industrial wastes, tomato pomace, sugar beet pulp, and olive oil cake with hemicellulose content of 31.42, 38.77, and 34.35% (w/w), and yielded 800 ± 60, 1320 ± 87, and 210 ± 10 mg total reducing sugars/100 g substrate, respectively. Such enzymes have different industrial applications for example bioethanol, juice industries, food processing, animal feed and papers industries. Kheiralla *et al.* (2018a) proved the concept and the feasibility of the production of different valuable products from tomato (*Solanum lycopersicum*) agro-industrial wastes (AW). Where, different sub-products and biofuels can be produced (Figure 26); lipids, protein and ash as reach sources of essential carotenoids, amino acids and inorganic minerals, respectively; lignin which has different industrial applications, holocellulose (cellulose and hemicellulose) for the production of bioethanol and single cell protein and

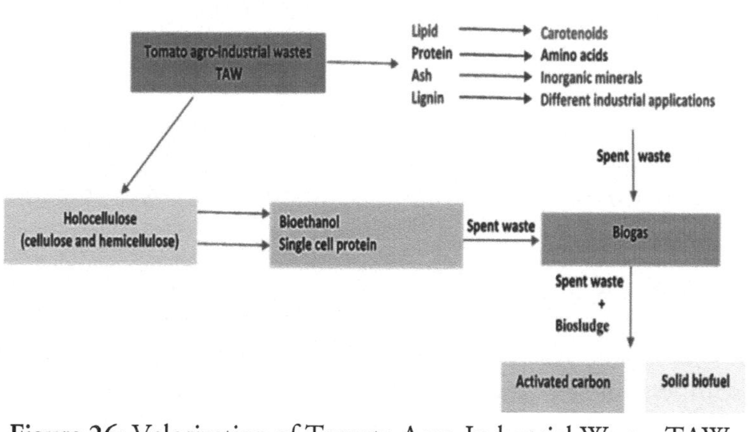

Figure 26: Valorization of Tomato Agro-Industrial Wastes TAW

the spent wastes obtained from all the aforementioned extraction and bio-refining processes can be used for biogas production. Moreover, to reach for the zero waste point; the spent wastes and bio-sludge after biogas fermentation, can be used for production of solid biofuel or activated carbon, which has different valuable applications.

Macroalgae are abundant, sustainable, with high carbohydrate content and little or no lignin. Thus, sugars can be easily extracted compared with lignocellulosic biomass, with a decrease in energy requirements and the overall cost. Macroalgae recycle CO_2 and it does not require arable land, fertilizers, pesticides or fresh water resources. Its cultivation is highly productive and has no negative impact on food supplies and fresh water resources. The spent wastes of macroalgae are already used in biogas and biofertilizer production (Ben Yahmed *et al.*, 2016). Different marine macroalgal strains, red *Jania rubens*, green *Ulva lactuca* and brown *Sargassum latifolium*, have been collected from Egyptian Mediterranean and Red Sea shores with carbohydrate content ranged between 12–20% (Soilman *et al.*, 2018). Those macroalgae were found to be also good source of lipids, proteins and inorganic minerals. Moreover, to catch the wave of third-generation bioethanol, Soliman eta l. (2018) proved that macroalgae can be considered as a promising feedstock, for the third-generation bioethanol, through a simple hydrothermal hydrolysis process of *S. latifolium* followed by solid state biological scarification using *Trichoderma asperellum* RM1. That produced total sugars of 510 mg/g macroalgal biomass. Further, the integration of that hydrothermal and fungal hydrolyses of the macroalgal biomass with a separate batch fermentation of the produced sugars using two *Saccharomyces cerevisiae* strains, produced approximately 0.29 g bioethanol/g total reducing sugars. Thus, using *S. latifolium* as feedstock for third-generation bioethanol would act as a triple fact solution; (1) bioremediation process for this ecosystem, as it causes eutrophication problem, (2) a source for renewable energy source and (3) economy savings as its presence affects the touristic activities.

Reaching to the point of zero waste via developing an economically viable wastewater management process using the spent waste biomass from bioethanol production process (i.e. lingo-cellulosic waste after the scarification step and yeast after the bioethanol fermentation step) is a vital research point. Farah and El-Gendy (2013) used the spent waste *Saccharomyces cerevisiae* from bioethanol fermentation process for the removal of Acid Red 14 (AR14) anionic dye. Further, El-Gendy *et al.* (2015f,g) used the spent waste *Saccharomyces cerevisiae* from bioethanol fermentation process for the removal of Reactive Red-84 dye and Basic Blue 41. The biosorbent was regenerated and reused efficiently for four successive times. The adsorbed dyes were completely photodegraded under UV-irradiation using TiO_2 NPs and TiO_2 hydrosol, respectively. El-Gendy and Nassar (2015b) proved the higher efficiency of spent waste *Saccharomyces cerevisiae* for phenol removal than the spent tea waste, recording 17.96 and 13.98 mg/g, respectively. The synergetic effect between nano-polyaniline (PANI) and spent waste *Saccharomyces cerevisiae* for the removal of anionic sulphonated Acid Red 14 (AR) dye from aqueous solution has been also reported (Ahmed *et al.*, 2016). The PANI/biomass recorded maximum adsorption capacity of 430 mg/g compared to PANI and biomass, which recorded 323 and 263 mg/g, respectively. Younis *et al.* (2014, 2015) reported the potential application of rice straw and sugarcane bagasse spent waste biomass residue from bioethanol production process as adsorbents for the removal of phenol and PAHs from petroleum refinery wastewater. Refuse derived fuels (RDF), from spent waste biomass produced as pellets or fluff, are the principal products of waste treatment processes. For example the calorific value of spent waste bagasse and rice straw after solid state fermentation would range from 11 to 14 MJ/kg. However, after the biosorption of such pollutants can reach to approximately 22–24 MJ/kg. Co-firing of RDFs with fossil fuels has become not only an industrial necessity, but also an option for decreasing the overall fuel costs. In another study, spent waste sugarcane bagasse from bioethanol production has been used as an ecofriendly adsorbent to remove diesel oil and kerosene spill (El-Gendy and Nassar, 2016). Such biomass expressed good biosorption capacity 44–86 and 78–110 mg/g over a wide range of pH and salinities. Thus it can be applied for bioremediation of oil spill in different type of water (i.e. fresh or saline). Moreover, the calorific value of the biomass after the biosorption of diesel oil and kerosene recorded $\approx$ 32.91 and 33.61 MJ/kg, respectively. The results recommend the applicability of the low cost, readily available, sustainable and environmentally friendly sugar-cane bagasse for enzymes production, bioethanol production, wastewater treatment, and production of a renewable solid biofuel, which would have positive impact on economy, energy, and environment.

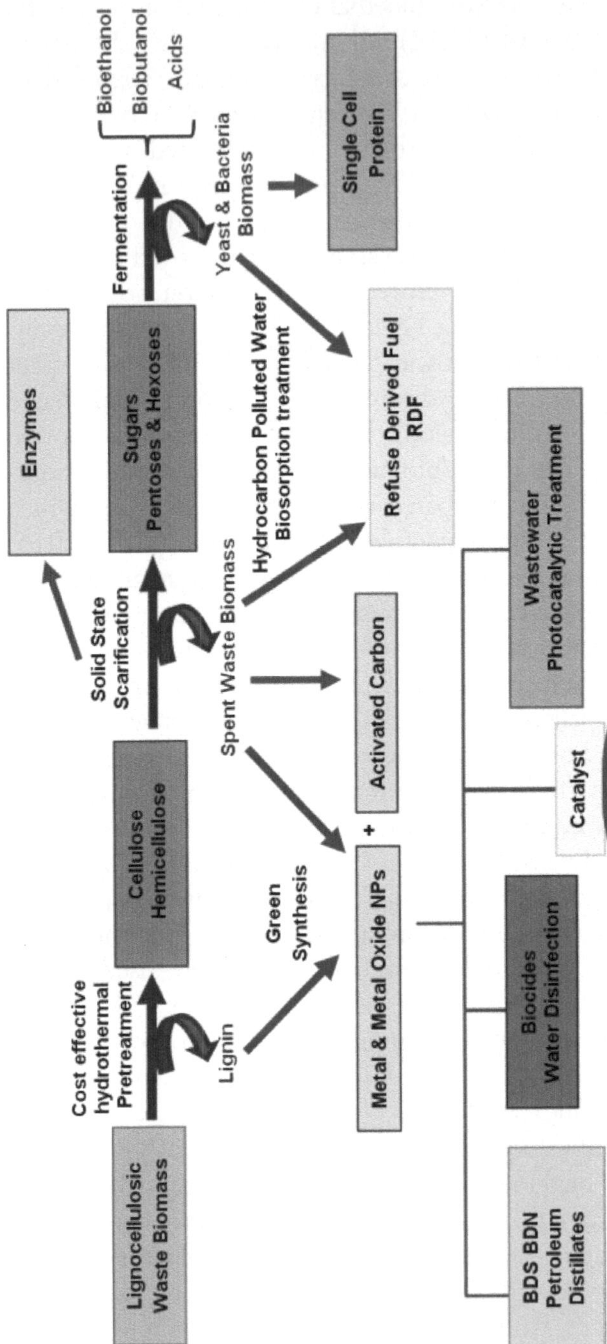

Figure 27: A Suggested Full Integrated Process to Valorize Lignocellulosic Waste Reaching to the Point of Zero Waste

CONCLUSION

It can be concluded from this chapter, that, although waste lingo-cellulosic wastes and agro-industrial wastes in general are thought to be problematic otherwise it can be considered as a hidden treasure that would solve a lot of problems. Figure 27 summarizes a suggested simple integrated process for valorization of lignocellulosic biomass (e.g. bagasse, orange peel, seaweed... etc.) into (1) enzymes with different applications in food processing, animal fodder, paper, cosmetics and pharmaceutical industries and (2) sugars for bioethanol and other biorefineries production. The spent wastes produced during such valorization can be valorized into (1) single cell protein as a valuable product and food supplement, (2) metal and metal oxides which have different applications as catalysts for biodiesel production, photocatalysts for photodegradation of different organo-pollutants and water treatment, biocides for water densification, and it can be also used in biodesulfurization and biodenitrogenation of petroleum and its fractions. The extract of such spent wastes can be also used as corrosion inhibitor and for mitigating of biocorrosion and biofouling. The spent waste itself can act as a good adsorbent for pollutants removal and production of solid refuse derived fuel with considerable calorific value. Moreover, it can be valorized into activated carbon which has different applications. Finally, all the spent wastes from all of the aforementioned suggested steps can be collected in a digester to produce biogas.

REFERENCES

[1] Aboelazayem, O., El-Gendy, N. Sh., Ashour, F.H. and Sadek, M.A. (2017). Complete analysis of castor oil methanolysis for biodiesel production. *10th International Conference on Sustainable Energy and Environmental Protection*, 27–30 June, Bled, Slovenia), Bioenergy and Biofuels, J. Krope, A. Ghani Olabi, D. Goričanec and S. Božičnik, pp. 45–55.

[2] Aboelazayem, O., El-Gendy, N. Sh., Ashour, F.H. and Sadek, M.A. (2018). Biodiesel production from castor oil in Egypt: Process optimisation, kinetic study, diesel engine performance and exhaust emissions analysis. *Energy*, 157: 843–852.

[3] Abo-State, M.A., El-Gendy, N. Sh., El-Temtamy, S.A., Mahdy, H.M. and Nassar, H.N. (2014a). Modification of basal salts medium for enhancing dibenzothiophene biodesulfurization by *Brevibacillus invocatus*

C19 and *Rhodococcus erythropolis* IGTS8. *World Applied Sciences Journal,* 30(2): 133–140.

[4] Abo-State, M.A., Ragab, A.M.E., EL-Gendy, N. Sh., Farahat, L.A. and Madian, H.R. (2013). Effect of different pretreatments on Egyptian sugar-cane bagasse saccharification and bioethanol production. *Egyptian Journal of Petroleum,* 22(1): 161–167.

[5] Abo-State, M.A., Ragab, A.M.E., EL-Gendy, N. Sh., Farahat, L.A. and Madian, H.R. (2014b). Bioethanol production from rice straw enzymatically saccharified by fungal isolates, *Trichoderma viride* F94 and *Aspergillus terreus* F98. Soft. 3: 19–29.

[6] Adam, F., Bertoncini, F., Dartiguelongue, C., Marchand, K., Thiebaut, D. and Hennion M. (2009). Comprehensive two-dimensional gas chromatography for basic and neutral nitrogen speciation in middle distillates. *Fuel.* 88: 938–946.

[7] Adlakha, J., Singh, P., Ram, S.K., Kumar, M., Singh, M.P., Singh, D., Sahai, V. and Srivastava, P. (2016). Optimization of conditions for deep desulfurization of heavy crude oil and hydrodesulfurized diesel by *Gordonia* sp. IITR100. *Fuel.* 184: 761–769.

[8] Agarwal, M., Dikshit, P.K., Bhasarkar, J.B., Borah, A.J. and Moholkar, V.S. (2016). Physical insight into ultrasound-assisted biodesulfurization using free and immobilized cells of *Rhodococcus rhodochorous* MTTC3552. *Biochemical Engineering Journal,* 112: 105–111.

[9] Agarwal, P. and Sharma, D.K. (2010). Comparative studies on the biodesulfurization of crude oil with other desulfurization techniques and deep desulfurization through integrated processes. *Energy and Fuels,* 24(1): 518–524.

[10] Ahmed, S.M., El-Dib, F.I., El-Gendy, N. Sh., Sayed, W.M. and El-Khodary, M. (2016). A kinetic study for the removal of anionic sulphonated dye from aqueous solution using nanopolyaniline and Baker's yeast. *Arabian Journal of Chemistry,* 9: S1721–S1728.

[11] Al-Degs, Y.S., El-Sheikh, A.H., Al-Bakain, R.Z., Newman, A.P. and Al-Ghouti, M.A. (2016). Conventional and upcoming sulfur-cleaning technologies for petroleum fuel: A review. *Energy Technology,* 6: 679–699.

[12] Ali, H.R., El-Gendy, N.S., Moustafa, Y.M., Mohamed, I., Roushdy, M.I. and Hashem, A.I. (2012). Degradation of asphaltenic fraction by locally isolated halotolerant bacterial strains. *ISRN Soil Science,* ID 435485. doi:10.5402/2012/435485.

[13] Ali, H.R., El-Gendy, N. Sh., El- Ezbewy, S., El-Gemeie, G.H., Moustafa, Y.M. and Roushdy, R.I. (2006). Assessment of polycyclic aromatic hydrocarbons contamination in water, sediment and fish of Temsah Lake, Suez Canal, Egypt. *Current World Environment,* 1(1): 11–22.

[14] Ali, H.R., Ismail, D.A. and El-Gendy, N. Sh. (2014). The Biotreatment of oil polluted seawater by biosurfactant producer halotolerant

Pseudomonas aeruginosa Asph2. Energy Sources, Part A, Part A: Recovery, *Utilization and Environmental Effects,* 36(13): 1429–1436.

[15] Ali, H.R., Nassar, H.N. and El-Gendy, N. Sh. (2017). Green synthesis of α-Fe_2O_3 using *Citrus reticulum* peels extract and water decontamination from different organic pollutants. Energy Sources, Part A Recover, *Utilization and Environmental Effects,* 39(13): 1425–1434.

[16] Ali, I.H. and Khan, M.I. (2017). Study of the inhibition effect of carbendazim towards carbon steel corrosion in 0.50 M HCl solutions. *International Journal Electrochemical Sciences,* 12: 2285–96.

[17] Al-Otaibi, M.S., Al-Mayouf, A.M., Khan, M., Mousa, A.A., Al-Mazroa, S.A. and Alkhathlan, H.Z. (2014). Corrosion inhibitory action of some plant extracts on the corrosion of mild steel in acidic media. *Arabian Journal of Chemistry,* 7(3): 340–346.

[18] Al-Senani, G.M. (2016). Corrosion inhibition of carbon steel in acidic chloride medium by *Cucumis sativus* (cucumber) peel extract. *International Journal of Electrochemical Science,* 11:291–302.

[19] Alves L., Paixão, S.M., Pacheco, R., Ferreira, A.F. and Silva, C.M. (2015). Biodesulphurization of fossil fuels: energy, emissions and cost analysis. *RSC Advances,* 5: 34047–34057.

[20] Alves, L. and Paixão, S.M. (2014). Enhancement of dibenzothiophene desulfurization by *Gordonia alkanivorans* Strain 1B using sugar beet molasses as alternative carbon source. *Applied Biochemistry and Biotechnology,* 172: 3297–3305.

[21] Alves, L., Marques, S., Matos, J., Tenreiro, R. and Gírio FM. (2008). Dibenzothiophene desulfurization by *Gordonia alkanivorans* strain 1B using recycled paper sludge hydrolyzate. *Chemosphere,* 70: 967–973.

[22] Amer, R., El-Gendy, N. Sh., Taha, T., Farag, S. and Abdel Fattah, Y. (2014). Biodegradation of crude oil and its kinetics using indigenous bacterial consortium isolated from contaminated area in Egyptian Mediterranean ecosystem. *Jökull Journal* 64: 42–58.

[23] Ansari, F. (2008). Use of magnetic nanoparticles to enhance biodesulfurization. A Ph.D Thesis. Cranfield University. England.

[24] Ansari, F., Grigoriev, P., Libor, S., Tothill, I. E. and Ramsden, J.J. (2009). DBT degradation enhancement by decorating *Rhodococcus erythropolis* IGST8 with magnetic Fe_3O_4 nanoparticles. *Biotechnology and Bioengineering,* 102(5): 1505–1512.

[25] Aribike, D.S., Susu, A.A., Nwachukwu, S.C.U. and Kareem, S.A. (2009). Microbial desulfurization of diesel by *Desulfobacterium aniline.* Academia Arena. 1(4): 11–17.

[26] Aribike, D.S., Susu, A.A., Nwachukwu, S.C.U. and Kareem, S.A. (2008). Biodesulfurization of kerosene by *Desulfobacterium indolicum. Nature and Science,* 6(4): 55–63.

[27] Atabani, A.E., Silitonga, A.S., Ong, H.C., Mahlia, T.M.I., Masjuki, H.H., Badruddin, I.F. and Fayaz, H. (2013). Non-edible vegetable oils: A critical evaluation of oil extraction, fatty acid compositions, biodiesel production, characteristics, engine performance and emissions production. *Renewable and Sustainable Energy Reviews,* 18: 211–245.

[28] Bahuguna, A., Lily, M.K., Munjal, A., Singh, R. and Dangwal, K. (2011). Desulfurization of dibenzothiophene (DBT) by a novel strain *Lysinibacillus sphaericus* DMT-7 isolated from diesel contaminated soil. *Journal of Environmental Sciences,* 23: 975–982.

[29] Bakr, M.M.Y. (2009). Occurrence and geochemical significance of carbazoles and xanthones in crude oil from the Western Desert, Egypt. *Journal of King Abdulaziz University Earth Sciences,* 20(2): 127–159.

[30] Balat, M. (2011). Production of bioethanol from lignocellulosic materials via the biochemical pathway: A review. *Energy Conversion and Management,* 52(2), 858–875.

[31] Bandyopadhyay, S., Chowdhury, R. and Bhattacharjee, C. (2013). Steady state performance of a bioreactor for production of near zero sulfur diesel (NZSD) and bio-surfactant. *Journal of Clean Energy Technologies,* 1(3): 189–193.

[32] Bardania, H., Raheb, J., Mohammad-Beigi, H., Rasekh, B. and Arpanaei, A. (2013). Desulfurization activity and reusability of magnetite nanoparticle-coated *Rhodococcus erythropolis* FMF and *R. erythropolis* IGTS8 bacterial cells. *Biotechnology and Applied Biochemistry,* 60: 323–329.

[33] Barros, F.F.C., Simiqueli, A.P.R., de Andrade, C.J. and Pastore, G.M. (2013). Production of enzymes from agroindustrial wastes by biosurfactant-producing strains of *Bacillus subtilis. Biotechnology Research International,* 103960. http://dx.doi.org/10.1155/2013/103960

[34] Belal, E.B. (2013). Bioethanol production from rice straw residues. *Brazilian Journal of Microbiology,* 44(1): 225–234.

[35] Ben Yahmed, N., Jmel, A.M., Ben Alaya, M., Bouallagui, H., Marzouki, N.M. and Smaali, I. (2016). A biorefinery concept using the green macroalgae *Chaetomorpha linum* for the coproduction of bioethanol and biogas. *Energy Conversion and Management,* 119: 257–265.

[36] Benedik, M.J., Gibbs, P.R., Riddle, R.R. and Willson, R.C. (1998). Microbial denitrogenation of fossil fuels. *Trends in Biotechnology,* 16: 390–395.

[37] Bhadra, B.N. and Jhung, S.H. (2019). Oxidative desulfurization and denitrogenation of fuels using metal-organic framework-based/-derived catalysts. *Applied Catalysis B: Environmental,* 259: 118021.

[38] Bhatia, S. and Sharma, D.K. (2010). Biodesulfurization of dibenzothiophene, its alkylated derivatives and crude oil by a newly isolated

strain of Pantoea agglomerans D23W3. *Biochemical Engineering Journal,* 50(3): 104–109.

[39] Bhatia, S. and Sharma, D.K. (2012). Thermophilic desulfurization of dibenzothiophene and different petroleum oils by *Klebsiella* sp. 13T. *Environmental Science and Pollution Research,* 19(8):3491–3497.

[40] Booth, G.H. and Tiller, A.K. (1968). Cathodic characteristics of mild steel in suspensions of sulphate-reducing bacteria. *Corrosion Science,* 8(8): 583–600.

[41] Bordoloi, N.K., Rai, S.K., Chaudhuri, M.K. and Mukherjee, A.K. (2014). Deep-desulfurization of dibenzothiophene and its derivatives present in diesel oil by a newly isolated bacterium *Achromobacter* sp. to reduce the environmental pollution from fossil fuel combustion. *Fuel Processing Technology,* 119: 236–244.

[42] Cambier, S., Røgeberg, M., Georgantzopoulou, A., Serchi, T., Karlsson, C., Verhaegen, S., Iversen, T.G., Guignard, C., *et al.* (2018). Fate and effects of silver nanoparticles on early life-stage development of zebrafish (Danio rerio) in comparison to silver nitrate. *Science of The Total Environment,* 610/611: 972–982.

[43] Castorena, G., Suarez, C., Valdez, I., Amador, G., Fernandez, L. and Le Borgne, S. (2002). Sulfur selective desulfurization of dibenzothiophene and diesel oil by newly isolated *Rhodococcus* sp. Strains. *FEMS Microbiology Letters,* 215(1): 157–161.

[44] Chandel, A.K., da Silva, S.S., Carvalho, W. and Singh, O.V. (2012). Sugarcane bagasse and leaves: foreseeable biomass of biofuel and bio-products. *Journal of Chemical Technology and Biotechnology,* 87(1): 11–20.

[45] Chang, J., Chang, Y.K., Cho, K.-S. and Chang, H.N. (2000). Desulfurization of model and diesel oils by resting cells of *Gordona* sp. *Biotechnology Letters,* 22: 193–196.

[46] Chang, J.H., Rhee, S.-K., Chang, Y.K. and Chang, H.N. (1998). Desulfurization of diesel oils by a newly isolated dibenzothiophene-degrading *Nocardia* sp. strain CYKS2. *Biotechnology Progress,* 14(6): 851–855.

[47] Chaubey, N., Singh, V.K. and Quraishi, M. A. (2015). Effect of some peel extracts on the corrosion behavior of aluminum alloy in alkaline medium. *International Journal of Industrial Chemistry,* 6:317–328.

[48] Chauhan, A.K., Ahmad, A., Singh, S.P. and Kumar, A. (2015). Biodesulfurization of benzonaphtothiophene by an isolated *Gordonia* sp. IITR100. *International Biodeterioration and Biodegradation,* 104: 105–111.

[49] Chen, H., Cai, Y., Zhang, W. and Li, W. (2009). Methoxylation pathway in biodesulfurization of model organosulfur compounds with *Mycobacterium* sp. *Bioresource Technology,* 100: 2085–2087.

[50] Chen, S., Zhao, C., Liu, Q., Zhang, X., Sun, S. and Zang, M. (2019). Biodesulfurization of diesel oil in oil–water two phase reaction system by *Gordonia* sp. SC-10. *Biotechnology Letters,* 41:547–554.

[51] Constanti, M., Giralt, J. and Bordons, A. (1996). Degradation and desulfurization of dibenzothiophene sulfone and other sulfur compounds by *Agrobacterium* NC501 and a mixed culture. *Enzyme and Microbial Technology,* 19: 214–219.

[52] Dai, Y., Shao, R., Qi, G. and Ding, B.B. (2014). Enhanced dibenzothiophene biodesulfurization by immobilized cells of Brevibacterium lutescens in n-octane–water biphasic system. *Applied Biochemistry Biotechnology*. 174(6): 2236–2244.

[53] Das, B. and Deka, S. (2019). A cost-effective and environmentally sustainable process for phycoremediation of oil field formation water for its safe disposal and reuse. *Science Reports.* 9:15232. https://doi.org/10.1038/s41598-019-51806-5

[54] Deriase, S.F. and El-Gendy, N. Sh. (2010). Modeling and simulation for degradation of different polyaromatic sulfur heterocyclic compounds by *Bacillus sphaericus* HN1 in a batch reactor, 4(2): 1–24.

[55] Deriase, S.F., El-Gendy, N. Sh. and Nassar, H.N. (2012). Enhancing biodegradation of dibenzothiophene by *Bacillus sphaericus* HN1 using factorial design and response surface optimization of medium components. Energy Sources, Part A: Recovery, *Utilization and Environmental Effects,* 34(22): 2073–2083.

[56] Deriase, S.F., Nassar, H.N. and El-Gendy, N. Sh. (2015). Modeling and simulation for biodegradation of mono-, binary- and tertiary-substrate batch systems of different polyaromatic sulfur heterocyclic compounds. *Petroleum Science and Technology,* 33(10): 1063–1076.

[57] Deriase, S.F., Younis, Sh.A. and El-Gendy, N. Sh. (2013). Kinetic evaluation and modeling for batch degradation of 2-hydroxybiphenyl and 2, 2'-dihdroxybipheny by *Corynebacterium variabilis* Sh42. *Desalination and Water Treatment,* 51: 4719–4728.

[58] Derikvand, P., Etemadifar, Z. and Biria, D. (2014). Taguchi optimization of dibenzothiophene biodesulfurization by *Rhodococcus erythropolis* R1 immobilized cells in a biphasic system. *International Biodeterioration and Biodegradation,* 86: 343–348.

[59] Diamantopoulos, N., Panagiotaras, D. and Nikolopoulos, D. (2015). Comprehensive review on biodiesel production using solid acid heterogeneous catalysts. *Journal of Thermodynamics and Catalysis,* 5: 143. doi:10.4172/2157-7544.1000143.

[60] Duarte G.F., Rosado A.S., Seldin L., de Araujo W. and van Elsas J.D. (2001). Analysis of bacterial community structure in sulfurous-oil-containing soils and detection of species carrying dibenzothiophene desulfurization (dsz) genes. *Applied Environmental Microbiology,* 67: 1052–1062.

[61] Eckart, V., Hieke, W., Bauch, J. and Gentzsch, H. (1981). Microbial desulfurization of petroleum and heavy petroleum fractions. 1. Studies on microbial aerobic desulfurization of Romashkino crude oil. *Chemical Abstracts*, 94: 142230q.

[62] Eckart, V., Hieke, W., Bauch, J. and Gentzsch, H. (1982). Microbial desulfurization of petroleum and heavy petroleum fractions. 3. Change in the chemical com position of fuel-D-Oil by microbial aerobic desulfurization. *Chemical Abstracts*, 97: 147259c.

[63] Eckart, V., Kohler, M. and Hieke, W. (1986). Microbial desulfurization of petroleum and heavy petroleum fractions. 5. Anaerobic desulfurization of Romashkino petroleum. *Zentralblatt für Mikrobiologie,* 141: 291–300.

[64] El-Gendy, N. Sh. (2006). Biodegradation potentials of dibenzothiophene by new bacteria isolated from hydrocarbon polluted soil in Egypt. Biosciences, *Biotechnology Research Asia,* 3(1a): 95–106.

[65] El-Gendy, N. Sh., Abidin, Z.A., Abu Amr, S.S. and Nassar, H.N. (2015e). Statistical optimization of alginate immobilization process of *Candida stauntonica* strain MY1 for bioethanol production. *International Journal of Current Microbiology and Applied Sciences,* 4(3): 397–420.

[66] El-Gendy, N. Sh., Abu Amr, S.S. and Abdul Aziz, H. (2014d). The optimization of biodiesel production from waste frying sunflower oil using a heterogeneous catalyst. Energy Sources, Part A: Recovery, *Utilization and Environmental effects,* 36(15): 1615–1625.

[67] El-Gendy, N. Sh., Abu Amr, S.S., El-Salamony, R.A. and Nassar, H.N. (2015f). Factorial design analysis for biosorption of Reactive Red-84 dye using fermentation spent waste biomass, biosorbent regeneration and desorbed dye photodegradation using TiO_2 nanoparticles. *Desalination and Water Treatment,* 56(12): 3403–3413.

[68] El-Gendy, N. Sh., Ali, A.A.., Abu Amr, S.S., Abdul Aziz, H. and Mohamed, A.S. (2016c). Application of D-optimal design and RSM to optimize the transesterification of waste cooking oil using natural and chemical heterogeneous catalyst. *Energy Sources, Part A: Recovery, Utilization and Environmental effects,* 38(13): 1852–1866.

[69] El-Gendy, N. Sh., Ali, H.R., El-Nady, M.M., Deriase, S.F., Moustafa, Y.M. and Roushdy, M.I. (2014b). Effect of different bioremediation techniques on petroleum biomarkers and asphaltene fraction in oil polluted sea water. *Desalination and Water Treatment,* 52(40–42): 7484–7494.

[70] El-Gendy, N. Sh. and Abo-State, M.A. (2008). Isolation, characterization and evaluation of *Staphylococcus gallinarum* NK1 as a degrader for dibenzothiophene, phenantherene and naphthalene. *Egyptian Journal of Petroleum,* 17(2): 75–91.

[71] El-Gendy, N. Sh. and Deriase, S.F. (2013). Statistical optimization of bio-diesel production from different types of waste cooking oils using

basic heterogeneous catalyst. *International Journal of Chemical and Biochemical Science,* 4: 79–88.

[72] El-Gendy, N. Sh. and Deriase, S.F. (2015). Waste eggshells for production of biodiesel from different types of waste cooking oil as waste recycling and a renewable energy process. *Energy Sources, Part A: Recovery, Utilization and Environmental effects,* 37(10): 1114–1124.

[73] El-Gendy, N. Sh. and Deriase, S.F. (2018). Application of statistical approaches to optimize the productivity of biodiesel and investigate the physicochemical properties of the bio/petro-diesel blends. In "High-Performance Materials and Engineered Chemistry" Eds. Torrens, F., Balköse, D. & Thomas, S., "Innovations in Physical Chemistry: Monograph Series" Eds. Haghi, A.K., Pogliani, L. & Ribeiro, A.C.F., *Apple Academic Press AAP Inc.,* NJ, USA. Chapter 7, Part II, pp. 161–239.

[74] El-Gendy, N. Sh. and Hamdy, A. (2014). Application of different bio-catalysts for bio-diesel production from waste cooking oil as a sustainable and green energy process", *Proceedings of the 24th Annual International Conference on Soil,* Water, Energy & Air, Volume 2, p. 2–37, published by the Association for Environmental Health and Sciences (AEHS) Foundation.

[75] El-Gendy, N. Sh. and Moustafa, Y.M. (2007). Environmental assessment of petroleum hydrocarbons contaminating Temsah Lake, Suez Canal, Egypt. *Oriental Journal of Chemistry,* 22(1): 11–26.

[76] El-Gendy, N. Sh. and Nassar, H.N. (2015a). Kinetic modeling of the bioremediation of diesel oil polluted seawater using pseudomonas aeruginosa NH1. Energy Sources, Part A: Recovery, *Utilization and Environmental Effects,* 37(11): 1147–1163.

[77] El-Gendy, N. Sh. and Nassar, H.N. (2015b). A comparative study for batch process of phenol biosorption onto different spent waste biomass: Kinetics and equilibrium. Energy Sources, Part A: Recovery, *Utilization and Environmental Effects,* 37: 1098–1109.

[78] El-Gendy, N. Sh. and Nassar, H.N. (2016). Study on the effectiveness of spent waste sugarcane bagasse for adsorption of different petroleum hydrocarbons water pollutants: Kinetic and equilibrium isotherm. *Desalination and Water Treatment,* 57(12): 5514–5528.

[79] El-Gendy, N. Sh. and Nassar, H.N. (2018). Biodesulfurization in Petroleum Refining. John Wiley & Sons, Inc., 111 River Street, Hoboken, NJ 07030, USA and Scrivener Publishing LLC, 100 Cummings Center, Suite 541J, Beverly, MA 01915, USA.

[80] El-Gendy, N. Sh. and Omran, B.A. (2019). Green Synthesis of Nanoparticles for Water Treatment" in "Nano and Bio-Based Technologies for Wastewater Treatment: Prediction and Control Tools for the Dispersion of Pollutants in the Environment" Elvis Fosso-

Kankeu (ed.), John Wiley & Sons, Inc., 111 River Street, Hoboken, NJ 07030, USA and Scrivener Publishing LLC, 100 Cummings Center, Suite 541J, Beverly, MA 01915, USA, pp. 205–264.

[81] El-Gendy, N. Sh. and Speight, J.G. (2015). Handbook of reûnery desulfurization. CRC Press, Taylor & Francis, Boca Raton, Florida, USA.

[82] El-Gendy, N. Sh., Deriase, S.F. and Hamdy, A. (2014e). The optimization of biodiesel production from waste frying corn oil using snails shells as a catalyst. Energy Sources, Part A: Recovery, *Utilization and Environmental Effects*, 36(6): 623–637.

[83] El-Gendy, N. Sh., Deriase, S.F., Hamdy A. and Abdallah, R.I. (2015a). Statistical optimization of biodiesel production from sunflower waste cooking oil using basic heterogeneous biocatalyst prepared from eggshells. *Egyptian Journal of Petroleum*, 24: 37–48.

[84] El-Gendy, N. Sh., Deriase, S.F. and Osman, D.I. (2014c). The optimization of bio-diesel production from waste frying oil using response surface methodology and the investigation of correlations for changes in basic properties of bio-petro-diesel blends. Energy Sources, Part A: Recovery, *Utilization and Environmental effects*, 36(5): 457–470.

[85] El-Gendy, N. Sh., El-Gharabawy, A.S.A., Abu Amr, S.S. and Ashour F.H. (2015c). Response surface optimization of an alkaline trancesterification of waste cooking oil. *International Journal of ChemTech Research,* 8(8): 385–398.

[86] El-Gendy, N. Sh., El-Salamony, R.A., Abu Amr, S.S. and Nassar, H.N. (2015g). Statistical optimization of Basic Blue 41 dye biosorption by *Saccharomyces cerevisiae* spent waste biomass and photo-catalytic regeneration using acid TiO_2 hydrosol. *Journal of Water Process Engineering,* 6: 193–202.

[87] El-Gendy, N. Sh., El-Salamony, R.A. and Younis, S.A. (2016a). Green synthesis of fluorapatite from waste animal bones and the photo-catalytic degradation activity of a new ZnO/green biocatalyst nano-composite for removal of chlorophenols. *Journal of Water Process Engineering,* 12: 8–19.

[88] El-Gendy, N. Sh., Farahat, L.A., Moustafa, Y.M., Shaker, N. and El-Temtamy, S.A. (2006). Biodesulfurization of crude and diesel oil by *Candida parapsilosis* NSh45 isolated from Egyptian hydrocarbon polluted sea water. *Bioscience Biotechnology Research Asia*, 3(1a): 5–16.

[89] El-Gendy, N. Sh., Hamdy, A., Fatthallah, N.A. and Omran, B.A. (2016b). Recycling of some domestic wastes to produce green corrosion inhibitors and biocides for sulfate reducing bacteria. Energy Sources, Part A: Recovery, *Utilization and Environmental Effects*, 38(24): 3722–3732.

[90] El-Gendy, N. Sh., Hamdy, A. and Abu Amr, S.S. (2014d). An investigation of biodiesel production from wastes of seafood restaurants.

International Journal of Biomaterials, Volume 2014, Article ID 609624. http://dx.doi.org/10.1155/2014/609624.

[91] El-Gendy, N. Sh., Hamdy, A. and Abu Amr, S.S. (2015b). Application of D-optimal design and RSM to optimize the transesterification of waste cooking oil using a biocatalyst derived from waste animal bones and Novozym 43. Energ. Sources Part A: Recovery, *Utilization and Environmental effects*, 37(11): 1233–1251.

[92] El-Gendy, N. Sh., Hamdy, A. and Omran, B.A. (2018a). Thermal and surface studies on the corrosion inhibition of petroleum pipeline by aqueous extract of *Allium cepa* skin under acidic condition. *Energy Sources, Part A: Recovery, Utilization, and Environmental Effects*, 40(8): 905–915.

[93] El-Gendy, N. Sh., Madian, H.R. and Abu Amr, S.S. (2013). Design and optimization of a process for sugarcane molasses fermentation by *saccharomyces cerevisiae* using response surface methodology. *International Journal of Microbiology,* Article ID 815631. http://dx.doi.org/10.1155/2013/815631.

[94] El-Gendy, N. Sh., Madian, H.R., Nassar, H.N. and Abu Amr S.S. (2015d). Response surface optimization of the thermal acid pretreatment of sugar beet pulp for bioethanol production using *Trichoderma viride* and *Saccharomyces cerevisiae. Recent Patents on Biotechnology*, 9(1): 50–62.

[95] El-Gendy, N. Sh., Moustafa, Y.M., Habib, S.A. and Ali, Sh. (2010). Evaluation of *Corynebacterium variabilis* Sh42 as a degrader for different polyaromatic compounds. *Journal of American Science*, 6(11): 343–356.

[96] El-Gendy, N. Sh., Moustafa, Y.M., Barakat, M.A.K. and Deriase, S.F. (2009). Evaluation of a bioslurry remediation of petroleum hydrocarbons contaminated sediments using chemical, mathematical and microscopic analysis. *International Journal of Environmental Studies*, 66(5): 563–579.

[97] El-Gendy, N. Sh., Nassar, H.N. and Abu Amr, S.S. (2014a). Factorial design and response surface optimization for enhancing a biodesulfurization process. *Petroleum Science and Technology*, 32(14): 1669–1679.

[98] El-Mahdi, A.M., Aziz, H.A., Abu Amr S.S., El-Gendy, N. Sh. and Nassar, H. (2015b). Performance of isolated *Kocuria* sp. SAR1 in light crude oil biodegradation. *Bioremediation and Biodegradation,* 6:303. doi:10.4172/2155-6199.1000303.

[99] El-Mahdi, A.M., Aziz, H.A., Abu Amr S.S., El-Gendy, N. Sh. and Nassar, H. (2016). Isolation and characterization of Pseudomonas sp. NAF1 and its application in biodegradation of crude oil. *Environmental Earth Sciences*, 75:380. doi.10.1007/s12665-016-5296-z

[100] El-Mahdi, A.M., Aziz, H.A., Abu Amr S.S., El-Gendy, N. Sh. and Nassar, H. (2015c). Performance of some new bacterial isolates on biodegradation of Libyan light crude oil using agro-industrial wastes as co-substrates. *Applied Mechanics and Materials*, 802: 496–500.

[101] El-Mahdi, A.M., Aziz, H.A., El-Gendy, N. Sh., Abu Amr, S.S. and Nassar, H.N. (2015a). Effectiveness of corn steep liquor as a natural low-cost material for the biodegradation of two types of Libyan crude oil. *Australian Journal of Basic and Applied Sciences*, 9(11): 494–505.

[102] El-Mahdi, A.M., Aziz, H.A., El-Gendy, N. Sh., Abu Amr, S.S. and Nassar, H.N. (2014). Optimization of Libyan crude oil biodegradation by using solid waste date as a natural low-cost material. *Journal of Bioremediation and Biodegradation,* 5:252. doi:10.4172/2155-6199.1000252.

[103] El-Sheshtawy, H.S., Khalil, N.M., Ahmed, W. and Abdallah, R.I. (2014). Monitoring of oil pollution at Gemsa Bay and bioremediation capacity of bacterial isolates with biosurfactants and nanoparticles. *Marine Pollution Bulletin*, 87: 191–200.

[104] Fadhil, A.M.A., Al-Jailawi, M.H. and Mahdi, M.S. (2014). Isolation and characterization of a new thermophilic, carbazole degrading bacterium (*Anoxybacillus rupiensis*) Strain Ir3 (JQ912241). *International Journal of Advanced Research*, 2:795–805.

[105] Fallon, R.D., Hnatow, L.L., Jackson, S.C. and Keeler, S.J. (2010). Method for identification of novel anaerobic denitrifying bacteria utilizing petroleum components as sole carbon source. US patent 7740063 B2.

[106] Farah, J.Y. and El-Gendy, N. Sh. (2013). Performance, kinetics and equilibrium in biosorption of anionic dye Acid Red 14 by the waste biomass of *Saccharomyces cerevisiae* as a low-cost biosorbent. *Turkish Journal of Engineering and Environmental Sciences*, 37: 146–161.

[107] Farooq, M., Ramli, A. and Naeem, A. (2015). Biodiesel production from low FFA waste cooking oil using heterogeneous catalyst derived from chicken bones. *Renewable Energy,* 76: 362–368.

[108] Feng, J., Zeng, Y., Ma, C., Cai, X., Zhang, Q., Tong, M., Yu, B. and Xu, P. (2006). The surfactant Tween 80 enhances biodesulfurization. Applied and Environmental. *Microbiology,* 72:7390–7393.

[109] Finnerty, W.R. (1993). Symposium on: Bioremediation and bio-processing presented before the division of petroleum chemistry, Inc. 205th National Meeting, American Chemical Society, Denver, Co., pp. 282–285.

[110] Folsom, B., Schieche, D., DiGarazia, P., Werner, J. and Palmer, S. (1999). Microbial desulfurization of alkylated dibenzothiophenes from a hydrodesulfurized middle distillate by *Rhodococcus erythropolis* I–19. *Applied and Environmental Microbiology,* 65(11): 4967–4972.

[111] Furuya, T., Ishii, Y., Noda, K., Kino, K. and Kirimura, K. (2003). Thermophilic biodesulfurization of hydrodesulfurized light gas oils by *Mycobacterium phlei* WU-F1. *FEMS Microbiology Letters,* 221: 137–142.

[112] Ghulamullah, K., Kazi, M., Salim, N., Wan, J.B., Hapipah, B.M.A., Fadhil, L.F. and Ghulam, M.K. (2015). Application of natural product

extracts as green corrosion inhibitors for metals and alloys in acid pickling processes—A review. *International Journal of Electrochemical Science,* 10:6120–6134.

[113] Girilal, M., Krishnakumar, V., Paramasivan, P., Fayaz, A.M. and Kalaichelvan, P.T. (2015). A comparative study on biologically and chemically synthesized silver nanoparticles induced heat shock proteins on fresh water fish *Oreochromis niloticus. Chemosphere,* 139: 461–468.

[114] González-Hernández, J.C. and Peña, A. (2002). Adaptation strategies of halophilic microorganisms and Debaryomyces hansenii (halophilic yeast). *Revista Latinoamericana de Microbiología,* 44(3–4): 137–156.

[115] Gorji, A. and Ghanei, R. (2014). A review on catalytic biodiesel production. *Journal of Biodiversity and Environmental Sciences,* 5(4): 48–59.

[116] Granados, M.I., Poves, M.D.Z., Alonso, D.M., Mariscal, R., Galisteo, F.C., Moreno-Tost, R., Santamaria, J. and Fierro, J.L.G. (2007). Biodiesel from sunflower oil by using activated calcium oxide. *Applied Catalysis B: Environmental,* 73: 317–326.

[117] Grossman, M.J., Lee, M.K., Prince, R.C., Garrett, K.K., George, G.N. and Pickering, I.J. (1999). Microbial desulfurization of a crude oil middle-distillate fraction: analysis of the extent of sulfur removal and the effect of removal on remaining sulfur. *Applied and Environmental Microbiology,* 65: 181–188.

[118] Grossman, M.J., Lee, M.K., Prince, R.C., Minak-Bernero, V., George, G.N. and Pickering, I.J. (2001). Deep desulfurization of extensively hydrodesulfurized middle distillate oil by *Rhodococcus* sp. strain ECRD-1. *Applied and Environmental Microbiology,* 67: 1949–1952.

[119] Gui, M.M., Lee, K.T. and Bhatia, S., 2008. Feasibility of edible oil vs. non-edible oil vs. waste edible oil as biodiesel feedstock. *Energy,* 33: 1646–1653.

[120] Gunam, I.B., Yaku, Y., Hirano, M., Yamamura, K., Tomita, F., Sone, T. and Asano, K. (2006). Biodesulfurization of alkylated forms of dibenzothiophene and benzothiophene by *Sphingomonas subarctica* T7b. *Journal of Bioscience and Bioengineering,* 101(4): 322–327.

[121] Guobin, S., Huaiying, Z., Weiquan, C., Jianmin, X. and Huizhou, L. (2005). Improvement of biodesulfurization rate by assembling nanosorbents on the surfaces of microbial cells. *Biophysical Journal,* 89(6): 58–60.

[122] Gurunathan, S., Han, J.W., Kwon, D-N. and Kim, J-H. (2014). Enhanced antibacterial and anti-biofilm activities of silver nanoparticles against Gram-negative and Gram-positive bacteria. Nanoscale Res. Lett. 9:373–390. Organization of Petroleum Exporting Countries OPEC, 2013. World Oil Outlook. Helferstorferstrasse 17 A-1010, Vienna, Austria.

[123] Hamdy, A. and El-Gendy, N. Sh. (2013). Thermodynamic, adsorption and electrochemical studies for corrosion inhibition of carbon steel by henna extract in acid medium. *Egyptian Journal of Petroleum,* 22: 17–25.

[124] Hamouda, H.I., Nassar, H.N., Madian, H.R., Abu Amr, S.S. and El-Gendy, N. Sh. (2015a). Statistical optimization of batch ethanol fermentation of sugarcane molasses by *Candida tropicalis* strain HSC-24. *International Journal of ChemTech Research,* 8(6): 878–889.

[125] Hamouda, H.I., Nassar, H.N., Madian, H.R., Abu Amr, S.S. and El-Gendy, N. Sh. (2015b). Response surface optimization of bioethanol production from sugarcane molasses by *Pichia veronae* strain HSC-22. *Biotechnology Research International,* Article ID 905792. http://dx.doi.org/10.1155/2015/905792.

[126] Hamouda, H.I., Nassar, H.N., Madian, H.R., El-Sayed, M.H., El-Ghamry, A.A. and El-Gendy, N. Sh. (2016). Isolation of fermentative microbial isolates from sugar cane and beet molasses and evaluation for enhanced production of bioethanol. *Energy Sources, Part A: Recovery, Utilization and Environmental Effects,* 38(15): 2170–2180.

[127] Han, L., Feng J., Zhang, S., Ma, Z., Wang, Y. and Zhang, X. (2012). Alkali pretreated of wheat straw and its enzymatic hydrolysis. *Brazilian Journal of Microbiology,* 43(1): 53–61.

[128] Hegazi, A.H., Andersson, J.T. and El-Gayar, M.S. (2003). Application of gas chromatography with atomic emission detection to the geochemical investigation of polycyclic aromatic sulfur heterocycles in Egyptian crude oils. Fuel Process. *Technol,* 85: 1–19.

[129] Helwani, Z., Othman, M.R., Aziz, N., Fernando, W.J.N. and Kim, J. (2009). Technologies for production of biodiesel on green catalytic techniques: A review. *Fuel Processing Technology,* 90: 1502–1514.

[130] Hirano, S.I., Haruki, M., Takano, K., Imanaka, T., Morikawa, M. and Kanaya, S. (2004). Gene cloning and in vivo characterization of a dibenzothiophene dioxygenase from *Xanthobacter polyaromaticivorans. Applied Microbiology and Biotechnology,* 69: 672–681.

[131] Huang T., Qiang L., Zelong W., Daojang Y. and Jiamin X. (2012). Simultaneous removal of thiophene and dibenzothiophene by immobilized *Pseudomonas delafieldii* R-8 cells. *The Chinese Journal of Chemical Engineering,* 20:47–51.

[132] Ibrahim, M.K., Salama, M.F., Abdel-Moghny, T. and El-Shahawy, M.R. (2004). A study on biodesulfurization of an Egyptian crude oil. II-valuation of microbial desulfurization capacity of some local bacterial isolates. *Egyptian Journal of. Biotechnology,* 17: 329–335.

[133] ICCT, International Council on Clean Transportation. US report, October 24, 2011.

[134] Irani, Z.A., Yazdian, F., Mohebali, G., Soheili, M. and Mehrnia, M.R. (2011). Determination of growth kinetic parameters of a desulfurizing

bacterium, *Gordonia alkanivorans* RIPI90A. *Chemical Engineering Transactions,* 24: 937–942.

[135] Ishihara, A., Dumegnil, F., Aoyagi, T., Nishikawa, M., Hosomi, M., Qian, E.W. and Kabe, Y. (2008). Degradation of carabzole by *Novosphingobium* sp. strain NIY3. *Journal of the Japan Petroleum Institute,* 51:174–179.

[136] Ishii, Y., Kozaki, S., Furuya, T., Kino, K. and Kirimura, K. (2005). Thermophilic biodesulfurization of various heterocyclic sulfur compounds and crude straight-run light gas oil fraction by a newly isolated strain *Mycobacterium phlei* WU-0103. *Current Microbiology,* 50: 63–70.

[137] Ismail, A.R., El-Henawy, S.B., Betiha, M.A., Abu Amr, S.A. El-Gendy, N. Sh., Azab, M.S. and Sedky, N.M. (2017a). Optimization of batch Novozym 435-catalyzed transesterification of waste cooking oil with methanol for biodiesel production in a solvent-free medium. Energy Sources, Part A: Recovery, *Utilization and Environmental Effects,* 39(9): 911–925.

[138] Ismail, A.R., El-Henawy, S.B., Younis, S.A., Betiha, M.A., Abu Amr, S.A. El-Gendy, N. Sh., Azab, M.S. and Sedky, N.M. (2017b). Optimization of a batch CaO-catalyzed transesterification of used domestic waste oil with methanol and elucidation of a mathematical correlation between biodiesel yield and percent conversion. Energy Sources, Part A: Recovery, *Utilization and Environmental Effects,* 39(10): 1013–1028.

[139] Ismail, A.R., El-Henawy, S.B., Younis, S.A., Betiha, M.A., El-Gendy, N. Sh., Azab, M.S. and Sedky, N.M. (2018). Statistical enhancement of lipase extracellular production by *Bacillus stratosphericus* PSP8 in a batch submerged fermentation process. *Journal of Applied Microbiology,* 125: 1076–1093.

[140] Jae, P.S., Lee, I-S., Chang, Y.K. and Lee, S.Y. (2003). Desulfurization of dibenzothiophene and diesel oil by metabolically engineered Escherichia coli. *Journal of Microbiology and Biotechnology,* 13(4): 578–583.

[141] Jairam, S., Kolar, P., Sharma-Shivappa, R., Osborn, e J.A. and Davis, J.P. (2012). KI-impregnated oyster shell as a solid catalyst for soybean oil transesterification. *Bioresource Technology,* 104: 329–335.

[142] Johnson, S., Monticello, D., Hazan, C. and Colin, J.M. (2000). Conversion of organosulfur compounds to oxyorganosulfur compounds for desulfurization of fossil fuels. US Patent No. 6, 071, 738.

[143] Kafayati, M., Raheb, J., Angazi, M., Alizadeh, S. and Bardania, H. (2013). The effect of magnetic Fe_3O_4 nanoparticles on the growth of genetically manipulated bacterium, *Pseudomonas aeruginosa* PTSOX4. *Iranian Journal of Biotechnology,* 111: 41–46.

[144] Kareem, S.A., Aribike, D.S., Nwachukwu, S.C. and Latinwo, G.K. (2012), Microbial desulfurization of diesel by *Desulfobacterium indolicum. Journal of Environmental Science and Engineering,* 54(1): 98–103.

[145] Kareem, S.A., Aribike, D.S., Susu, A.A. and Nwachukwu, S.C.U. (2016). Anaerobic biodesulfurization of kerosene part I: Identifying a capable microorganism. *Chemical Engineering and Process Techniques,* 2(2): 1028.

[146] Karimi, E., Jeffryes, C., Yazdian, F., Akhavan Sepahi, A., Hatamian, A., Rasekh, B. and Ashrafi, S. J. (2017). DBT desulfurization by decorating *Rhodococcus erythropolis* IGTS8 using magnetic Fe_3O_4 nanoparticles in a bioreactor. *Engineering in Life Sciences,* 17(5): 528–535.

[147] Kayser, K.J. and Kilbane, J.J. II (2004). Method for metabolizing carbazole in petroleum, US Patent 6943006.

[148] Kedziora, A., Speruda, M., Krzyzewska, E., Rybka, J., Łukowiak, A. and Bugla-Płosko´nska, G. (2018). Similarities and differences between silver ions and silver in nanoforms as antibacterial agents. *International Journal of Molecular Sciences,* 19: 444–461.

[149] Khan, T.M.Y., Atabani, A.E., Badruddin, I.A., Badarudin, A., Khayoon, M.S and Triwahyono, S. (2014). Recent scenario and technologies to utilize non-edible oils for biodiesel production. *Renewable and Sustainable Energy Reviews,* 37: 840–851.

[150] Kheiralla, Z.H., El-Gendy, N. Sh., Ahmed, H.A., Shaltout, T.H. and Hussein, M.M.D. (2018b)One-factor-at-a-time (OFAT) optimization of hemicellulases production from *Fusarium moniliforme* in submerged fermentation. Energy Sources, Part A: Recovery, *Utilization and Environmental Effects,* 40(15): 1877–1885.

[151] Kheiralla, Z.H., El-Gendy, N. Sh., Ahmed, H.A., Shaltout, T.H. and Hussein, M.M.D. (2018a). Upgrading of tomato (*Solanum lycopersicum*) agroindustrial wastes. *Journal of Microbial and Biochemical Technology,* 10(2): 46–48.

[152] Kilbane, J.J. (2006). Microbial biocatalyst developments to upgrade fossil fuels. *Current Opinion in Biotechnology,* 17: 305–314.

[153] Kim, B.Y., Kim, H.Y., Kim, T.S. and Park, D.H. (1995). Selectivity of desulfurization activity of *Desulfovibrio desulfuricans* M6 on different petroleum products. *Fuel Processing Technology,* 43: 87–94.

[154] Kim, H.Y., Kim, T.S. and Kim, B.H. (1990a). Degradation of organic compounds and the reduction of dibenzothiophene to biphenyl and hydrogen sulûde by *Desulfovibrio desulfuricans* M6. *Biotechnology Letters,* 12:761–764.

[155] Kim, T.S., Kim, H.Y. and Kim, B.H. (1990b). Petroleum desulfurization by *Desulfovibrio desulfuricans* M6 using electrochemically supplied reducing equivalent. *Biotechnology Letters,* 12: 757–760.

[156] Kim, Y.J., Chang, J.H., Cho, K.S., Ryu, H.W. and Chang, Y.K. (2004). A physiological study on growth and dibenzothiophene (DBT) desulfurization characteristics of *Gordonia* sp. CYKS1. *Korean Journal of Chemical Engineering,* 21: 436–441.

[157] Kirkwood, K.M., Andersson, J.T., Fedorak, P.M., Foght, J.M. and Gray, M.R. (2007). Sulfur from benzothiophene and alkyl benzothiophenes supports growth of *Rhodococcus* sp. strain J7H1. *Biodegradation,* 18: 541–549.

[158] Kitashov, Y.N., Nazarov, A.V., Zorya, E.I. and Muradov, A.V. (2019). Alternative methods for the removal of sulfur compounds from petroleum fractions. *Chemistry and Technology of Fuels and Oils,* 55: 584–589.

[159] Knightes, C.D. and Peters, C.A. (2006). Multi-Substrate biodegradation kinetics for binary and complex mixtures of polycyclic aromatic hydrocarbons. *Environmental Toxicology and Chemistry,* 25:1746–1756.

[160] Kohler M., Genz I.L., Schicht B. and Eckart V. (1984). Microbial desulfurization of petroleum and heavy petroleum fractions. 4. Anaerobic degradation of organic sulfur compounds. *Zentralblatt für Mikrobiologie,* 139: 239–247.

[161] Kumar, A., Some, D.K and Kiriamiti, K.H. (2014). Pretreatment of CaO catalyst for transesterification of Croton megalocarpus oil. *Journal of Sustainable Research in Engineering,* 1(3): 57–62.

[162] Kumar, K.S. and Kathireswari, P. (2016). Biological synthesis of Silver nanoparticles (AgNPs) by Lawsonia inermis (Henna) plant aqueous extract and its antimicrobial activity against human pathogen. *International Journal of Current Microbiology and Applied Sciences,* 5: 926–937.

[163] Kumar, R.S., Sureshkumar, K. and Velraj, R. (2015). Optimization of biodiesel production from *Manilkara zapota* (L.) seed oil using Taguchi method. Fuel. 140: 90–96.

[164] Kumar, S.A. and Pushpa, A. (2012). Saccharification by Fungi and ethanol production by bacteria using lignocellulosic materials. *International Research Journal of Pharmacy,* 3, 411–414.

[165] Kumari, A., Mahapatra, P., Garlapati, V.K. and Banerjee R. (2009). Enzymatic transesterification of Jatropha oil. *Biotechnology for Biofuels,* 2(1). DOI: 10.1186/1754-6834-2-1

[166] Kunze, W. (2014). Technology Brewing and Malting, 5th revised English ed. Versuchs-und Lehranstalt für Brauerei (VLB), Berlin.

[167] Labana, S., Pandey, G. and Jain, R.K. (2005). Desulfurization of dibenzothiophene and diesel oils by bacteria. *Letters in Applied Microbiology,* 40(3): 159–163.

[168] Larentis, A.L., Sampaio, H.C.C., Carneiro, C.C., Martins, O.B. and Alves, T.L.M. (2011). Evaluation of growth, carbazole biodegradation and anthranilic acid production by Pseudomonas stutzeri. *Brazilian Journal of Chemical Engineering,* 28:37–44.

[169] Lateef, S.A., Ajumobi, O.O. and Onaiz, S.A. (2019). Enzymatic desulfurization of crude oil and its fractions: A mini review on the recent progresses and challenges. *Arabian Journal for Science and Engineering,* 44: 5181–5193.

[170] Lee, K.X. and Valla, J.A. (2019). Adsorptive desulfurization of liquid hydrocarbons using zeolite-based sorbents: a comprehensive review. *Reaction Chemistry and Engineering,* 4: 1357–1386.

[171] Li, F., Zhang, Z., Feng, J., Cai, X. and Ping, X. (2007a). Bio-desulfurization of DBT in tetradecane and crude oil by a facultative thermophilic bacterium *Mycobacterium goodii* X7B. *Journal of Biotechnology,* 127: 222–228.

[172] Li, F.L., Xu, P., Ma, C.Q., Luo, L.L. and Wang, X.S. (2003). Deep desulfurization of hydrodesulfurization-treated diesel oil by a facultative thermophilic bacterium *Mycobacterium* sp. X7B. *FEMS Microbiology Letters,* 223: 301–307.

[173] Li, G., Ma, T., Zhang, Q., Wang, C. and Chen, Q. (2007b). Sulfur-selective desulfurization of dibenzothiophene and diesel oil by newly isolated *Rhodococcus erythropolis* NCC1. *Chinese Organic Chemistry,* 25:400–405.

[174] Li, J., Yang, Z., Li, S., Jin, Q. and Zaho, J. (2020). Review on oxidative desulfrrization of fuel by supported heteropolyacid catalysts. *Journal of Industrial and Engineering Chemistry,* 82: 1–16.

[175] Li, M., Zheng, Y., Chen, Y. and Zhu, X. (2014). Biodiesel production from waste cooking oil using a heterogeneous catalyst from pyrolyzed rice husk. *Bioresource Technology,* 154: 345–348.

[176] Li, W. and Jiang, X. (2013). Enhancement of bunker oil biodesulfurization by adding surfactant. *World Journal of Microbiology and Biotechnology,* 29(1): 103–108.

[177] Li, W., Tang, H., Liu, Q., Xing, J., Li, Q., Wang, D., Yang, M., Li, X. and Liu, H. (2009). Deep desulfurization of diesel by integrating adsorption and microbial method. *Biochemical Engineering Journal,* 44: 297–301.

[178] Li, Y., Du X., Wu C., Liu X., Wang X. and Ping X. (2013). An efficient magnetically modified microbial cell biocomposite for carbazole biodegradation. *Nanoscale Research. Letters,* 8: 522.

[179] Li, Y-G., Li, W.-L., Huang, J.-X., Xiong, X.C., Gao, H.S., Xing, J.M. and Liu, H.Z. (2008). Biodegradation of carbazole in oil/water biphasic system by a newly isolated bacterium *Klebsiella* sp. LSSE-H2. *Biochemical. Engineering Journal,* 41:166–170.

[180] Liaquata, A.M., Masjukia, H.H., Kalama, M.A., Varmana, M., Hazrata, M.A, Shahabuddin M. and Mofijur, M. (2012). Application of blend fuels in a diesel engine. *Energy Procedia,* 14: 1124–1133.

[181] Liu, X.Q., Liu, H.Z., Xing, J.M., Guan, Y.P., Ma, Z.Y., Shan, G.B. and Yang, C.L. (2003). Preparation and characterization of superparamagnetic functional polymeric microparticles. *China Particuol Journal,* 1: 76–79.

[182] Liu, Y., Bisson, T.M., Yang, H. and Xu, Z. (2010). Recent developments in novel sorbents for flue gas clean up. *Fuel Processing Technology,* 91(10): 1175–1197.

[183] Luque, R., Lovett, J.C., Datta, B., Clancy, J., Campelo, J.M., Luque, R., Lovett, J.C., Datta, B., Clancy J. and Campelo, J.M. (2010). Biodiesel as feasible petrol fuel replacement: A multidisciplinary overview. *Energy and Environmental Science,* 3: 1706–1721.

[184] Lyons, L.A., Codina, O., Post, R.M. and Rutledge, D.E. (1988). Evaluation of new molluscicide for alleviating macrofouling by Asiatic clam. *American Power Conference,* Chicago, IL, Pp. 18–20.

[185] Maass, D., Todescato, D., Moritz, D.E., Vladimir Oliveira, J., Oliveira, D., Ulson de Souza, A.A., Guelli Souza, S.M.A. (2015). Desulfurization and denitrogenation of heavy gas oil by *Rhodococcus erythropolis* ATCC 4277. *Bioprocess and Biosystems Engineering,* 38(8): 1447–1453.

[186] Madian, H.R., EL-Gendy, N. Sh., Farahat, L.A., Abo-State, M.A. and Ragab, A.M.E. (2012). Fungal hydrolysis and saccharification of rice straw and ethanol production. *Biosciences Biotechnology Research Asia,* 9(2): 467–476.

[187] Maghsoudi, S., Vossoughi, M., Kheirolomoom, A., Tanaka, E. and Katoh, S. (2001). Biodesulfurization of hydrocarbons and diesel fuels by Rhodococcus sp. strain P32C1. *Biochemical Engineering Journal,* 8:151–156.

[188] Malik, P., Shankar, R., Malik, V., Sharma, N. and Mukherjee, T.K. (2014). Green chemistry based benign routes for nanoparticle synthesis. *Journal of nanoparticle research,* 1–14.

[189] Martinez, I., Santos, V.E., Alcon, A. and Garcia-Ochoa, F. (2015). Enhancement of the biodesulfurization capacity of *Pseudomonas putida* CECT5279 by co-substrate addition. *Process Biochemistry,* 50: 119–124.

[190] Mazubert, A., Poux, M. and Aubin, J. (2013). Intensified processes for FAME production from waste cooking oil: A technological review. *Chemical Engineering Journal,* 233: 201–223.

[191] Meira, M., Quintella, C.M., Ribeiro, E.M.O., Silva, H.R.G. and Guimarães, A.K. (2016). Overview of the challenges in the production of biodiesel. *Biomass Conversion and Biorefinery,* 5(3): 321–329.

[192] Mingfang, L., Zhongxuan, G., Jianmin, X., Huizhou, L. and Jiayong, C. (2003). Microbial desulfurization of model and straight-run diesel oils. *Journal of Chemical Technology and Biotechnology,* 78: 873–876.

[193] Mogollon, L., Rodriguez, R., Larrota, W., Ortiz, C. and Torres, R. (1998). Biocatalytic removal of nickel and vanadium from petro-porphyrins and asphaltenes. *Apllied Biochemistry and Biotechnology,* 70/72: 765–777.

[194] Mohebali, G. and Ball, A.S. (2016). Biodesulfurization of diesel fuels—past, present and future perspectives. *International Journal of biodeterioration and Biodegradation,* 110: 163–180.

[195] Mostafa, S.S.M. and El-Gendy N. Sh. (2017). Evaluation of fuel properties for microalgae Spirulina platensis bio-diesel and its blends

with Egyptian petro-diesel. *Arabian Journal of Chemistry,* 10: S2040–S2050.

[196] Moustafa, Y.M.M., El-Gendy, N. Sh., Farahat, L.A., Abo-State, M.A. and El-Temtamy S.A. (2006). Biodesulfurization of Ras Badran crude oil and its constituents with special emphasis on its asphaltene fraction. *Egyptian Journal of Petroleum,* 15(1): 21–30.

[197] Mužic, M. and Sertić-Bionda, K. (2013). Alternative processes for removing organic sulfur compounds from petroleum fractions. *Chemical and Biochemical Engineering,* Q. 27(1): 101–108.

[198] Nadalig, T., Raymond, N., Nimatuzahroh, M., Budzinsk, H. and Bertrand, J.C. (2002). Degradation of phenanthrene, methylphenanthrenes and dibenzothiophene by a *Sphingomonas* strain 2mpII. *Applied Microbiology and Biotechnology,* 59(1): 79–85.

[199] Nasab, N.A., Kumleh, H.H., Kazemzad, M., Panjeh, F.G. and Davoodi-Dehaghani, F. (2015). Improvement of desulfurization performance of *Rhodococcus erythropolis* IGTS8 by assembling spherical mesoporous silica nanosorbents on the surface of the bacterial cells. *Journal of Applied Chemical Research,* 9(2): 81–91.

[200] Nassar, H.N. (2010). Potentials of polycyclic aromatic sulfur heterocycles (PASHs) degrading microorganisms isolated from different Egyptian hydrocarbon polluted sites. MSc Thesis, Al-Azhar University.

[201] Nassar, H.N. (2015). Development of biodesulfurization process for petroleum fractions using nano-immobilized catalyst; Ph.D. Degree; Al-Azhar University, Cairo, Egypt.

[202] Nassar, H.N., Deriase, S.F. and El-Gendy, N. Sh. (2016). Modeling the relationship between microbial biomass and total viable count for a new bacterial isolate used in biodesulfurization of petroleum and its fractions. *Petroleum Science and Technology,* 34(11–12): 980–985.

[203] Nassar, H.N., Deriase, S.F. and El-Gendy, N. Sh. (2017a) Statistical optimization of biomass production and biodesulfurization activity of *Rhodococcus erythropolis* HN2. *Petroleum Science and Technology,* 35(20): 1951–1959.

[204] Nassar, H.N., Deriase, S.F. and El-Gendy, N. Sh. (2017b). Kinetic modeling and simulation of a batch biodesulfurization process by *Rhodococcus erythropolis* HN2. *Petroleum Science and Technology,* 35(20): 1995–2002.

[205] Nassar, H.N., El-Gendy, N. Sh., Abo-State, M.A., Mostafa, Y.M., Mahdy, H.M. and El-Temtamy, S.A. (2013). Desulfurization of dibenzothiophene by a novel strain *Brevibacillus invocatus* C19 isolated from Egyptian coke. *Biosciences Biotechnology Research Asia,* 10(1): 29–46.

[206] Nazari, F., Kefayati, M.E. and Raheb, J. (2017). The study of biological technologies for the removal of sulfur compounds. *Journal of Sciences, Islamic Republic of Iran,* 28(3): 205–219.

[207] Nehlsen, P.J. (2005). Developing Clean Fuels: Novel Techniques for Desulfurization. Ph.D. Thesis, Princeton University.

[208] Niju, S., Begum, K.M.M.S. and Anantharaman, N. (2014). Modification of egg shell and its application in biodiesel production. *Journal of Saudi Chemical Society,* 18: 702–706.

[209] Nuhoglu, A. and Yalcin, B. (2005). Modelling of phenol removal in a batch reactor, *Process Biochemistry,* 40: 1233–1239.

[210] Nunes, A.A., Franca, A.S. and Oliveria, L.S. (2009). Activated carbons from waste biomass: An alternative use for biodiesel production solid residues. *Bioresource Technology,* 100: 1786–1792.

[211] Obayori, O.S., Adebusoye, S.A., Ilori, M.O., Oyetibo, G.O., Omotayo, A.E. and Amund, O.O. (2010). Effects of corn steep liquor on growth rate and pyrene degradation by *Pseudomonas* strains. *Current Microbiology,* 60(6): 407–411.

[212] Ogbonna, N., Matthew, M.C. and Okechukwu, O.D. (2011). Inhibition of copper corrosion by acid extracts of *Gnetum africana* and *Musa acuminate* peel. *International Journal of Multidisciplinary Sciences and Engineering,* 5:5–9.

[213] Omran, B.A., Fatthallah, N.A., El-Gendy, N. Sh., El-Shatoury, E.H. and Abouzeid, M.A. (2013). Green Biocides against sulphate reducing bacteria and macrofouling organisms. *Journal of Pure and Applied Microbiology,* 7(3): 2219–2232.

[214] Omran, B.A., Nassar, H.N., Fatthallah, N.A., Hamdy, A., El-Shatoury, E.H. and El-Gendy, N. Sh. (2018a).Characterization and antimicrobial activity of silver nanoparticles mycosynthesized by *Aspergillus brasiliensis. Journal of Applied Microbiology,* 125: 370–382.

[215] Omran, B.A., Nassar, H.N., Fatthallah, N.A., Hamdy, A., El-Shatoury, E.H. and El-Gendy, N. Sh., 2018b. Waste upcycling of *Citrus sinensis* peels as a green route for the synthesis of silver nanoparticles. Energy Sources, Part A Recover, *Utilization and Environmental Effects,* 40(2): 227–236.

[216] Omran, B.A., Nassar, H.N., Younis, S.A., El-Salamony, R.A., Fatthallah, N.A., Hamdy, A., El-Shatoury, E.H. and El-Gendy, N. Sh. (2019a). Novel mycosynthesis of cobalt oxide nanoparticles using *Aspergillus brasiliensis* ATCC 16404—optimization, characterization and antimicrobial activity. *Journal of Applied Microbiology,* 128: 438–457.

[217] Omran, B.A., Nassar, H.N., Younis, S.A., Fatthallah, N.A., Hamdy, A., El-Shatoury, E.H. and El-Gendy, N. Sh. (2019b). Physiochemical properties of *Trichoderma longibrachiatum* DSMZ 16517—synthesized silver nanoparticles for the mitigation of halotolerant sulphate-reducing bacteria. *Journal of Applied Microbiology,* 126: 138–154.

[218] Ouada, S.B., Ali, R.B., Leboulanger, C., Zaghden, H., Choura, S., Ouada, H.B. and Sayadi, S. (2018). Effect and removal of bisphenol A

by two extremophilic microalgal strains (Chlorophyta). *Journal of Applied Phycology,* 30: 1765–1776.

[219] Pacheco, M.A., Lange, E.A., Pienkos, P.T., Yu, L.Q., Rouse, M.P., Lin, Q. and Linguist, L.K. (1999). Recent advances in biodesulfurization of diesel fuel. NPRA Annual Meeting, March 21–23, San Antonio, Texas. *National Petrochemical and Refiners Association*, paper AM-99-27, pp. 1–26.

[220] Paixão, S.M., Arez, B.F., Roseiro, J.C. and Alves, L. (2016). Simultaneously saccharification and fermentation approach as a tool for enhanced fossil fuels biodesulfurization. *Journal of Environmental Management,* 182: 397–405.

[221] Panda, K.K., Achary, V.M.M., Krishnaveni, R., Padhi, B.K., Sarangi, S.N., Sahu, S.N. and Panda, B.B. (2011). In vitro biosynthesis and genotoxicity bioassay of silver nanoparticles using plants. *Toxicology in Vitro,* 25: 1097–1105.

[222] Paramjeet, S., Manasa, P. and Korrapati, N. (2018). Biofuels: Production of fungal-mediated ligninolytic enzymes and the modes of bioprocesses utilizing agro-based residues. *Biocatalysis and Agricultural Biotechnology,* 14: 57–71.

[223] Patel, S. (2012). Threats, management and envisaged utilizations of aquatic weed *Eichhornia crassipes*: an overview. *Reviews in Environmental Science and Bio/Technology,* 11(3): 249–259.

[224] Peng, B. and Zhou, Z. (2016). Study on growth characteristic and microbial desulfurization activity of the bacterial strain MP12. *Biochemical Engineering Journal,* 112: 202–207.

[225] Pichtel, J. (2016). Oil and gas production wastewater: Soil contamination and pollution prevention. *Applied and Environmental Soil Science.* 24. https://doi.org/10.1155/2016/2707989.

[226] Pifferi, P., Lanzarini, G. and Malleuzzi, D. (1990). The desulfurization of heavy oils and residua. EU Patent 401922.

[227] Porto, B., Maass, D., Oliveira, J.V., de Oliveira, D., Yamamoto, C.I., Ulson de Souzaa, A.A. and Ulson de Souza, S.M.A.G. (2017). Heavy gas oil biodesulfurization *by Rhodococcus erythropolis* ATCC 4277: optimized culture medium composition and evaluation of low-cost alternative media. *Journal of Chemical Technology and Biotechnology,* 92: 2376–2382.

[228] Raafat, T., El-Gendy, N. Sh., Farahat, L., Kamel, M. and ElShafy, E.A. (2007). Bioremediation of industrial wastes of oil refineries as an environmental solution for water pollution. *Eurasian Chemistry and Technology Journal,* 9(2): 153–162.

[229] Rabelo, S.C., Carrere, H., Maciel Filho, R. and Costa, A.C. (2011). Production of bioethanol, methane and heat from sugarcane bagasse in a biorefinery concept. *Bioresource Technology,* 102: 7887–7895.

[230] Rahpeyma, S.S., Mohammadi, M. and Rehab, J. (2017). Biodesulfurization of dibenzothiophene by two bacterial strains in cooperation with Fe_3O_4, ZnO and CuO nanoparticles. *Journal of Microbial and Biochemical Technology*, 9(2): 587–591.

[231] Ray, D.K., Mueller, N.D., West, P.C. and Foley, J.A. (2013). Yield trends are insufficient to double global crop production by 2050. PLoS ONE 8(6): e66428. doi:10.1371/journal.pone.0066428.

[232] Rhee, S.K., Chang, J.H., Chang, Y.K. and Chang, H.N. (1998). Desulfurization of dibenzothiophene and diesel oil by a newly isolated *Gordona* strain, CYKS1. *Applied and Environmental Microbiology*, 64(6): 2327–2331.

[233] Rhee, S.K., Lee, K.S., Chung, J.C. and Lee, S.T. (1997). Degradation of pyridine by *Nocardioides* sp. strain OS4 isolated from the oxic zone of a spent shale column. *Canadian Journal of Microbiology*, 43(2): 205–209.

[234] Roig, B., Mnif, W., Hassine, A.I.H., Zidi, I., Bayle, S., Bartegi, A. and Thomas, O. (2013). Endocrine disrupting chemicals and human health risk assessment: A critical review. *Critical Reviews of Environmental Science and Technology*, 43(21): 2297–2351.

[235] Ryabov, V.D. (2009). Khimiya nefti i gaza (Oil and gas chemistry). Moscow: Forum.

[236] Sadare, O.O., Obazu, F and Daramola, M.O. (2017). Biodesulfurization of petroleum distillates—current status, opportunities and future challenges. *Environments*, 4: 85.

[237] Saed, D., Nassar, H.N., El-Gendy, N. Sh., Zaki, T., Moustafa, Y.M. and Badr, I.H.A. (2014). Enhancement of pyrene biodegradation by assembling MFe3O4 nano-sorbents on the surface of microbial cells. Energy Sources, Part A, Part A: Recovery, *Utilization and Environmental Effects*, 36:1931–1937.

[238] Sagardia, F., Rigau, J.J., Martinez, L.A., Fuentes, F. and Lopez, C. (1975). Degradation of benzothiophene and related compounds by a soil Pseudomonas in an oil-aqueous environment. *Journal of Applied Microbiology*, 29: 722–725.

[239] Salama, M.F., Ibrahim, M.K. and El-Shahawy, M.R. (2004). A study on biodesulfurization of an Egyptian crude oil. I-evaluation of biodesulfurization capacity of *Rhodococcus rhodochrous* ATCC 53968. *Egyptian Journal of Biotechnology*, 17: 321–328.

[240] Santos, S.C., Alviano, D.S., Alviano, C.S., Padula, M., Leitao, A.C., Martins, O.B., Ribeiro, C.M., Sassaki, M.Y., Matta, C.P., Bevilaqua, J., Sebastian, G.V. and Seldin, L. (2006). Characterization of *Gordonia* sp. strain F.5.25.8 capable of dibenzothiophene desulfurization and carbazole utilization. *Applied Microbiology and Biotechnology*, 71: 355–362.

[241] Santos, S.C., Alviano, D.S., Alviano, C.S., Padula, M., Leitao, A.C., Martins, O.B., Ribeiro, C.M., Sassaki, M.Y., Matta, C.P. and Bevilaqua,

J. (2005). Characterization of *Gordonia* sp. strain F.5.25.8 capable of dibenzothiophene desulfurization and carbazole utilization. *Applied Microbiology and Biotechnology,* 66: 1–8.

[242] Saratha, R., Priya, S.V. and Thilagavathy, P. (2009). Investigation of Citrus aurantiifolia leaves extract as corrosion inhibitor for mild steel in 1 M HCL. *Journal of Chemistry,* 6(3):785–795.

[243] Setti, L., Farinelli, P., Di Martino, S., Frassinetti, S., Lanzarini, G. and Pifferi, P.G. (1999). Developments in destructive and non-destrucitve pathways for selective desulfurizations in oil biorefining processes. *Applied Microbiology and Biotechnology,* 52: 111–117.

[244] Setti, L., Rossi, M., Lanzarini, G. and Pifferi, P.G. (1993). The effect of n-alkanes in the degradation of dibenzothiophene and of organic sulfur compounds in heavy oil by a *Pseudomonas* sp. *Biotechnology Letters,* 14(6): 515–520.

[245] Shah, S., Sharma, S. and Gupta, M.N. (2003). Enzymatic trancesterification for biodiesel production. *Indian Journal of Biochemistry and Biophysics,* 40(6): 392–399.

[246] Shalaby, E.A. and El-Gendy, N. Sh. (2012). Two steps alkaline transesterification of waste cooking oil and quality assessment of produced biodiesel. *International Journal of Chemical and Biochemical Sciences,* 1: 30–35.

[247] Shan G.B., Xing J.M., Guo C., Liu H.Z. and Chen J.Y. (2005a). Biodesulfurization using *Pseudomonas delaûeldii* in magnetic polyvinyl alcohol beads. *Letters in Applied Microbiology,* 40: 30–36.

[248] Shan G.B., Xing J.M., Zhang H. and Liu, H.Z. (2005b). Biodesulfurization of dibenzothiophene by microbial cells coated with magnetite nanoparticles. *Applied and Environmental Microbiology,* 71(8): 4497–4502.

[249] Shan, G.B., Xing, J.M., Luo, M.F., Liu, H.Z. and Chen, J.Y. (2003). Immobilization of *Pseudomonas delafieldii* with magnetic polyvinyl alcohol beads and its application in biodesulfurization. *Biotechnology Letters,* 25, 1977–1981.

[250] Shang, H., Du, W., Liu, Z. and Zhang, H. (2013). Development of microwave induced hydrodesulfurization of petroleum streams: A review. *Journal of Industrial and Engineering Chemistry,* 19: 1061–1068.

[251] Silva, E.J., Rocha e Silva, N.M.P., Rufino, R.D., Luna, J.M., Silva, R.O. and Sarubbo, L.A. (2014b). Characterization of a biosurfactant produced by *Pseudomonas cepacia* CCT6659 in the presence of industrial wastes and its application in the biodegradation of hydrophobic compounds in soil. *Colloids and Surfaces B: Biointerfaces,* 117: 36–41.

[252] Silva, T.P., Paixão, S.M., Roseiro, J.C. and Alves, L. (2014a). Jerusalem artichoke as low-cost fructose-rich feedstock for fossil fuels desulphurization by a fructophilic bacterium. *Journal of Applied Microbiology,* 118: 609–618.

[253] Silva, T.P., Paixão, S.M., Teixeira, A.V., Roseiro, J.C. and Alves, L. (2013). Optimization of low sulfur carob pulp liquor as carbon source for fossil fuels biodesulfurization. *Journal of Chemical Technology and Biotechnology*, 88: 919–923.

[254] Singh, A. (2012). How specific microbial communities benefit the oil industry: Biorefining and bioprocessing for upgrading petroleum oil. In: Applied microbiology and molecular biology in oilfield systems. Whitby, C., Skovhus, T.L. (eds.), pp. 121–178.

[255] Singh, G. B., Srivastava, S., Gupta, S. and Gupta, N. (2011a). Evaluation of carbazole degradation by *Enterobacter* sp. Isolated from hydrocarbon contaminated soil. *Rec. Res. Sci. Technol*, 311:44–48.

[256] Singh, G.B., Gupta, S., Srivastava, S. and Gupta, N. (2011b). Biodegradation of carbazole by newly isolated *Acinetobacter* spp. *Bulletin of Environmental Contamination and Toxicology*, 87: 522–526.

[257] Singh, N., Chakraborty, R. and Gupta, R.K. (2018). Mutton bone derived hydroxyapatite supported TiO_2 nanoparticles for sustainable photocatalytic applications. *Journal of Environmental Chemical Engineering*, 6(1): 459–467.

[258] Singh, P. (2015). Improved biodesulfurization of persistent organosulfur compounds. PhD Thesis, Department of Biochemical Engineering and Biotechnology, Indian Institute of Technology Delhi, New Delhi, India.

[259] Singh, S., Khanna, S., Moholkar, V. and Goyal, A. (2014). Screening and optimization of pretreatments for *Parthenium hysterophorus* as feedstock for alcoholic biofuels. *Applied Energy*, 129: 195–206.

[260] Slinn, M. and Kendall, K. (2009). Developing the reaction kinetics for a biodiesel reactor. *Bioresource Technology*, 100: 2324–2327.

[261] Soleimani, M., Bassi, A. and Margaritis, A. (2007). Biodesulfurization of refractory organic sulfur compounds in fossil fuels. *Biotechnology Advances*, 25: 570–596.

[262] Soliman, R.M., Younis, S.A., El-Gendy, N. Sh., Mostafa, S.S.M., El-Temtamy, S.A. and Hashim, A.I. (2018). Batch bioethanol production via the biological and chemical saccharification of some Egyptian marine macroalgae. *Journal of Applied Microbiology*, 125: 422—440.

[263] Sorate, K.A. and Bhale, P.V. (2015). Biodiesel properties and automotive system compatibility issues. *Renewable and Sustainable Energy Reviews*, 41: 777–798.

[264] Speight, J.G. (2014). The Chemistry and Technology of Petroleum 5th Edition. CRC Press, Taylor & Francis Group, Boca Raton, Florida, USA.

[265] Speight, J.G. and El-Gendy, N. Sh. (2017). Introduction to Petroleum Biotechnology. Gulf Professional Publishing is an imprint of Elsevier, Cambridge, MA, USA.

[266] Srinivasaraghavan, K., Sarma, P.M. and Lal, B. (2006). Comparative analysis of phenotypic and genotypic characteristics of two desulfurizing

bacterial strains, *Mycobacterium phlei* SM120-1 and *Mycobacterium phlei* GTIS10. *Letters in Applied Microbiology,* 42(5): 483–489.

[267] Srivastava, P. and Kumar, A. (2008). Biodesulfurization of petroleum fractions for a cleaner environment. 28(3): 8–12.

[268] Srivastava, V.C. (2012). An evaluation of desulfurization technologies for sulfur removal from liquid fuels. *RSC Advances,* 2: 759–783.

[269] Sulaiman, S. and Amin, M.H.M. (2016). Fish bone-derived catalyzed methanolysis of waste cooking oil. *Bulletin of Chemical Reaction Engineering and Catalysis,* 11(2): 245–249.

[270] Tan, Y.H., Abdullah, M.O. and Nolasco-Hipolito, C. (2015). The potential of waste cooking oil-based biodiesel using heterogeneous catalyst derived from various calcined eggshells coupled with an emulsification technique: a review on the emission reduction and engine performance" *Renewable and sustainable Energy review,* 47: 589–603.

[271] Tanaka, Y., Matsui, T., Konishi, J. and Maruhashi, K. (2002). Biodesulfurization of benzothiophene and dibenzothiophene by a newly isolated *Rhodococcus* strain. *Applied Microbiology and Biotechnology,* 59: 325–328.

[272] Thavasi, R., Jayalakshmi, S. and Banat, I.M. (2011). Application of biosurfactant produced from peanut oil cake by Lactobacillus delbrueckii in biodegradation of crude oil. *Bioresource Technology,* 10(23): 3366–3372.

[273] Tilstra, L., Eng, G., Olson, G.J. and Wang, F.W. (1992). Reduction of sulphur from polysulphidic model compounds by the hyperthermophilic *Archaebacterium Pyrococcus furiosus.* Fuel. 71: 779–783.

[274] Tissot, B.B. and Welte, D.H. (1984). Petroleum Formation and Occurrence. 2nd revised and Enlarged Ed., Springer-Verlag, Berlin Heidelberg New York Tokyo.

[275] Torkamani, S., Shayegan, J., Yaghmaei, S. and Alemzadeh, I. (2008a). Study of a newly isolated thermophilic bacterium capable of Kuhemond heavy crude oil and dibenzothiophene biodesulfurization following 4S pathway at 60°C. *Journal of Chemical Technology and Biotechnology,* 83(12): 1689–1693.

[276] Torkamani, S., Shayegan, J., Yaghmaei, S. and Alemzadeh, I. (2008b). Study of the first isolated fungus capable of heavy crude oil biodesulfurization. *Industrial and Engineering Chemistry Research,* 47(19): 7476–7482.

[277] van Hamme, J.D., Fedorak, P.M., Foght, J.M., Gray, M.R. and Dettman, H.D. (2004). Use of novel fluorinated organosulfur compound to isolate bacteria capable of carbon-sulfur bond cleavage. *Applied and Environmental Microbiology,* 70: 1487–1493

[278] Vane, L.M., Alvarez, F.R., Rosenblum, L. and Govindaswamy, S. (2012). Efficient ethanol recovery from yeast fermentation broth with integrated

distillation–membrane process. *Industrial and Engineering Chemistry Research,* 52(3): 1033–1041.

[279] Vashist, D. and Ahmad, M. (2014). Statistical analysis of diesel engine performance for castor and jatropha biodiesel-blended fuel. *International Journal of Automotive and Mechanical Engineering,* 10: 2155–2169.

[280] Vazquez-Duhalt, R., Torres, E., Valderrama, B. and Le Borgne S. (2002). Will biochemical catalysis impact the petroleum refining industry? *Energy and Fuels,* 16: 1239–1250.

[281] Voloshchuk, A., Krasilnkova, O., Serbrykova, N., Artamonova, S., Aliev, A., Khozina, E. and Kiselev, V. (2009). Selective adsorption of organic sulfur-containing compounds from diesel fuel using type Y-zeolite and γ-aluminum oxide. *Protection of Metals and Physical Chemistry of Surfaces,* 25: 512–517.

[282] Wang, X., Gai, Z., Yu, B., Feng, J., Xu, C., Yuan, Y., Deng, Z. and Xu, P. (2007). Degradation of CAR by microbial cells immobilized in magnetic gellan gum gel beads. *Applied and Environmental Microbiology,* 73: 6421–6428.

[283] Wang, X., Wang, X., Liu, M., Bu, Y., Zhang, J., Chen, J. and Zhao, J. (2015). Adsorption-synergic biodegradation of diesel oil in synthetic seawater by acclimated strains immobilized on multifunctional materials. *Marin Pollution Bulletin,* 92: 195–200.

[284] Wen, Z., Yu, X., Tu, S.T., Yan, J. and Dahlquist, E. (2010). Synthesis of biodiesel from vegetable oil with methanol catalyzed by Li-doped magnesium oxide catalysts. *Applied Energy,* 87: 743–748.

[285] Xiong, X, Xing, J., Li, X., Bai, X., Li, W., Li, Y. and Liu, H. (2007). Enhancement of biodesulfurization in two-liquid systems by heterogeneous expression of *Vitreoscilla* hemoglobin. *Applied Environmental Microbiology,* 73: 2394–2397.

[286] Yamada, K.O., Morimoto, M. and Tani, Y. (2001). Degradation of dibenzothiophene by sulfate-reducing bacteria cultured in the presence of only nitrogen gas. *Journal of Bioscience and Bioengineering,* 91: 91–93.

[287] Yang J., Hu Y., Zhao D., Wang S., Lau P.C.K. and Marison I.W. (2007). Two-layer continuous-process design for the biodesulfurization of diesel oils under bacterial growth conditions. *Biochemical Engineering Journal,* 37(2): 212–218.

[288] Younis, S.A., El-Gendy, N. Sh. and Moustafa, Y.M. (2013). A study on bio-treatment of petrogenic contamination in El-Lessan area of Damietta River Nile Branch, Egypt. *International Journal of Chemical and Biochemical Sciences,* 4: 112–124.

[289] Younis, S.A., El-Gendy, N. Sh. and Nassar, H.N. (2020). Biokinetic aspects for biocatalytic remediation of xenobiotics polluted seawater. *Journal of Applied Microbiology,* 129: 319–334.

[290] Younis, S.A., El-Gendy, N. Sh., El-Azab, W.I. and Moustafa, Y.M. (2015). Kinetic, isotherm and thermodynamic studies of polycyclic aromatic hydrocarbons biosorption from petroleum refinery wastewater using spent waste biomass. *Desalination and Water Treatment,* 56(11): 3013–3023.

[291] Younis, S.A., El-Gendy, N. Sh., El-Azab, W.I., Moustafa, Y.M. and Hashem, A.I. (2014). The biosorption of phenol from petroleum refinery wastewater using spent waste biomass. Energy Sources, Part A: Recovery, *Utilization and Environmental Effects,* 36(23): 2566–2578.

[292] Yu, B., Xu, P., Shi, Q. and Ma, C. (2006a). Deep desulphurization of diesel oil and crude oils by a newly isolated *Rhodococcus erythropolis* strain. *Applied and Environmental. Microbiology,* 72: 54–58.

[293] Yu, B., Xu, P., Zhu, S., Cai, X., Wang, Y., Li, L., Li, F., Liu, X. and Ma, C. (2006b). Selective biodegradation of S and N heterocycles by recombinant *Rhodococcus erythropolis* strain containing carbazole dioxygenase. *Applied and Environmental Microbiology,* 72(3): 2235–2238.

[294] Zakaria, B.S. Nassar, H.N., Saed, D. and El-Gendy, N. Sh. (2015b). Enhancement of carbazole denitrogenation rate using magnetically decorated Bacillus clausii BS1. *Petroleum Science and Technology,* 33(7): 802–811.

[295] Zakaria, B.S., Nassar, H.N., Abu Amr, S.S. and EL-Gendy, N. Sh. (2015a). Applying factorial design and response surface methodology to enhance microbial denitrogenation by tween 80 and yeast extract. *Petroleum Science and Technology,* 33(8): 880–892.

[296] Zakaria, B.S., Nassar, H.N., EL-Gendy, N. Sh., ElTemtamy, S.A. and Sherif, S.M. (2016). Denitrogenation of carbazole by a novel strain *Bacillus Clausii* BS1 isolated from Egyptian Coke, Energy Sources, Part A: Recovery, *Utilization and Environmental Effects,* 38(13): 1840–1851.

[297] Zaki T., Saed D., Aman D., Younis, S.A. and Moustafa Y.M. (2013). Synthesis and characterization of MFe_3O_4 sulfur nanoadsorbants. *Journal of Sol-Gel Science and Technology,* 65: 269–276.

[298] Zhang T., Li W., Chen V., Tang H., Li Q., Xing J. and Liu H. (2011). Enhanced biodesulfurization by magnetic immobilized Rhodococcus erythropolis LSSE8 assembled with nano-γ-AlO. *World Journal of Microbiology and Biotechnology,* 27:299–305.

[299] Zhang, H., Liu, Q.F., Li, Y., Li, W., Xiong, X., Xing, J. and Liu, H. (2008). Selection of adsorbents for in-situ coupling technology of adsorptive desulfurization and biodesulfurization. *Science in China Series B-Chemistry,* 51(1): 69–77.

[300] Zhang, H., Shan, G., Liu, H. and Xing, J. (2007b). Surface Modification of γ-Al_2O_3 Nanoparticles with gum Arabic and its applications in adsorption and biodesulfurization. *Surface and Coatings Technology,* 201: 6917–6921.

[301] Zhang, Q., Tong, M.Y., Li, Y.S., Gao, H.J. and Fang, X.C. (2007a). Extensive desulfurization of diesel by *Rhodococcus erythropolis*. *Biotechnology Letters,* 29:123–127.

[302] Zhang, T. and Fang, H.H.P. (2001). Phylogenetic diversity of SRB-rich marine biofilm. *Applied Microbiology and Biotechnology,* 57: 437–440.

[303] Zhang, Y., Dube, M.A., McLean, D.D. and Kates, M. (2003). Biodiesel production from waste cooking oil: 2. Economic assessment and sensitivity analysis. *Bioresource Technology,* 90: 229–240.

[304] Zheng, Y., Lee, C., Yu, C., Cheng, Y-S., Zhang, R. and Jenkins, B.M. (2013). Dilute acid pretreatment and fermentation of sugar beet pulp to ethanol. *Applied Energy,* 105: 1–7.

[305] Zobell C.E. (1953). Process of removing sulfur from petroleum hydrocarbons and apparatus. US Patent 2, 641, 564.

Effect of Air Polycyclic Aromatic Hydrocarbons in Food Chain: Direct and Indirect Exposure

Kavita Verma and Vartika Mathur*
Animal Plant Interactions Lab, Department of Zoology,
Sri Venkateswara College, University of Delhi, Delhi–110021, India
*E-mail: vmathur@svc.ac.in

ABSTRACT: Polycyclic aromatic hydrocarbons (PAH) are ubiquitous, semi-volatile and persistence organic pollutants that are highly resistant to degradation and considered as major carcinogenic pollutants present in the environment. In the last few decades, air PAH concentration has increased rapidly due to various anthropogenic activities. In higher trophic levels, majority of PAH exposure takes place directly through respiration, dry or wet deposition, whereas indirect exposure takes place due to decreased PAH assimilation at lower trophic level, which transfers to the higher animals through food chain. This study discusses the exposure and transfer of PAH to different trophic levels, their effect on the food chain and some of their bioremediation methods. Plants are known to accumulate around 40% of the PAH present in the environment. In most of the plants, PAH get absorbed from air by deposition in their waxy cuticle and affect plant germination, morphology, photosynthesis and physiology. PAH accumulation ultimately leads to phytotoxicity. These drastic changes can also influence microbial communities associated with plants. In humans and animals, the exposure takes place generally by inhalation of polluted air, direct dermal contact and through ingestion of PAH affected food. This leads to carcinogenic, morphogenic and mutagenic effects. Methods for removal of PAH from environment include photo-oxidation, chemical oxidation and biological degradation. Animals, plants and microbes have the ability to metabolize PAH through various enzymatic activities. However, the actual mechanism is still unclear. Microorganisms such as Pseudomonas, Micrococcus and Staphylococcus strains utilize PAH as a carbon source and degrading them via salicylate pathway or P-450 monooxygenase activity, thus, they are being utilized for bioremediation of PAH. However, standardization and effectiveness of these techniques is still unclear. Consequently, assessment of the risks and hazards of PAH exposure to the various trophic levels has generated a need to prioritize its mitigation and remediation.

Keywords: Phytotoxicity, Bioaccumulation, Carcinogenic, Benzo[a] pyrene, Bioremediation, P-450 monooxygenase.

INTRODUCTION

Air pollution is the fifth leading risk factor for deaths in the world according to *State of Global Air report (2019)*. PAH is the class of air pollutants known to be responsible for one of the highest numbers of deaths in the world (WHO, 2012). Increased anthropogenic activities, such as burning of fossil fuels, mining, incomplete combustion of organic fuels, power and heat generation processes, residential burning (stoves, gas and oil burners) and crop residue burning, have led to higher exposure of PAH in the ecosystem, which contributes to diseases in humans and other organisms. These are primarily formed by pyrolysis (decomposition under high temperature) and pyro-synthesis (synthesis resulting from heating) due to the incomplete combustion of organic compounds (Ravindra *et al.*, 2008). PAH are present both in gaseous as well as particulate phase and are distributed in air, soil and water. Their concentration in air is greater in winter as compared to summer. Out of the 500 PAH known so far, 16 have been described as priority PAH by the Environmental Protection Agency (EPA) based on their abundance and toxicity (Bostrom *et al.*, 2002) as listed in Table 1. They are persistent and ubiquitous in nature (Santodonato *et al.*, 1981) following their direct exposure (through air) or indirect through food chain (feeding on PAH contaminated food). Due to their volatile nature, they tend to act both at the point of exposure and target tissue or organs. Most of the low molecular weight PAH (such as naphthalene and anthracene) are non-carcinogenic but highly toxic to lower organisms due to their high volatility and concentration. On the contrary, the high molecular weight PAHs are carcinogenic to higher organisms due to their relatively low volatility and lipid solubility (Shen *et al.*, 2013). Of these, Benzo[*a*]pyrene (BaP) is considered as the most carcinogenic and is generally present in the highest concentration therefore, it is taken as an indicator of PAH in the environment.

Due to their carcinogenic, toxic and mutagenic properties, PAH equally affect all life forms including plants, animals, human and aquatic organisms. Plants accumulate PAH in their cuticle (Rey-Salgueiro *et al.,* 2008). Bioaccumulation of PAH in plants, is determined by their lipophilicity and volatility. PAH with low water solubility, high lipophilicity, high polarity and less degradative nature are transferred to higher trophic levels. PAH via plants get transferred to second trophic level leading to increased chances of their transfer to higher trophic level (Liang *et al.*, 2007). Although, during inter-trophic transfer, majority of PAH are metabolized by various microbes and plants resulting in their bio-dilution, the quantity transferred to the next trophic level can also affect the consumer organism (Dhananjyan and Murlidharan, 2012). In humans, PAH exposure causes respiratory and pulmonary diseases, low reproducibility, neurodegenerative disorders, mental impairment, geno-toxic effects, carcinogenicity as well as physical disability in all age groups. PAH are also potent immune-suppressants in humans causing interference in host resistance, humoral immunity and immune system development.

This review summarizes the properties, sources, fate and distribution of PAH, their transfer in food chain and effects on higher trophic levels. It also discusses the methods of biodegradation and phytoremediation of air PAH in a sustainable manner.

Properties, Distribution and Fate of PAH

PAH are also referred to as Polycyclic Organic Matter (POM) or Polynuclear Aromatic Hydrocarbons (PNAs). Chemically, they are non-polar, uncharged, inert, semi-volatile and persistent organic environmental pollutants, composed of 2–7 fused aromatic rings (Skupinska *et al.,* 2004; Kaushik *et al.*, 2006). Benzene rings are present in angular, linear and cluster arrangements with substituted or unsubstituted configuration. The general properties of PAH include lipophilicity, light sensitivity, conductivity, fluorescence, non-corrosiveness, heat resistance, high boiling and melting point and either being colorless or white or pale yellow in color (Hussein *et al.,* 2016). Properties such as low vapor pressure and low water solubility decrease with increase in molecular weight of PAH, whereas resistance to oxidation and reduction properties increases with increasing molecular weight. The low molecular weight PAH (2 or 3 ringed) are relatively more soluble, volatile and have less affinity for biological surfaces, making them more toxic. In contrast, the high molecular weight PAH (4 to 7 ringed) are more carcinogenic, due to their less volatility and low solubility. PAH have characteristic UV absorbance spectra due to their aromatic structure. The carcinogenic and mutagenic properties are dependent on degree of non-planarity (Baum *et al.,* 1978). *International Agency for Research on Cancer* (IARC) monographs program has reviewed 60 PAH based on their carcinogenicity. For example, BaP is considered as group 1 known human carcinogen. Acyclopenta[*cd*]pyrene, dibenz[*a,h*]anthracene and dibenzo[*a,l*]pyrene are classified as probably carcinogenic to humans (Group 2A). Benz[*j*]aceanthrylene, benz[*a*]anthracene, benzo[*b*] fluoranthene, benzo[*j*]fluoranthene, benzo[*k*]fluoranthene, benzo[*c*] phenanthrene, chrysene, dibenzo[*a,h*]pyrene, dibenzo[*a,i*]pyrene, indeno[1,2,3-*cd*]pyrene and 5-methylchrysene, are classified as possibly carcinogenic to humans (Group 2B). Rest 45 PAH could not be classified as carcinogenic to humans (Group 3), because of limited or inadequate experimental evidence (IARC, 2010). PAH undergoes a photochemical reaction in ambient air and is converted into water soluble compounds making them highly toxic to various life forms (Yu *et al.,* 2002). For example, photochemical reaction of nitro PAH binds it to DNA, thus disrupting it and consequently, leads to the formation of more mutagenic nitroquinone (Schauer *et al.,* 2004).

PAH are widely distributed in the environment and are found in air, water as well as soil in different compositions. However, due to their semi volatile nature, their major concentration is present in air. The fate of the air PAH depends on its physico-chemical reaction, photo-chemical transformation, wet and dry deposition and interaction with other pollutants (Delgado*et al.*, 2010). Depending on their property (i.e. vapor pressure and volatility), nature (i.e. origin and structure) and atmospheric conditions (i.e. temperature and humidity), PAH either exist in vapor phase or are adsorbed on particulate matter (Wang *et al.,* 2013; Zhang and Tao, 2009). Low molecular weight PAH (MW $<$ 200), which are characterized by; high vapor pressure, high volatility and more toxic, are present in gaseous form and present throughout the year. In contrast, high molecular weight PAH (MW $>$ 250), which are characterized by; low vapor pressure, low volatility and less toxic, are present in particulate form in some atmospheric conditions (Kaushik and Haritash, 2006; Singh *et al.*, 2008) and transported over long distance before accumulation to soil, water, plants and other living organisms. Thus, PAH predominantly exist in gas phase in summer and as particulate phase associated in winter (Mohanraj *et al.*, 2012). Adsorption of PAH also depend on humidity and type of suspended particulate matter (such as soot, dust, pollen and fly ash) (Zhang and Tao, 2009). Photo-oxidation, a process of removal of PAH from atmosphere, may convert them to other toxic, carcinogenic and persistent products such as quinine, which may then enter into food chain (Edward, 1983).

Table 1: Water Solubility, Carcinogenicity and Structure of PAH

PAH	*Water Solubility*	*Carcinogenicity*	*Structure*
Naphthalene	31.7	Weak Carcinogenic	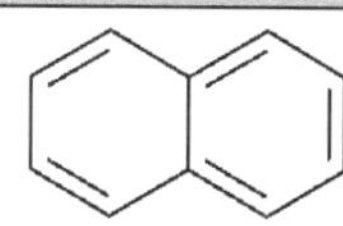
Acenaphtylene	3.57	Weak Carcinogenic	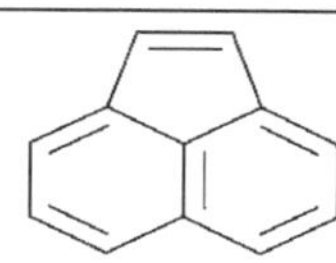
Acenaphthene	3.42	Weak Carcinogenic	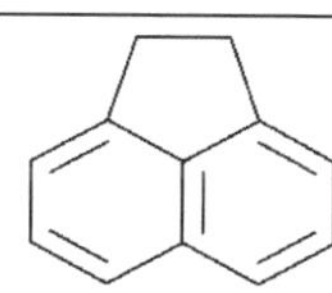
Fluorene	1.98	Weak Carcinogenic	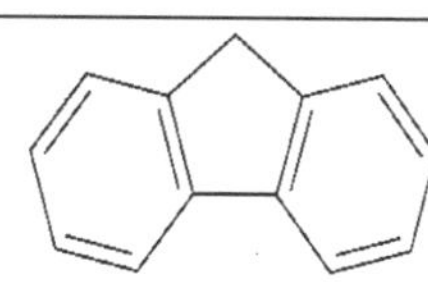
Phenanthrene	1.29	Highly Carcinogenic	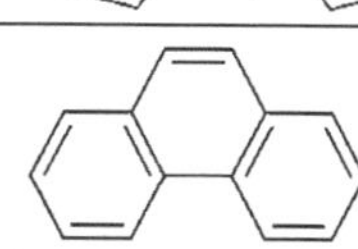
Anthracene	0.07	Weak Carcinogenic	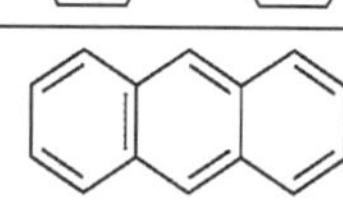
Fluoranthene	0.26	Weak Carcinogenic	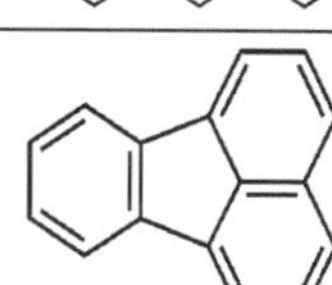

PAH	*Water Solubility*	*Carcinogenicity*	*Structure*
Pyrene	0.135	Highly Carcinogenic	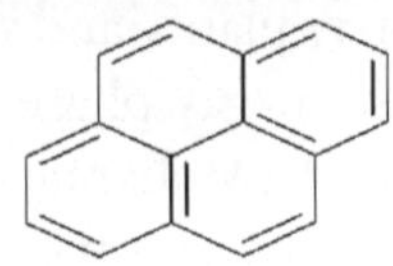
Benzo[*a*] anthracene	0.011	Highly Carcinogenic	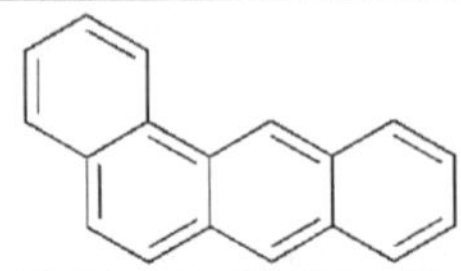
Chrysene	0.002	Moderately Carcinogenic	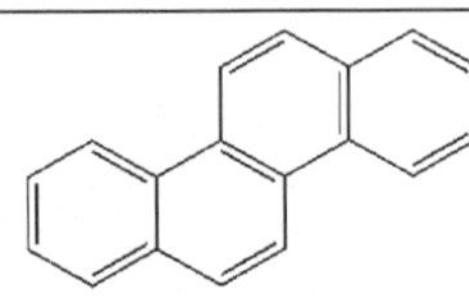
Benzo[*b*] fluoranthene	0.0015	Moderately Carcinogenic	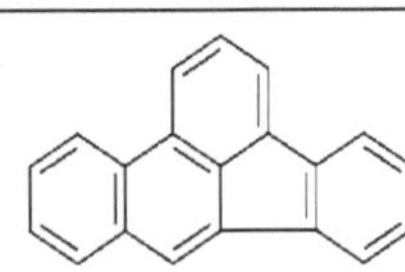
Benzo[*k*] fluoranthrene	0.0008	Moderately Carcinogenic	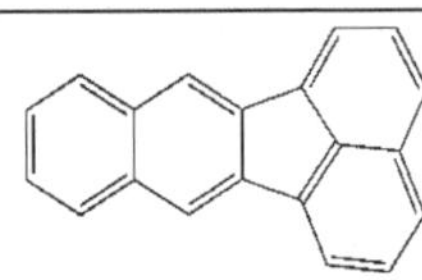
Benzo[*a*] pyrene	< 0.0001	Highly Carcinogenic	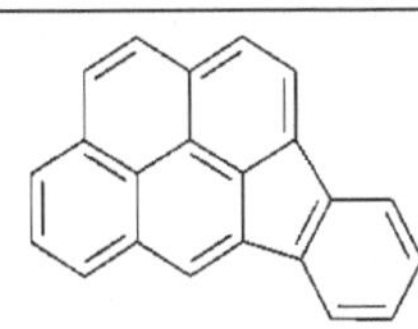
Indeno[1,2,3-*cd*]pyrene	< 0.001	Highly Carcinogenic	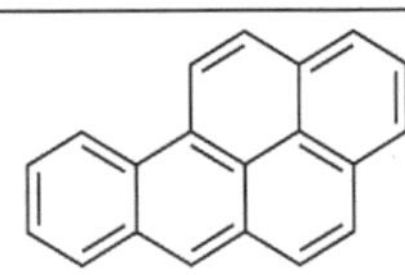
Dibenzo[*a,h*] anthracene	< 0.001	Moderately Carcinogenic	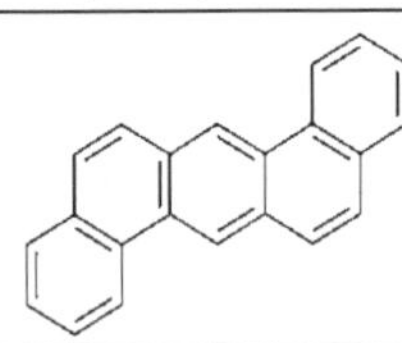
Benzo[*ghi*] perylene	< 0.001	Moderately Carcinogenic	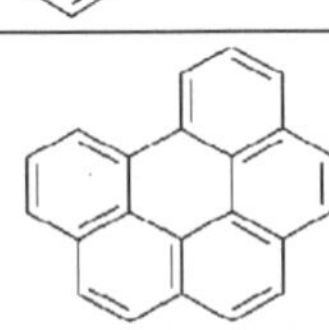

Indian Scenario about Air PAH

Most of the Indian cities recorded very unhealthy (201–300 µg/m^3) and hazardous (>300 µg/m^3) air quality over the last few years (AAIQ, 2018). Owing to the deteriorating air quality, Central Pollution Control Board categorized a total of 102 Indian cities that failed to attain Indian air quality standards (CPCB, 2017). Global burden of disease report (2017) stated that over 1.24 million deaths occurred due to long term exposure to indoor and outdoor air pollution in India in that year. Anthropogenic activities such as mining, burning of fossil fuel, crop residue burning and residential burning have majorly contributed in making India the second largest emitter of PAH (90,000 tonnes per year) in the world (Dybing *et al.,* 2013). Incomplete combustion of solid bio-fuel (such as wood, biomass pellets, crop residues and coal), especially in Indian rural households, contributes to majority amount (92.5%) of air PAH, with about 1.56 Gg/y of BaP emission (International Energy Agency, 2006). According to Global burden of disease 2019 report, incomplete combustion of biomass fuels especially in North Indian rural households have increased the load of PAH related respiratory diseases. One of the major reasons of these statistics is the dependence of the country's energy requirements on petroleum, firewood, straw and animal biomass fuels. Furthermore, India contributes about 3.6% of high molecular weight PAH emission, which is greater than the average global emission (Zhang and Tao, 2009). Exposure to black carbon from rural households of India along with ambient air exposure leads to 1.6 million premature deaths and 49 million disability adjusted life years (Global burden of Disease, 2010). These data indicate, India is currently in an alarming state of air quality and PAH emissions that are contributing to major health issues in the world and needs better remediation approaches.

Effect of PAH on Plants

Plants grown in areas with high Air Quality Index (AQI) have 100X higher PAH concentration compared to plants grown in unpolluted areas (Santodonato *et al.,* 1980). High quantity of air PAH is accumulated by the broad, hairy leaves or those with high surface to volume ratio (Böhme *et al.,* 1999). Different atmospheric conditions, such as humidity and temperature, as well as plant physiology, such as relative lipid content, leaf shape and well-structured wax crystals, play a role in the accumulation of PAH (Kömp and McLachlan, 1997).

Factors such as wet and dry deposition of gas phase and particulate phase PAH and equilibrium partitioning between gaseous phase and vegetation are also responsible for deposition of PAH on plants (McLachlan, 1999). PAH accumulation leads to increase in various antioxidants enzymes such as monodehydroascorbate reductase, MDHAR; dehydroascorbate reductase, DHAR; glutathione peroxidase, GPX; guaicol peroxidase, GOPX and glutathione-S-transferase, GST and some non-enzymatic compounds such as; ascorbic acid, ASH; glutathione, GSH; phenolic compounds, alkaloids, non-protein amino acids and a-tocopherols (Gill *et al.*, 2010). PAH are known to cause an array of phytotoxic effects in plants (Li *et al.,* 2008). For example, in rice, they cause decreased chlorophyll content, biomass and oxidative stress (Li *et al.,* 2008). In *Arabidopsis thaliana,* PAH accumulation leads to induction of leaf lesions, poor growth and development (Alkio *et al.*, 2005). PAH accumulated in plants are transferred to animals through food chain.

Effect of PAH on Animals

Although animals are constantly exposed to PAH through direct dermal contact with air containing PAH, or inhalation, the major source of exposure remains contaminated food (Kim *et al.*, 2013). PAH predominantly affect immune system, skin, haemopoietic system, kidney and mammary glands (Eisler, 1987) and also cause low egg count, low gametes number, low reproducibility, different antioxidants level, effect on physiology and morphology of animals. PAH and their metabolites can induce cancerous and non-cancerous tumor formation in most of the body tissue of animals (Kaushik *et al.,* 2006). A study showed that 7, 12-dimethylbenz(α)anthracene proved to be carcinogenic for mammary gland of female rat and BaP can cause leukemia, stomach tumor skin adenoma in rats feeding on PAH affected food grains (Dipple,1985).

Air PAH are known to have lethal and nonlethal effects on eggs of birds such as thinning of egg shell, breakage of shell and death of eggs. PAH caused cancerous tumors in air sacs of crane that proved fatal for them (Kaushik *et al.,* 2006). Many microbes (such as *Pseudomonas, Micrococcus)* use PAH as their carbon source, but increasing PAH degrading microbes lead to substantial decline in non-degrading microbes (Kawasaki *et al.,* 2012).

Effect of PAH on Humans

Humans are exposed to PAH via direct contact (inhalation of PAH polluted air or dermal contact) or indirectly by consumption of plants and animals contaminated by PAH (Omodara *et al.*, 2014). About 88–98% PAH exposure in humans may be attributed to food such as grains and smoked food (Kluska, 2003). Metabolism of PAH in humans takes place in liver and kidney and they are removed through feces and urine. PAH presence in human is diagnosed and quantified by using metabolites such as 1-hydroxypyrene, hydroxynaphthalene and hydroxyl phenanthrene present in urine and are used as bio-markers (Elovaara *et al.*, 2006). Active metabolite production leads to DNA binding and carcinogenicity (Naz, 1999; USEPA, 1993). Dietary exposure to PAH early in life by children may lead to more serious effects than the same exposure later in life (Perera *et al.,* 2012). Pregnant females exposed to PAH can result in cancer and low birth weight of fetus (Dejmek *et al.*, 2000). Exposure to PAH in womb resulted in higher risk of developing depression, anxiety and attention deficiencies later in life in children (Perera *et al.*, 2012). Moreover, even minute quantities of PAH can cause short term symptoms such as eye irritation, vomiting, nausea as well as long term effects such as immunotoxicity, low birth rate, liver damage, kidney problem, cataract, jaundice, decreased fecundity and cancer (Abdel-Shafy and Mansour, 2016). Phenanthrene (3 ringed PAH) causes mild allergy, photosensitize the skin, inhibitor of gap junction intracellular communication and mutagenic to bacterial community in humans (Mastrangela *et al.*, 1997; Weis *et al.,* 1998). Not only PAH, but also their reactive metabolites such as epoxides and dihydrodiolare the major health concern as they can bind to DNA and cellular protein, causing malformation, tumors, cancer and mutations in humans (Armstrong *et al.*, 2004).

Carcinogenicity of all other PAH are measured w.r.t BaP (5 ringed PAH).Incremental lifetime cancer risk (ILCR) is a probabilistic risk model approach to quantitatively estimate the exposure risk for three age groups of adult, children and infant. It can be calculated to estimate the human health risk as follows:

$$\text{ILCR} = \text{Exposure } (\mu\text{g/kg/day}) \times \text{Cancer slope factor } (\mu\text{g/kg/day})$$

Metabolization and Bioremediation of PAHs

Plants, animals and humans use various enzymes such as cytochrome P450, epoxide hydrolase, glutathione transferase, UDP-glucuronosyl transferase for PAH metabolization. The process involves their activation by cytochrome P-450, followed by sequential oxygenation and hydration steps, catalyzed by a cytochrome P-450 isozyme. The resultant product (Epoxides or phenols) is either detoxified to form glutathione conjugate and glucoronides or further oxidized to produce diol-epoxide (carcinogenic) (Stegeman *et al.*, 2001).

$$\underset{\text{(Inert)}}{\text{B}[a]\text{P}} \xrightarrow{\text{CYP1A1}} \underset{\text{Oxide}}{\text{BP-7, 8}} \xrightarrow{\text{Epoxide-hydroxylase}} \underset{\text{Diol}}{\text{BP 7, 8}} \xrightarrow{\text{P 450}} \underset{\text{(carcinogenic)}}{\text{BP-7, 8 Diol-9, 10 Oxide}}$$

Plants have evolved intricate mechanisms to respond to various biotic and abiotic stresses (Fujita *et al.,* 2006). Phytoremediation is the Plants remove pollutants from air, water and soil by a process known as phytoremediation (Sharma *et al.*, 2019). Phytoremediation of PAH using transgenic plants is very well studied (Bisht *et al.*, 2015; Abdel-Shafy and Mansour, 2018). Plants use glutathione-S-transferase to protect against PAH phytotoxicity and hence their over expression maybe an efficient approach for remediation (Paskova *et al.,* 2006). Genes (*ndo, nah* and *dox)* from animals, microbes and plants are genetically modified and transferred in plants to enhance their ability to degrade PAH (Doty *et al.,* 2008). Mammalian gene encoding P-450 cytochrome is responsible for the metabolization of PAH. However, when this gene was transferred to plants system, the rate of PAH degradation was observed to decrease (Stegeman *et al.*, 1991). Bio-remediation using endophytes (microbes living in plant tissue) has been a recent approach for removal of PAH from the environment (Sarma *et al.,* 2019). Endophytes help in plant growth, provide plant resistant to various abiotic and biotic stresses (Doty *et al.,* 2008). They reside in-between the plant cell and adapted to plant's environment. Hence, using such microbes as bioremediation method to degrade PAH is a recent advancement in biological research. In plants, colony of PAH degrading endophytes increases as they use PAH as carbon source. Also, during abiotic stress, plants recruit more PAH degrading endophytes (Siciliano *et al.,* 2001). However, this increase in specific type of endophytes is also dependent on type of pollutant present in environment, type of plant species in which endophytes grows (Khan *et al.,* 2014). In recent years, use of engineered PAH degrading endophytes for phytoremediation is being explored (Sun *et al.,* 2014). For example, endophytic fungus *Ceratobasidum stevensii*, have shown novel capability to degrade phenanthrene (Dai *et al.*, 2010). *Phanerochaete chrysosporium* (white rot fungi) have been reported to be involved in PAH degradation by producing ligninolytic enzymes (Paszczynski and Crawford, 1995).

FUTURE ASPECTS

Further research is required to understand the relationship between increase in certain type of endophytic population and PAH degradation. The possibility of culturing these microbes on a large scale and their commercial application needs to be explored. Research efforts for plants transgenic and endophytic degradation need to be strengthened. These approaches may be considered for mitigating PAH from not only soil, but also air and water.

REFERENCES

[1] Abdel-Shafy, H.I. and Mansour, M.S. (2016). A review on polycyclic aromatic hydrocarbons: source, environmental impact, effect on human health and remediation. *Egyptian Journal of Petroleum, 25*(1), 107–123.

[2] Abdel-Shafy, H.I. and Mansour, M.S. (2018). Phytoremediation for the elimination of metals, pesticides, PAHs and other Pollutants from wastewater and soil. In *Phytobiont and Ecosystem Restitution* (pp. 101–136). Springer, Singapore.

[3] Alkio M., Tabuchi T.M., Wang X. and Colon-Carmona, A. (2005). Stress responses to polycyclic aromatic hydrocarbons in Arabidopsis include growth inhibition and hypersensitive response-like symptoms. *Journal of Experimental Botany*, 56(421), 2983–2994.

[4] Armstrong, B.G., Hutchinson, E., Unwin, J., and Fletcher, T., Lung cancer risk after exposure to polycyclic aromatic hydrocarbons: A review and meta-analysis. *Environ Health Perspect,* 2004; 112(9):970–8.

[5] Artic Monitoring and Assessment Program (AMAP) (1998). The arctic monitoring and assessment program: arctic pollution issues: a state of the arctic environment report. Oslo, Norway: Arctic Monitoring and Assessment Program. Bach PB, Kelley MJ, Tate.

[6] Baum, E.J. (1978). Occurrence and surveillance of polycyclic aromatic hydrocarbons. *Polycyclic hydrocarbons and cancer,* 1, 45–70.

[7] Bisht, S., Pandey, P., Bhargava, B., Sharma, S., Kumar, V. and Sharma, K.D. (2015). Bioremediation of polyaromatic hydrocarbons (PAHs) using rhizosphere technology. *Brazilian Journal of Microbiology*, 46(1), 7–21.

[8] Böhme, F., Welsch-Pausch, K. and McLachlan, M. (1999). Uptake of airborne semivolatile organic compounds in agricultural plants: Field measurements of interspecies variability. *Environmental Science and Technology* 33 (11): 1805–1813. Behling G., Koller.

[9] Boström, C.E., Gerde, P., Hanberg, A., Jernström, B., Johansson, C., Kyrklund, T., and Westerholm, R. (2002). Cancer risk assessment, indicators and guidelines for polycyclic aromatic hydrocarbons in the ambient air. *Environmental health perspectives, 110* (suppl 3), 451–488.

[10] Dai, C.C., Tian, L.S., Zhao, Y.T., Chen, Y. and Xie, H. (2010). Degradation of phenanthrene by the endophytic fungus *Ceratobasidumstevensii* found in *Bischofiapolycarpa*. Biodegradation, 21(2), 245–255.

[11] Dejmek J., Solansk, I. and Benes, I., *et al.* (2000). The impact of polycyclic aromatic hydrocarbons and fine particles on pregnancy outcome. *Environ Health Perspect,* 108: 1159–64.

[12] Delgado Saborit, J.M., Aquilina, N., Baker, S., Harrad, S., Meddings, C. and Harrison, R.M. Determination of atmospheric particulate-phase policyclic aromatic hydrocarbons from low volume air samples. *Anal Methods*, 2010;2: 231–42.

[13] Dhananjayan, V. and Muralidharan, S. (2012). Polycyclic aromatic hydrocarbons in various species of fishes from Mumbai harbour, India and their dietary intake concentration to human. *International Journal of Oceanography*, 2012.

[14] Dipple, A. (1985). Polycyclic aromatic hydrocarbon carcinogenesis: an introduction.

[15] Doty, S.L., Xin, G., Oakley, B., Kang, J.W., Singleton, G., Vajzovic, A. and Staley, J.T. (in press). Diazotrophicendophytes of native poplar and willow. Symbiosis.

[16] Doty, S.L. (2008). Enhancing phytoremediation through the use of transgenics and endophytes. New Phytologist, 179(2), 318–333.

[17] Dybing, E., Schwarze, P.E., Nafstad, P., Victorin, K. and Penning, T.M. (2013). Polycyclic aromatic hydrocarbons in ambient air and cancer. In Air Pollution and Cancer (Vol. 161). *International Agency for Research on Cancer*, WHO Press Geneva.

[18] Edwards, N.T. (1983). Polycyclic Aromatic Hydrocarbons (PAH's) in the Terrestrial Environment—A Review 1. *Journal of Environmental Quality, 12*(4), 427–441.

[19] Eisler, R. (1987). Polycyclic aromatic hydrocarbon hazards to fish, wildlife and invertebrates: A synoptic review. Biological Report 85 (1.11). *Contaminant Hazard Reviews Report, 11*, 55.

[20] Elovaara, E., Mikkola, J., Mäkelä, M., Paldanius, B. and Priha, E. (2006). Assessment of soil remediation workers' exposure to polycyclic aromatic hydrocarbons (PAH): Biomonitoring of naphthols, phenanthrols and 1-hydroxypyrene in urine. *Toxicology letters, 162*(2–3), 158–163.

[21] Environment protection agency (EPA), (2015). https://www.epa.gov/sites/production/files/2015-10/documents/ace3_criteria_air_pollutants.pdf

[22] Franzaring, J. and van der Eerden, L.J. (2000). Accumulation of airborne persistent organic pollutants (POPs) in plants. *Basic and Applied Ecology, 1*(1), 25–30.

[23] Fujita, M., Fujita, Y., Noutoshi, Y., Takahashi, F., Narusaka, Y., Yamaguchi-Shinozaki, K. and Shinozaki, K. (2006). Crosstalk between abiotic and biotic stress responses: A current view from the points of convergence in the stress signaling networks. *Current opinion in plant biology, 9*(4), 436–442.

[24] Gill, S.S. and Tuteja, N. (2010). Reactive oxygen species and antioxidant machinery in abiotic stress tolerance in crop plants. *Plant physiology and biochemistry, 48*(12), 909–930.

[25] Global atmospheric emissions of polycyclic aromatic hydrocarbons from 1960 to 2008 and 806.

[26] https://aqicn.org/map/asia (2018).

[27] Hussein, R.A., Al-Ghanim, K.A., Abd-El-Atty, M.M. and Mohamed, L.A. (2016). Contamination of Red Sea Shrimp (Palaemonserratus) with Polycyclic Aromatic Hydrocarbons: a Health Risk Assessment Study. *Polish Journal of Environmental Studies, 25*(2).

[28] IARC (2010). Some non-heterocyclic polycyclic aromatic hydrocarbons and some related exposures. *IARC Monogr Eval Carcinog Risks Hum*, 92: 1–853. PMID:21141735

[29] Kameda, T. Atmospheric chemistry of policy clicaromatic hydrocarbons and related compounds. *J Health Sci.,* 2011;57(6):504–11.

[30] Kaushik, C.P. and Haritash, A.K. (2006). Polycyclic aromatic hydrocarbons (PAHs) and environmental health. *Our earth,* 3(3), 1–7.

[31] Kawasaki, A., Watson, E.R. and Kertesz, M.A. (2012). Indirect effects of polycyclic aromatic hydrocarbon contamination on microbial communities in legume and grass rhizospheres. *Plant and soil,* 358(1–2), 169–182.

[32] Kim, K.H., Jahan, S.A., Kabir, E. and Brown, R.J. (2013). A review of airborne polycyclic aromatic hydrocarbons (PAHs) and their human health effects. *Environment international, 60*, 71–80.

[33] Kluska, M. (2003). Soil contamination with polycyclic aromatic hydrocarbons in the vicinity of the ring road in Siedlce city. *Polish Journal of Environmental Studies, 12*(3), 309–314.

[34] Kömp, P. and McLachlan, M. (1997) Interspecies variability of the plant/air partitioning of polychlorinated biphenyls. *Environmental Science and Technology,* 31: 2944–2948.

[35] Li, J.H., Gao, Y., Wu, S.C., Cheung, K.C., Wang, X.R. and Wong, M.H. (2008). Physiological and biochemical responses of rice (Oryza sativa L.) to phenanthrene and pyrene. *International journal of phytoremediation,* 10(2), 106–118.

[36] Liang, Y., Tse, M.F., Young, L. and Wong, M.H. (2007). Distribution patterns of polycyclic aromatic hydrocarbons (PAHs) in the sediments and fish at Mai Po Marshes Nature Reserve, Hong Kong. *Water Research, 41*(6), 1303–1311.

[37] Mastrangela, G. *et al.* (1997). Polycyclic aromatic hydrocarbons and cancer in man. *Environ. Health Perspect.,* 104, 1166–1170.

[38] McLachlan, M.S. (1999). Framework for the interpretation of measurements of SOCs in plants. *Environmental Science and Technology,* 33: 1799–1804.

[39] Mohanraj, R., Dhanakumar, S. and Solaraj, G. Polycyclic aromatic hydrocarbons bound to PM2.5 in urban Coimbatore, India with emphasis on source apportionment. *Sci. World J.,* 2012. http://dx.doi.org/10.1100/2012/980843. [Article ID 980843].

[40] Naz, R.K. (1999). Endocrine disruptors, effects on male and female reproductive systems. Boca Raton (FL): CRC Press LLC.

[41] Omodara, N.B., Amoko, J.S. and Ojo, B.M. (2014). Polycyclic aromatic hydrocabons (PAHs) in the environment, sources, effects and reduction risks. *Sky J. Soil Sci. Environ Man,* 3:96–101.

[42] Pašková, V., Hilscherová, K., Feldmannová, M. and Bláha, L. (2006). Toxic effects and oxidative stress in higher plants exposed to polycyclic aromatic hydrocarbons and their N heterocyclic derivatives. *Environmental Toxicology and Chemistry: An International Journal, 25*(12), 3238–3245.

[43] Paszczynski, A. and Crawford, R.L. (1995). Potential for bioremediation of xenobiotic compounds by the white-rot fungus *Phanerochaetechrysosporium*. Biotechnol Prog, 11:368–379.

[44] Perera, F.P., Tang, D., Wang, S., *et al.* (2012). Prenatal polycyclic aromatic hydrocarbon (PAH) exposure and child behavior at age 6–7 years. *Environ Health Perspect,* 120:921–6.

[45] Ravindra, K., Sokhi, R. and Van Grieken, R. (2008). Atmospheric polycyclic aromatic hydrocarbons: source attribution, emission factors and regulation. *Atmospheric Environment,* 42(13), 2895–2921.

[46] Rey-Salgueiro, L., Martínez-Carballo, E., García-Falcón, M.S. and Simal-Gándara, J. (2008). Effects of a chemical company fire on the occurrence of polycyclic aromatic hydrocarbons in plant foods. *Food Chemistry, 108*(1), 347–353.

[47] Santodonato, J., Howard, P. and Basu, D. (1981). Health and ecological assessment of polynuclear aromatic hydrocarbons. *Journal of environmental pathology and toxicology,* 5(1), 1.

[48] Santodonato, J., Howard, P. and Basu, D. (1981). Health and ecological assessment of polynuclear aromatic hydrocarbons. *Journal of environmental pathology and toxicology, 5*(1), 1.

[49] Sarma, H., Nava, A.R. and Prasad, M.N.V. (2019). Mechanistic understanding and future prospect of microbe-enhanced phytoremediation of polycyclic aromatic hydrocarbons in soil. *Environmental technology & innovation, 13*, 318–330.

[50] Schauer, C., Niessner, R. and Pöschl, U. (2004). Analysis of nitrated polycyclic aromatic hydrocarbons by liquid chromatography with fluorescence and mass spectrometry detection: Air particulate matter, soot and reaction product studies. *Anal Bioanal. Chem.*, 378: 725–736.

[51] Sharma, G., Guleria, R. and Mathur, V. (2020). Differences in plant metabolites and microbes associated with Azadirachtaindica with variation in air pollution. *Environmental Pollution, 257*, 113595.

[52] Shen, H., Huang, Y., Wang, R., Zhu, D., Li, W., Shen, G. and Chen, H. (2013). Global atmospheric emissions of polycyclic aromatic hydrocarbons from 1960 to 2008 and future predictions. *Environmental science & technology, 47*(12), 6415–6424.

[53] Siciliano, S.D., Fortin, N., Mihoc, A., Wisse, G., Labelle, S., Beaumier, D., Ouellette, D., Roy, R., Whyte, L.G., Banks, M.K. *et al.* (2001). Selection of specific endophytic bacterial genotypes by plants in response to soil contamination. *Applied and Environmental Microbiology*, 6: 2469–2475.

[54] Singh, K.P., Malik, A., Kumar, R., Saxena, P. and Sinha, S. (2008). Receptor modeling for source apportionment of polycyclic aromatic hydrocarbons in urban atmosphere. *Environmental monitoring and assessment, 136*(1–3), 183–196.

[55] Šišović, A., Vađić, Ž., Šega, K., Bešlić, I. and Vađić, V. (2005). Comparison between PAH mass concentrations measured in PM10 and PM2. 5 particle fractions. *Bulletin of environmental contamination and toxicology, 75*(1), 121–126.

[56] Skupinska, K., Misiewicz, I. and Kasprzycka-Guttman, T. (2004). Polycyclic aromatic hydrocarbons: physicochemical properties, environmental appearance and impact on living organisms. *Acta Pol Pharm*, 61(3), 233–240.

[57] *State of global report, 2019.* https://www.stateofglobalair.org/sites/default/files/soga_2019_report.pdf

[58] Stegeman, J.J. and Lech, J.J. (1991). Cytochrome P-450 monooxygenase systems in aquatic species: carcinogen metabolism and biomarkers for carcinogen and pollutant exposure. *Environmental health perspectives, 90*, 101–109.

[59] Stegeman, J.J. *et al.* (2001). Cytochrome P450 1A expression in mid water fishes: potential effects of chemical contaminants in remote oceanic zones. *Environ. Sci. Technol.*, 35, 54–62.

[60] Sun, K., Liu, J., Jin, L. and Gao, Y. (2014). Utilizing pyrene-degrading endophytic bacteria to reduce the risk of plant pyrene contamination. *Plant and soil, 374*(1–2), 251–262.

[61] Sverdrup, L.E., Krogh, P.H., Nielsen, T., Kjær, C. and Stenersen, J. (2003). Toxicity of eight polycyclic aromatic compounds to red clover *(Trifoliumpratense)*, ryegrass (*Loliumperenne*) and mustard *(Sinapsisalba)*. Chemosphere, 53(8), 993–1003.

[62] USEPA (1993). Provisional guidance for quantitative risk assessment of polycyclic aromatic hydrocarbons. Cincinnati (OH): Environmental Criteria and Assessment Office. Final Draft.ECAO-CIN-842. March 1993.

[63] Wang, Y., Tian, Z., Zhu, H., Cheng, Z., Kang, M., Luo, C., *et al.*, Polycyclic aromatic hydrocarbons (PAHs) in soils and vegetation near an e-waste recycling site in South China: concentration, distribution, source and risk assessment. *Sci. Total Environ,* 2012; 439: 187–93.

[64] Weis, L.M. *et al.* (1998). Bay and baylike regions of polycyclic aromatic hydrocarbons were potent inhibitors of gap functional intercellular communication. *Environ. Health Perspect.* 106, 17–22.

[65] Yu, H. (2002). Environmental carcinogenic polycyclic aromatic hydrocarbons: photochemistry and phototoxicity. *Journal of Environmental Science and Health, Part C, 20*(2), 149–183.

[66] Zhang, Y. and Tao, S. (2009). Global atmospheric emission inventory of polycyclic aromatic hydrocarbons (PAHs) for 2004. *Atmospheric Environment,* 43(4), 812–819.

Reducing Carbon Dioxide Emissions through the Production of Synthetic Natural Gas

Radwa A. El-Salamony
Egyptian Petroleum Research Institute, Nasr City 11727, Cairo, Egypt
E-mail: radwa2005@hotmail.com

ABSTRACT: Carbon dioxide gas in the atmosphere causes a maintaining temperature that makes the Earth habitable, by trapping heat to create a natural greenhouse effect. Today the level of carbon dioxide is higher than at any time in human history. Scientists widely agree that Earth's average surface temperature has already increased since the 1880s and that human-caused increases in carbon dioxide and other heat-trapping gases are extremely likely to be responsible. Today, climate change is the term scientists use to describe the complex shifts, driven by greenhouse gas concentrations which are now affecting our planet's weather and climate systems. Climate change encompasses not only the rising average temperatures we refer to as global warming but also extreme weather events, shifting wildlife populations and habitats, rising seas and a range of other impacts. Recently, because of the increasing demand for natural gas and the increasing of greenhouse gases, interests have focused on the production of synthetic natural gas (SNG), which is suggested as an important future energy carrier. Among the catalytic reactions of CO_2, hydrogenation of CO_2, the so called methanation reaction, is a suitable technique for the fixation of CO_2. This technique can be used to convert exhausted CO_2 into methane (CH_4), which can be recycled for use as a fuel or a chemical as well as by contributing to the reduction of CO_2 emissions. The basic aim of this chapter is to summarize the recent progress in CO_2 methanation as a particularly promising technique for reducing CO_2 emissions and producing energy carrier or chemical.

Keywords: Greenhouse Gas, CO_2 Hydrogenation, Methanation, Catalytic Reaction, Synthetic Natural Gas (SNG).

INTRODUCTION

CO_2 emissions increased dramatically due to human activities, including power generation, transportation, deforestation and fossil fuels (Eckle *et al.,* 2011; Pastor-Pérez *et al.,* 2018). The CO_2 recycling has become an important topic in recent years and a lot of researches have been conducted (Borgschulte *et al.,* 2013; Hwang *et al.,* 2013). As CO_2 is one of the main contributors to greenhouse effect and hence to climate change, there is a growing interest in its use as a feed-stock in chemical processes (Hwang *et al.,* 2013; Pandey *et al.,* 2018), wherein CO_2 capture and storage (CCS) and CO_2 chemical conversion have attracted wide interests. It is important to convert CO_2 to fuels or raw materials which must be also easily transportable. CH_4 is suitable for this because it benefits from the existing infrastructure for transport and storage of natural gas (Xu *et al.,* 2017; Thomas *et al.,* 2018). However, natural gas is considered a clean source of energy compared to burning other fossil fuels due to the reduced emissions of CO_2 (Spivey and Hutchings, 2014). For example, it is viewed as one of the cleanest solutions to power the transportation industry and, due to its operational flexibility, can be more easily paired with renewable resources for delivering electricity at peak times. Demand for natural gas has grown over the past few decades and is expected to continue over the coming years.

In addition, methane formation from CO_2 is a simpler reaction which can generate methane under atmospheric pressure and low temperature (Galletti *et al.,* 2010; Beuls *et al.,* 2012). It is called the Sabatier reaction and could be applied in industry, provided that hydrogen is generated from renewable sources (Aziz *et al.,* 2015a). Also, compared with CO methanation, CO_2 methanation is also favorable in thermo-dynamics. In order to obtain a high CH_4 yield at 1 atm and low temperature. However, the eight-electron reduction of CO_2 to CH_4 by hydrogen is difficult to achieve because of significant kinetic barriers (Thampi *et al.,* 1987). The CH_4 selectivity below 500°C in the CO_2 methanation system is relatively higher than that in CO methanation. CO_2 methanation (Eq. (1)) is a strong exothermic reaction and thermodynamically favored at low temperature (Zhang *et al.,* 2020). However, CO_2 reduction process owns high kinetic barriers, making this process difficult to achieve and CO_2 is so stable that difficult to be activated (Liu *et al.,* 2012).

$$CO_2 + 4H_2 \rightarrow CH_4 + 2H_2O \ \Delta H_{298} = -165 \text{ kJ/mol} \qquad \dots (1)$$

Therefore, CO_2 activation and conversion at low temperature is a critical challenge. At a relative low temperature (200–250°C) the products are mainly CH_4 and H_2O. Rising the temperature above 450°C results in the increase in the CO by-product, due to reversed water gas shift reaction (Eq. (2)),

$$CO_2 + H_2 \rightarrow H_2O + CO \; \Delta H_{298} = 41.2 \text{ kJ mol}^{-1} \quad \ldots (2)$$

and meanwhile, unreacted CO_2 and H_2 also increase, along with a decrease in the CH_4 product. Since CO_2 methanation is also a strongly exothermic reaction, increasing temperature is unfavorable for this reaction. However, when the temperature exceeds about 550°C, the mole fraction of CO_2 reaches its maximum and then decreases because the reversed water gas shift reaction dominates. The calculation also reveals that no significant carbon deposition is generated under that condition (Zhang *et al.,* 2020).

The mechanisms proposed for CO_2 methanation according to the studies from experimental works and theoretical calculations can be divided into two categories. The first one involves that CO_2 will be converted to CO by reverse water gas shift (RWGS) reaction followed with CO methanation (Gao *et al.,* 2012), as listed in Eq. (2) and Eq. (3), respectively.

$$CO + 3H_2 \rightarrow CH_4 + H_2O \; \Delta H298 \text{ K} = -206.1 \text{ kJ mol}^{-1} \quad \ldots (3)$$

The other mechanism is direct hydrogenation of CO_2, which is related with the formation of intermediate such as formate, $-CH_2OH$, $-CH_3$ and then CH_4. Where, CO is considered as a byproduct of CO_2 methanation, not an intermediate for CO_2 methanation (Pan *et al.,* 2014).

CO_2 methanation has been studied extensively using different types of metals and supports (Karelovic and Ruiz, 2013). The CO_2 methanation reaction is exothermic and an eight-electron process is required to reduce the fully oxidized carbon to CH_4 with an appropriate catalyst. Ni (Ocampo *et al.,* 2011; Takano *et al.,* 2011; Zhi *et al.,* 2011; da Silva *et al.,* 2012), Ru (Sharma *et al.,* 2011) and Rh (Karelovic and Ruiz, 2013) based catalysts have been proved to express a good catalyst activity in CO_2 methanation and nickel based catalysts have been extensively investigated under varying experimental conditions for its low price and high activity. Currently, main supports of the catalysts are activated carbon (AC) (Swalus *et al.,* 2012), Al_2O_3 (Beuls *et al.,* 2012), ZrO_2 (da Silva *et al.,* 2012), molecular sieve, $Ni/Ce_xZr_{1-x}O_2$ (Cai *et al.,* 2013), SiC (Zhi *et al.,* 2011) and CeO_2 (Tada *et al.,* 2012).

EFFECT OF CATALYSTS

Effect of Metal Types

The presence of carbon dioxide and water molecules as impurities in crude natural gas decreases the quality of natural gas. Recently, the catalytic treatment of this toxic and acidic gas has become a promising technique by converting CO_2 to methane gas in the presence of H_2S gas; thus, enhancing the methane production and creating an environmentally friendly approach to the purification of natural gas. Abu Bakar *et al.* (2012) prepared a series of catalysts based on nickel oxide using the wetness impregnation technique, followed by calcination at 400°C. Pd/Ru/Ni(2 : 8 : 90)/Al_2O_3 catalyst was revealed as the most potential catalyst and achieved 43.60% of CO_2 conversion, with 6.82% of methane formation at 200°C. This catalyst had the highest percentage of 52.95% CO_2 conversion and yielded 39.73% methane at a maximum temperature of 400°C. In the presence of H_2S in the gas stream, the conversion dropped to 35.03%, with 3.64% yield of methane at a reaction temperature of 400°C. However, this catalyst achieved 100% H_2S desulfurization at 140°C and remained constant until the reaction temperature of 300°C.

The response surface methodology (RSM) was used to evaluate the effects of catalyst preparation parameter (Ru loading, calcination temperature and catalyst loading) toward the attainment of optimum preparation conditions of Ru/Mn/Fe–Al_2O_3 in CO_2/H_2 methanation at 270°C (Ab Halim *et al.*, 2015). The obtained model equation showed high coefficient of determination (R^2 = 0.9986). The optimum conditions for Ru loading, calcination temperature and catalyst loading were 5.5 wt %, 1010°C and 5 g, respectively. Verification experiment was performed at the optimum conditions and the experiment value (95.0%) closely agreed with predicted value (96.6%). This indicated that RSM is an accurate and applicable method to optimize the catalyst preparation parameters and to obtain the maximal conversion of CO_2 over Ru/Mn/Fe–Al_2O_3 catalyst.

The strategy of adding large amount of metal oxide promoters (15 wt%) to nickel supported on γ-Al_2O_3 micro-spheres catalyst was presented by (Guilera *et al.*, 2019). The addition of CeO_2, La_2O_3, Sm_2O_3, Y_2O_3 and ZrO_2 was clearly beneficial, as the corresponding metal-oxide promoted catalysts exhibited higher catalytic performance than Ni/Al_2O_3. This increase of catalytic activity is attributed to the higher amount of CO_2 adsorbed on the catalyst. Among the selected promoters, La_2O_3 showed the highest catalytic activity due to the enhancement of nickel reducibility, nickel dispersion and the presence of moderate basic sites. In addition, $Ni–La_2O_3/Al_2O_3$ was stable for one week, while the unpromoted catalyst exhibited a slight decline in its activity. The effects of CeO_2 content and reaction temperature on the performance of the $Ni–CeO_2/Al_2O_3$ catalyst with a nickel content of 15 wt% were studied in detail for methanation reaction (Liu *et al.,* 2012). The results showed that the catalytic performance was strongly dependent on the CeO_2 content in $Ni–CeO_2/Al_2O_3$ catalysts and that the catalysts with 2 wt% CeO_2 had the highest catalytic activity among the tested ones. Preliminary stability test of 120 h on stream over the $Ni–CeO_2/Al_2O_3$ catalyst at 350°C revealed that the catalyst was much better than the unpromoted one. Ni–Ce–Zr mixed oxides were prepared through one-pot hydrolysis of mixed metal nitrates with ammonium carbonate for CO_2 methanation (Nie *et al.,* 2017). The effects of Ce/Zr molar ratio and Ni content on catalysts' physical and chemical properties, reduction degree of Ni^{2+} and catalytic properties were systematically investigated. The results showed that Zr could lower metallic Ni particle sizes and alter interaction between Ni and supports, resulting in enhancements in the catalytic activity for methanation. The Ni–Ce–Zr catalyst containing 40 wt% Ni and Ce/Zr molar ratio of 9:1 exhibited the optimal catalytic properties, with 96.2% CO_2 conversion and almost 100% CH_4 selectivity at a low temperature of 275°C. During the tested period of 500 h, CO_2 conversion and CH_4 selectivity over Ni–Ce–Zr catalyst kept constant under 300°C.

The crucial problems for the catalysts of CO_2 methanation are the low activity at low temperature and deactivation caused by metal sintering. In order to overcome the problems or to improve the shortages, a new scheme has been put forward by loading $LaNi_{1-x}Mo_xO_3$ with perovskite-type structure on SiO_2 (Li *et al.,* 2019). After reduction, Ni nanoparticles, MoO_x and La_2O_3 would be all stay together and highly dispersed on SiO_2 ($Ni/MoO_x–La_2O_3/SiO_2$). Through effectively combining MoO_x which is active for the reaction of reverse water gas shift and Ni which can catalyze CO methanation, the resultant catalyst exhibited attractive good performance for CO_2 methanation, especially showing very good resistance to metal sintering. Ni–La_2O_3/SiO_2 catalyst without adding Mo was investigated for comparison. Since many metallic ions can enter into the lattice of a perovskite-type oxide, therefore, many combined catalysts for sequential reactions may be designed via this scheme.

The effects of Ni loading and water vapor on the properties of Ni/mesoporous silica nanoparticles (MSN) and CO_2 methanation were studied (Aziz *et al.,* 2015b). X-ray diffraction, N_2 adsorption–desorption and pyrrole-adsorbed infrared (IR) spectroscopy results indicated that the increasing Ni loading (1–10 wt%) decreased the crystallinity, surface area and basic sites of the catalysts. The activity of CO_2 methanation followed the order of 10Ni/MSN > 5Ni/MSN > 3Ni/MSN > 1Ni/MSN. These results showed that the balance between Ni and the basic-site concentration is vital for the high activity of CO_2 methanation. All Ni/MSN catalysts exhibited a high stability at 623 K for more than 100 h. The presence of water vapor in the feed stream induced a negative effect on the activity of CO_2 methanation. The water vapor decreased the carbonyl species concentration on the surface of Ni/MSN, as evidenced by CO + H_2O-adsorbed IR spectroscopy. The response surface methodology experiments were designed with face-centered central composite design (FCCCD) by applying 24 factorial points, 8 axial points and 2 replicates, with one response variable (CO_2 conversion). The Pareto chart indicated that the reaction temperature had the largest effect for all responses. The optimum CO_2 conversion was predicted from the response surface analysis as 85% at an operating treatment time of 6 h, reaction temperature of 614 K, gas hourly space velocity (GHSV) of 69105 ml/g_{cat}/h and H_2/CO_2 ratio of 3.68.

A series of mesoporous supported nickel-based catalysts on nanocrystalline alumina carrier promoted with various metals (Fe, Co, Zr, La and Cu) were prepared and employed in carbon dioxide methanation reaction (Moghaddam *et al.,* 2018). The results showed that among the prepared catalysts, Ni–La/Al_2O_3 and Ni–Fe/Al_2O_3 possessed the highest surface area and the largest pore volume, respectively. Likewise, there was a slight decrease in the pore volume and the average pore diameters of the promoted samples. The temperature programmed reduction (TPR) results depicted that the incorporation of the promoters enhanced the reducibility of the catalysts and shifted the reduction of NiO species to a lower reduction temperature. The

CO_2 conversions of all promoted catalysts except Cu promoted sample were higher peculiarly at low temperatures compared to those attained for the unpromoted catalyst. Ni-Fe/Al_2O_3 catalyst exhibited the best catalytic performance of 70.63% CO_2 conversion and 98.87% CH_4 selectivity at 350°C and showed high stability and desirable resistance against sintering.

Effect of Support

The effect of CeO_2 loading amount of Ru/CeO_2/Al_2O_3 on CO_2 methanation activity and CH_4 selectivity was studied (Tada *et al.,* 2014). The CO_2 reaction rate was increased by adding CeO_2 to Ru/Al_2O_3 and the order of CO_2 reaction rate at 250°C is Ru/30%CeO_2/Al_2O_3 > Ru/60%CeO_2/Al_2O_3 > Ru/CeO_2 > Ru/Al_2O_3. With a 30% CeO_2 loading of Ru/CeO_2/Al_2O_3, partial reduction of CeO_2 surface was promoted and the specific surface area was enlarged. Furthermore, it was observed using FTIR technique that intermediates of CO_2 methanation, such as formate and carbonate species, reacted with H_2 faster over Ru/30% CeO_2/Al_2O_3 and Ru/CeO_2 than over Ru/Al_2O_3. These could result in the high CO_2 reaction rate over CeO_2—containing catalysts. As for the selectivity to CH_4, Ru/30%CeO_2/Al_2O_3 exhibited high CH_4 selectivity compared with Ru/CeO_2, due to prompt CO conversion into CH_4 over Ru/30% CeO_2/Al_2O_3.

The novel nickel-based catalysts with a nickel content of 12 wt% were prepared with the zirconia-alumina composite as the supports (Cai *et al.,* 2011). The new carriers, ZrO_2 improved alumina, were synthesized by three methods, i.e., impregnation-precipitation, co-precipitation and impregnation method. The catalytic properties of these catalysts were investigated in the methanation of carbon dioxide. The new catalysts showed higher catalytic activity and better stability than Ni/γ-Al_2O_3. Furthermore, as a support for new nickel catalyst, the ZrO_2–Al_2O_3 composite prepared by the impregnation-precipitation method was more efficient than the other supports in the methanation of carbon dioxide. The highly dispersed zirconium oxide on the surface of γ–Al_2O_3 inhibited the formation of nickel aluminate-like phase, which was responsible for the better dispersion of Ni species and easier reduction of NiO species, leading to the enhanced catalytic performance of corresponding catalyst.

Rice husk ash-alumina (RHA–Al_2O_3) supports were prepared by impregnation of rice husk ash with an aluminum nitrate solution and were then used to prepare supported nickel catalysts (Ni/RHA–Al_2O_3) by incipient wetness impregnation method (Chang *et al.*, 2003). These catalysts with high surface area and mesopores structure are advantageous for chemical reactions. The interaction between nickel and support is very strong. Consequently, the small nickel oxide crystallites were formed with good dispersion and more than one species, such as NiO or $NiAl_2O_4$-like ones, were detected. Moreover, 500°C is suitable for hydrogenation of CO_2 with maximum yield and selectivity of CH_4. Furthermore, the hydrogenation tests showed that Ni/RHA–Al_2O_3 catalysts performed better than Ni/SiO_2–Al_2O_3 ones.

The effect of support type on CO_2 methanation activity and CH_4 selectivity was studied over Ni/CeO_2, Ni/α–Al_2O_3, Ni/TiO_2 and Ni/MgO catalysts (Tada *et al.*, 2012). Ni/CeO_2 exhibited high CO_2 conversion close to the equilibrium one from 300°C and above with CH_4 selectivity of ca. 1 under a large GHSV = 10,000 h^{-1} at CO_2: H_2 = 1: 4. It was found by H_2 consumption estimated from TPR results that not only Ni species but also CeO_2 was reduced at 600°C. CO_2–TPD demonstrated the amount of CO_2 adsorbed on Ni/CeO_2 was much larger than that on Ni/α–Al_2O_3. From these results, CO_2 reduction to CO over Ni/CeO_2 can be enhanced at low temperatures compared to Ni/α–Al_2O_3.

Also, CO_2 methanation was investigated over cobalt catalysts supported on different supports such as γ-Al_2O_3, SiO_2, TiO_2, CeO_2 and ZrO_2 (Le *at al.*, 2017). Among them, the cobalt catalyst supported on the high-surface-area CeO_2 was determined to be the most active for CO_2 methanation. The high Ni dispersion and an intimate contact between Ni and ceria appeared to be responsible for the high catalytic activity for CO_2 methanation. Furthermore, strong CO_2 adsorption on CeO_2 is beneficial for CO_2 methanation (Le *et al.*, 2017). This Co–CeO_2 also showed the stable catalytic activity even after an exposure to high-temperature reaction conditions. Whereas; the selectivity for methane was 100% with different CO_2 conversions over Ni/La_2O_3 catalyst due to the formation of lanthanum oxycarbonate ($La_2O_2CO_3$) which can play an important role in the activation of CO_2 (Huanling *et al.*, 2010). On the other hand, Muroyama *et al.* (2016) suggested that the moderate basic sites should have a positive effect on the catalytic activity for CO_2 methanation. The species formed on the surface during CO_2 methanation was compared for Ni/Y_2O_3, Ni/Al_2O_3 and Ni/La_2O_3. For Ni/Al_2O_3, the CO species on Ni surface was specifically confirmed, indicating that CO_2 methanation should proceed via the formation of CO intermediate. Only carbonates were formed over Ni/La_2O_3 even under the methanation condition. In contrast, over Ni/Y_2O_3 the carbonate-like species were converted to formate species by introducing H_2 to the CO_2 atmosphere, followed by the decomposition of formate species in the H_2 atmosphere. The promotion of the decomposition of formate species over the Ni/Y_2O_3 catalyst would be responsible for its high catalytic activity. Although the reaction mechanism of CO_2 methanation over Ni catalysts has not been elucidated sufficiently, the mechanism should be different depending on the catalysts. Thus, further investigations on the reaction mechanism as well as the search for new materials are requested for the development of highly-active catalysts.

Various supports (ZSM-5, SBA-15, MCM-41, Al_2O_3 and SiO_2) with various mesoporous structures were introduced to fabricate nickel-based catalysts for the CO_2 methanation by incipient wetness impregnation method (DOI: 10.1021/acs.energyfuels.7b03826). Ni/ZSM-5 catalyst displayed the most active catalytic properties, followed with Ni/SBA-15, Ni/Al_2O_3, Ni/SiO_2 and Ni/MCM-41 catalysts. The excellent catalytic property of Ni/ZSM-5 catalyst was resulted from the basic property and the synergistic effect between nickel metal and support. The reactivity of the reaction intermediate monodentate formate in Ni/ZSM-5 catalyst was more active than that of bidentate formate species as identified by in situ infrared spectroscopy. The Ni/ZSM-5 catalyst performed with excellent stability and no deactivation occurred up to 100 h.

FACTORS AFFECTING THE METHANATION REACTION

Effect of Reaction Temperature

The effect of the temperature on the CO_2 conversions was studied under methanation conditions (Cai et 2011). At lower temperatures (< 210°C), these catalysts were almost not active. At higher temperatures, the Ni/ZrO_2–Al_2O_3—Imp catalyst presented the best performance over the whole investigated temperature range and the CO_2 conversion ranged from 1.3% at 210°C to 69.8% at 360°C.

Panagiotopolou and Kodarides (2007) founded that the platinum catalyst is inactive within the temperature range of 200–400°C, since temperatures higher than 450°C are required in order to achieve conversion above 20%. Erdohelyi *et al.* (2004) reported that the rate of methane formation was suddenly higher in the CO_2 hydrogenation reaction on Rh/TiO_2 at higher temperatures of 400°C, around 75% selectivity for CH_4 formation and CO was also formed from the reaction. The addition of palladium into the Ru/Ni (10:90)/Al_2O_3 catalyst to form Pd/Ru/Ni (2:8:90)/Al_2O_3 catalyst coincidentally enhanced catalytic activity for the conversion of CO_2 of the prepared catalysts. It showed 37.94%, 43.60% and 45.77% of CO_2 conversion at reaction temperatures of 100°C, 200°C and 300°C, respectively, while 52.95% CO_2 conversion was achieved at the maximum studied temperature of 400°C (Abu Bakar *et al.,* 2011). The catalytic performance of the Ni/MoO_x–La_2O_3/SiO_2 catalysts was investigated within the temperature range of 250–550°C (Li *et al.*, 2019). From 250 to 400°C, CO_2 conversion was gradually increased, while upon reaching the temperature 400°C. Further, the conversion declined rapidly with the temperature increase, due to the fact that CO_2 methanation belongs to a strong exothermic reaction, following with the limitation of thermodynamic equilibrium (Gao *et al.*, 2012).

The CO_2 conversion increases rapidly as the reaction temperature increases up to 500°C and remains constant between 500 and 600°C, followed by gradual increase in conversion with reaction temperature over Ni/RHA–Al_2O_3 catalyst (Chang *et al.,* 2011). The CH_4 yield increases as the reaction temperature increases up to a maximum (500°C) and then decreases. This can be explained from the viewpoint of thermodynamics. Due to the highly exothermicity of methanation reaction; the excessive heat will result in inactivation of the catalyst and will affect the thermodynamic equilibrium. Therefore, when the temperature rises to such an extent, hydrogenation becomes limited due to the occurrence of thermodynamic equilibrium. A higher reaction temperature promotes side reactions and implies a lower CH_4 selectivity. Therefore, reaction proceeding under a suitable temperature can promote the yield of methane. Obviously, a reaction temperature of 500°C is the optimum condition for hydrogenation of CO_2 over Ni/RHA-Al_2O_3 catalyst.

Furthermore; CO_2 conversions over Ni/a–Al_2O_3, Ni/MgO and Ni/TiO_2 were gradually increased as the temperature was raised to ca.

450°C (Tada *et al.,* 2012). On the other hand, CO_2 conversion over Ni/CeO_2 exhibited a rapid increase up to 350°C and then approached the equilibrium one. CH_4 selectivity of Ni/α–Al_2O_3 and Ni/MgO was ca. 0.9 and 0.7, respectively, at 250°C and slightly increased up to 450°C. The rest of converted CO_2 existed as CO. As for Ni/CeO_2 and Ni/TiO_2, almost no CO was detected at the reactor outlet in the temperature range between 250 and 400°C. CO is a key intermediate involved in determining the selectivity of final product, therefore, CO converted from CO_2 on Ni/CeO_2 was promptly converted into CH_4, leading to the depletion of CO, namely high CH_4 selectivity close to 1 at low temperatures.

Effect of Pressure

According to le Chatelier's principle, the CO_2 methanation is favored at elevated pressures material. The effect of pressure on equilibrium conversion of CO_2 was calculated (Gao *et al.,* 2012). As can be seen, in the typical temperature range of 200–500°C, increasing the pressure is effective up to a certain point and further increase of pressure is less effective. A pressure of 10 to 30 atm is considered mild in terms of stress on the catalyst and should therefore not cause sintering for the catalyst. Carbon deposition theoretically does not occur if the H_2/CO_2 ratio is equal to or higher than the stoichiometric ratio. On the other hand, thermodynamic simulations performed by Jürgensen *et al.* (2015) found that elevated pressure increased the temperature at which carbon deposition would occur from 365°C at 1 bar to 515°C at 11 bar. Additionally, the selectivity for methane was improved at elevated pressure in both studies. Catalyst evaluation is mostly conducted at atmospheric pressure. However, pressure is of great importance in terms of optimizing the process. Further, temperatures above 500°C can cause sintering and increase carbon deposition, leading to catalyst deactivation. Thus, temperature control is vital as the exothermic methanation reaction will result in apparent temperature increase in large-scale operations (Jürgensen *et al.,* 2015).

Effect of Space Velocity

For the analysis of the effect of space velocity on Ni–Co/CeO_2–ZrO_2 catalyst performance, the reaction conditions were tested at the same temperatures and H_2/CO_2 ratio at space velocities of 6250, 12500 and

25000 ml/g_{cat}/h (Pastor-Pérez *et al.*, 2018). The reduced space velocity appears to enhance the catalytic activity as the 6250 mL/g/h sample achieves CH_4 selectivity's consistently around 98% at 300°C. On the other hand, an increase in space velocity at 25000 ml/g_{cat}/h; caused a decreased in the catalytic activity, CH_4 selectivity reaches a maximum of 82%. The CO_2 conversion versus reaction temperature for the three different space velocities investigated; an overall trend of slightly increasing CO_2 conversion can be seen when space velocity is reduced. On the other hand, an increase in space velocity (25,000 ml/g_{cat}/h) caused a decreased in the CO_2 conversion level, reaching 60%.

Abate *et al.* (2016) investigated the effect of gas hour space velocity (GHSV) on a Ni/Al_2O_3–TiO_2–CeO_2–ZrO_2 catalyst at temperatures from 250 to 400°C. The feed gas consisted of a H_2/CO_2 ratio of 4 and was diluted in 87.5% N_2. The GHSV was varied between 20000, 25000 and 30000 h^{-1}. At temperatures above 350°C the CO_2 conversion and CH_4 yield was found to be almost identical and close to equilibrium, which is apparently limited by thermodynamics. However, substantially higher CO_2 conversion and methane yield was observed at temperatures below 350°C as the GHSV was reduced. This is expected because the reaction is operated below thermodynamic equilibrium and a higher GHSV will result in lower CO_2 conversions. The CO selectivity data is quite close at different GHSVs, while a higher GHSV will give a slightly higher CO yield, which could be explained by the reduced chance for the hydrogenation of CO.

CO_2 methanation over sponge Ni was investigated (Tada *et al.*, 2017). When CO_2 methanation was carried out using sponge Ni without any pretreatment, the sponge Ni exhibited a CO_2 conversion of 83% at 250°C under a high space velocity (0.11 mol$_{CO_2}$/g_{cat}/h). We think that the sponge Ni is a promising new catalyst for CO_2 methanation because it showed the high activity even under the high GHSV and we can design a small plug flow reactor compared to a conventional reactor, resulting in a low manufacturing cost for the reactor. The high activity can be derived from the great number of crystal defects of face-centered cubic fcc-Ni in the sponge Ni. On the other hand, with high-temperature pretreatment, the sponge Ni lost its activity in CO_2 methanation as well as the surface defect sites. Thus, the activity loss can be explained by the disappearance of the surface defect sites by the high-temperature pretreatment (Tada *et al.*, 2017).

CO_2/H_2 Molar Ratio

The H_2:CO_2 ratio strongly influences the final product. Low ratios tend to provide larger amounts of high molecular mass products while at higher ratios, more methane is produced. The ideal H_2:CO_2 mole ratio creating an atmosphere relevant for methanation and leading to better selectivity and higher methane yield is generally agreed to be 3: 1 up to 4: 1 (Tsuji *et al.*, 1996; Graçca *et al.*, 2014). When the reaction proceeds with H_2 and CO_2 in the ratio of 4:1, more than 95% of the formed hydrocarbon is methane (Karn *et al.*, 1965).

When H_2/CO_2 is equal to 2, CO_2 conversion of only about 50–70% can be obtained at either 1 atm or 30 atm, maximum CH_4 selectivity of 73% and 88% is achieved at 1 atm and 30 atm, respectively and abundant carbon deposition (up to 50%) is found below 500°C at both 1 and 30 atm. But, when H_2/CO_2 ratio is equal to or more than 4, carbon deposition is not found (Gao *et al.*, 2012).

Thus, in order to obtain a high CH_4 yield and avoid carbon deposition, H_2/CO_2 ratio should not be lower than 4 even at 30 atm.

METHANATION REACTION MECHANISMS

The reaction pathways of CO_2 methanation are divided into two main categories. The first one proposes the conversion of CO_2 to CO via the reverse water gas shift reaction (RWGS) reaction followed with CO methanation (Eckle *et al.*, 2011; Borgschulte *et al.*, 2013), respectively and its subsequent reaction to methane through the same pathway as CO methanation. The second pathway proposes direct CO_2 methanation (Sharma *et al.*, 2011) which is related with the formation of intermediate such as formate species.

Methanation is a catalytic reaction depending on the nature of the metal serving as the catalyst; either CO or CH_4 is the main product of the reaction (Lapidus *et al.*, 2007).

Weatherbee and Bartholomew (1982) stated that the first step of CO_2 hydrogenation is dissociative adsorption to hydrogen atoms, CO_{ad} and oxygen atoms. Adsorbed CO can either dissociate to carbon and oxygen atoms or desorb. The next steps include the hydrogenation of adsorbed carbon and carbene intermediates to methane (Baraj *et al.*, 2016).

Even when the CO_2 methanation occurs via CO formation, it does not necessarily mean that CO formation should proceed through the reverse water gas shift reaction. Jacquemin *et al.* (2010) performed CO_2 methanation using a rhodium-based catalyst, Rh/γ–Al_2O_3; it was observed that the first step in CO_2 methanation is its dissociative adsorption to form CO_{ad} and O_{ad} on the surface of the catalyst. Oxidization of rhodium occurring during the reaction confirmed that CO_2 is dissociated on the surface of the catalyst and that the catalyst is oxidized by the O_{ad} species. In a related work, Beuls *et al.* (2014) performed CO_2 methanation using Rh/γ–Al_2O_3 catalyst and DRIFT measurements. It was confirmed that CO_2 dissociation is responsible for the oxidation of Rh.

Pan *et al.* (2014) studied CO_2 methanation on nickel supported ceria-zirconia (Ni/$Ce_{0.5}Zr_{0.5}O_2$) by FTIR spectroscopic measurements. It was observed that the main reaction intermediates of methanation were formate species. These formate species were believed to be derived from the hydrogenation of carbonates formed on active sites of the support, which were a result of surface oxygen sites and surface oxygen vacancies of Ce^{3+}, Ce^{4+} or Zr as well as hydroxyl surface sites of Ce^{4+} or Zr. Minor CO amounts were detected; however, CO was considered only as a by-product of the reaction and by no means as an intermediate.

Aldana *et al.* (2013) studied CO_2 methanation on the classic nickel catalyst supported on silica (Ni/SiO_2) and nickel catalysts supported on ceria-zirconia (Ni/CZ). It was observed that when using the Ni/CZ catalyst, CO_2 was adsorbed on sites of medium basicity forming large amounts of carbonate species. As the reaction proceeded it was assumed that carbonate species were reduced into formate species since an increase in the amount of the later, which were initially not present, was observed and then increased up to 523 K before sharply decreasing, which corresponds with the beginning of methane formation. However, when using a Ni/SiO_2 catalyst, a lower amount of carbonate species was observed due to the weaker basicity of the support. Even though formate species were observed, no evidence correlating their presence with the catalytic activity was found. Carbonyl species were detected indicating the dissociation of CO_2 on the metal surface. The overall performance of the Ni/CZ catalyst was better than that of the Ni/SiO_2 catalyst indicating that basic sites are very important for the CO_2 methanation.

CONCLUSION

Hydrogenation of CO_2, the so called methanation reaction, is a suitable technique for the fixation of CO_2. This technique can be used to convert exhausted CO_2 into methane (CH_4), which can be recycled for use as a fuel or a chemical as well as by contributing to the reduction of CO_2 emissions. CH_4 is suitable for this because it benefits from the existing infrastructure for transport and storage of natural gas. However, natural gas is considered a clean source of energy compared to burning other fossil fuels due to the reduced emissions of CO_2. CO_2 methanation has been studied extensively using different types of metals and supports. Nickle-based catalysts showed high activity and selectivity. To meet these requirements, the catalyst support of inert carrier should provide sufficient surface area for the metal to disperse. It has emerged that the support plays a very active role in the interaction between the nickel and the support. The nickel compounds on different support surfaces result in different extents of what are generally called "metal-support effects". This implies that catalysts with different characteristics exhibit different performances toward activity and selectivity for a methanation process. The CH_4 yield increases as the reaction temperature increases up to a maximum (500°C) and then decreases. The reaction of CO_2 hydrogenation to CH_4 is highly exothermic. The excessive heat will result in inactivation of the catalyst and will affect the thermodynamic equilibrium. Therefore, when the temperature rises to such an extent, hydrogenation becomes limited due to the thermodynamic equilibrium. Also; increasing the pressure is effective up to a certain point and further increase of pressure is less effective. The H_2: CO_2 ratio strongly influences the final product. Thus, in order to obtain a high CH_4 yield and avoid carbon deposition, H_2/CO_2 ratio should not be lower than 4.

REFERENCES

[1] Ab Halim, A.Z., Ali, R., Azelee W. and Abu Bakar, W. (2015). CO_2/H_2 methanation over M*/Mn/Fe-Al_2O_3 (M*: Pd, Rh and Ru) catalysts in natural gas; optimization by response surface methodology-central composite design. *Clean Techn. Environ. Policy*, 17:627–636.

[2] Abate, S., Mebrahtu, C., Giglio, E., Deorsola, F., Bensaid, S. (2016). Catalytic Performance of γ-Al_2O_3–ZrO_2–TiO_2–CeO_2 Composite Oxide

Supported Ni-Based Catalysts for CO_2 Methanation. *Ind. Eng. Chem. Res.*, 55: 4451–4460.

[3] Abu Bakar, W.A.W., Ali, R., Toemen, S. (2012). Catalytic methanation reaction over supported nickel–ruthenium oxide base for purification of simulated natural gas. *Scientia Iranica C*, 19(3): 525–534.

[4] Aziz, M.A.A., Jalil, A.A., Triwahyono, S. and Saad, M.W.A. (2015b). CO_2 methanation over Ni-promoted mesostructured silica nanoparticles: Influence of Ni loading and water vapor on activity and response surface methodology studies, *Chem. Enginee. J.*, 260: 757–764.

[5] Aziz, M.A.A., Jalil, A.A., Triwahyono, S., Ahmad, A. (2015a). CO_2 methanation over heterogeneous catalysts: recent progress and future prospects, *Green Chem.*, 17: 2647–2663.

[6] Baraj, E., Vagaský, S., Hlinčík, T., Ciahotný, K. and Tekáč, V. (2016). Reaction mechanisms of carbon dioxide methanation, *Chem. Papers*, 70 (4): 395–403.

[7] Beuls, A., Swalus, C., Jacquemin, M., Heyen, G., Karelovic, A., Ruiz, P. (2012). Methanation of CO_2: Further insight into the mechanism over Rh/γ-Al_2O_3 catalyst. *Appl. Catal.*, B, 2: 113–114.

[8] Borgschulte, A., Galladant, N., Probst, B., Suter, R., Callini, E., Ferri, D., Arroyo, Y., Erni, R., Geerlings, H. and Züttel, A. (2013). Sorption enhanced CO_2 methanation. *Phys. Chem. Chem. Phys.*, 15: 9620–9625.

[9] Burger, T., Koschany, F., Thomys, O., Köhler, K. and Hinrichsen, O. (2018). CO_2 methanation over Fe- and Mn-promoted co-precipitated Ni-Al catalysts: Synthesis, characterization and catalysis study, *Appl. Catal. A: Gen.*, 558: 44–54.

[10] Cai, M., Wen, J., Chu, W., Cheng, X. and Lib, Z. (2011). Methanation of carbon dioxide on Ni/ZrO_2–Al_2O_3 catalysts: Effects of ZrO_2 promoter and preparation method of novel ZrO_2–Al_2O_3 carrier. *J. Natural Gas Chem.*, 20(3): 318–324.

[11] Cai, W., Zhong, Q., Zhao, Y. (2013). Fractional-hydrolysis driven formation of non-uniform dopant concentration catalyst nanoparticles of Ni/CexZr1•$x$$O_2$ and its catalysis in methanation of CO_2. *Catal. Comm.*, 39: 30–34.

[12] Chang, F.-W., Kuo, M.-S., Tsay, M.-T. and Hsieh, M.-C. (2003). Hydrogenation of CO_2 over nickel catalysts on rice husk ash-alumina prepared by incipient wetness impregnation. *Appl. Catal. A: Gen.*, 247(2): 309–320.

[13] da Silva, D.C.D., Letichevsky, S., Borges, L.E.P. and Appel, L.G. (2012). The Ni/ZrO_2 catalyst and the methanation of CO and CO_2. *Int. J. Hydrogen Energy*, (37) 11: 8923–8928.

[14] Eckle, S., Anfang, H.-G. and Behm, R.J. (2018). What drives the selectivity for CO methanation in the methanation of CO_2-rich reformate gases on supported Ru catalysts? *Appl. Catal. A: Gen.*, 391(1–2): 325–333.

[15] Eckle, S., Anfang, H.G., Behm, R.J. (2011). Reaction intermediates and side products in the methanation of CO and CO_2 over supported Ru catalysts in H_2-rich reformate gases. *The J. Phys. Chem. C.,* 115: 1361–1367.

[16] Erdohelyi, A., Fodor, K. and Szailer, T. (2004). Effect of H2S on the reaction of methane with carbon dioxide over supported Rh catalysts. *Appl. Catal. B,* 53: 153–160.

[17] Galletti, C., Specchia, S., Saracco, G. and Specchia, V. (2010). CO-selective methanation over Ru-γ-Al_2O_3 catalysts in H_2-rich gas for PEM FC applications, *Chem. Eng. Sci.,* 65: 590.

[18] Gao, J., Wang, Y., Ping, Y., Hu, D., Xu, G., Gu, F., *et al.* (2012). A thermodynamic analysis of methanation reactions of carbon oxides for the production of synthetic natural gas, *RSC Adv.*, 2: 2358–2368.

[19] Graçca, I., González, L.V., Bacariza, M.C., Fernandes, A., Henriques, C., Lopes, J.M. and Ribeiro, M.F. (2014). CO_2 hydrogenation into CH4 on NiHNaUSY zeolites. *Appl. Catal. B.*, 147: 101–110.

[20] Huanling, S., Jian, Y., Jun, Z. and Lingjun, C. (2010). Methanation of Carbon Dioxide over a Highly Dispersed Ni/La_2O_3 Catalyst. *Chinese J. Catalysis,* 31(1): 21–23.

[21] Hwang, S., Hong, U.G., Lee, J., *et al.* (2013). Methanation of carbon dioxide over mesoporous Ni–Fe–Al_2O_3 catalysts prepared by a coprecipitation method: effect of precipitation agent. *J. Ind. and Engin. Chem.,* 19(6): 2016–2021.

[22] Jacquemin, M., Beuls, A., Ruiz, P. (2010). Catalytic production of methane from CO_2 and H_2 at low temperature: Insight on the reaction mechanism. *Catal. Today,* 157: 462–466.

[23] Jaimedel, J.G., Andreina, V., José, A., Díaz, A. and Andreu, T. (2019). Metal-oxide promoted Ni/Al_2O_3 as CO_2 methanation micro-size catalysts, *J. CO_2 Utilization,* 30: 11–17.

[24] Jürgensen, L., Ehimen, E.A., Born, J. and Holm-Nielsen, J.B. (2015). Dynamic biogas upgrading based on Sabatier process: Thermodynamic and dynamic process simulation. *Biores. Technol,* 178: 323–329.

[25] Karelovic and Ruiz, P. (2013). Mechanistic study of low temperature CO_2 methanation over Rh/TiO_2 catalysts. *J. Catal.,* 301: 141–153.

[26] Karn, F.S., Shultz, J.F. and Anderson, R.B. (1965). Hydrogenation of carbon monoxide and carbon dioxide on supported ruthenium catalysts at moderate pressures. *Ind. & Eng. Chem. Product Res. and Develop.,* 4: 265–269.

[27] Lapidus, A.L., Gaiadai, N.A., Nekrasov, N.V., Tishkova, L.A., Agafonov, Y.A. and Myshenkova, T.N. (2007). The mechanism of carbon dioxide hydrogenation on copper and nickel catalysts. *Petrol. Chem.,* 47: 75–82.

[28] Le, T.A., Kim, M.S., Lee, S.H. and Park, E.D. (2017). CO and CO_2 Methanation over Supported Cobalt Catalysts. *Top Catal.,* 60:714–720.

[29] Li, S., Guo, S., Gong, D., Kang, N., Fang, K.-G. and Liu, Y. (2019). Nano composite composed of MoOx–La_2O_3–Ni on SiO_2 for storing hydrogen into CH_4 via CO_2 methanation, *Int. J. Hydrogen energy*, 44: 1597–1609.

[30] Liu, H., Zou, X., Wang, X., Lu, X., Ding, W. (2012). Effect of CeO_2 addition on Ni/Al_2O_3 catalysts for methanation of carbon dioxide with hydrogen. *J. Nat. Gas Chem.*, 21: 703–707.

[31] Moghaddam, S.V., Rezaei, M., Meshkani, F. and Daroughegi, R. (2018). Carbon dioxide methanation over Ni-M/Al_2O_3 (M: Fe, CO, Zr, La and Cu) catalysts synthesized using the one-pot sol-gel synthesis method, *Int. J. Hydrogen Energy*, 43(34): 16522–16533.

[32] Muroyama, H., Tsuda, Y., Asakoshi, T., Masitah, H., Okanishi, T., Matsui, T. and Eguchi, K. (2016). Carbon dioxide methanation over Ni catalysts supported on various metal oxides. *J. Catal.,* 343: 178–184.

[33] Nie, W., Zou, X., Chen, C., Wang, X., Ding, W. and Lu, X. (2017). Methanation of Carbon Dioxide over Ni–Ce–Zr Oxides Prepared by One-Pot Hydrolysis of Metal Nitrates with Ammonium Carbonate, Catalysts, 7(4): 104.

[34] Ocampo, F., Louis, B., Kiennemann, A. and Roger, A.C. (2011). CO_2 methanation over Ni-Ceria-Zirconia catalysts: effect of preparation and operating conditions," *IOP Conference Series: Mate. Sci. and Eng.*, 19(1): Article ID 012007.

[35] Pan, Q., Peng, J., Wang, S. and Wang, S.D. (2014). In situ FTIR spectroscopic study of the CO_2 methanation mechanism on Ni/Ce0.5Zr0.5O2, *Catal. Sci. Technol.,* 4: 502–209.

[36] Panagiotopolou, P. and Kondarides, D.I. (2007). A comparative study of the water–gas shift activity of Pt catalysts supported on single (MOx) and composite (MOx/Al_2O_3, MOx/TiO_2) metal oxide carriers. *Catal. Today,* 127(1–4): 319–329.

[37] Pandey, D., Ray, K., Bhardwaj, R., Bojja, S., Chary, K.V.R. and Deo, G. (2018). Promotion of unsupported nickel catalyst using iron for CO_2 methanation. *Int. J. Hydrogen Energy,* 43: 4987–5000.

[38] Pastor-Pérez, L., LeSaché, E., Jones, C., Gu,S., Arellano-Garcia, H. and Reina. T.R. (2018). Synthetic natural gas production from CO_2 over Ni-x/CeO_2–ZrO_2 (x = Fe, Co) catalysts: Influence of promoters and space velocity. *Catal. Today,* 317: 108–113.

[39] Sharma, S., Hu, Z.P., Zhang, P., McFarland, E.W., Metiu, H. (2011). CO_2 methanation on Ru-doped ceria. *J. Catal.*, 278: 297–309.

[40] Spivey, J.J. and Hutchings, G. (2014). Catalytic aromatization of methane. *Chem. Soc. Rev.,* 43: 792–803.

[41] Swalus, C., Jacquemin, M., Poleunis, C., Bertrand, P. and Ruiz, P. (2012). CO_2 methanation on Rh/γ-Al_2O_3 catalyst at low temperature: '*in situ*' supply of hydrogen by Ni/activated carbon catalyst. *Appl. Catal. B: Envir.*, 125: 41–50.

[42] Tada, S., Ikeda, S., Shimoda, N., Honma, T., Takahashi, M., Nariyuki, A. and Satokawa, S. (2017). Sponge Ni catalyst with high activity in CO_2 methanation. *Int. J. Hydrogen Energy,* 42(51): 30126–30134.

[43] Tada, S., James, O., Kikuchi, R., Haned, T. and Kameyam, H. (2014). Promotion of CO_2 methanation activity and CH4 selectivity at low temperatures over $Ru/CeO_2/Al_2O_3$ catalysts. *Int. J. Hydrogen Energy,* 39(19): 10090–10100.

[44] Tada, S., Shimizu, T., Kameyama, H., Haneda, T. and Kikuchi, R. (2012). Ni/CeO_2 catalysts with high CO_2 methanation activity and high CH_4 selectivity at low temperatures. *Int. J. Hydrogen Energy*, 37(7): 5527–5531.

[45] Takano, H., Izumiya, K., Kumagai, N. and Hashimoto, K. (2011). The effect of heat treatment on the performance of the Ni/(Zr-Sm oxide) catalysts for carbon dioxide methanation. *Appl. Surf. Sci.,* 257(19): 8171–8176.

[46] Thampi, K.R., Kiwi, J. and Gratzel, M. (1987). Methanation and photo-methanation of carbon dioxide at room temperature and atmospheric pressure, *Nature*, 327: 506–508.

[47] Tsuji, M., Kodama, T., Yochida, T., Kitayama, Y., Tamaura, Y. (1996). Preparation and CO_2 methanation activity of an ultrafine Ni(II) ferrite catalyst. *J. Catal.,* 164: 315–321.

[48] Ussa Aldana, P.A., Ocampo, F., Kobl, K., Louis, B., Thibault- Starzyk, F., Daturi, M., Bazin, P., Thomas, S. and Roger, A.C. (2013). Catalytic CO_2 valorization into CH_4 on Ni based ceria-zirconia. Reaction mechanism by operando IR spectroscopy. *Catal. Today*, 215, 201–207.

[49] Weatherbee, G.D. and Bartholomew, C.H. (1982). Hydrogenation of CO_2 on group VIII metals: II. Kinetics and mechanism of CO_2 hydrogenation on nickel. *J. Catal.*, 77: 460–472.

[50] Xu, L., Wang, F., Chen, M., Nie, D., Lian X., Lu, Z., Chen, H., Kan Zhang, K., Pengxiang, G. (2017). CO_2 methanation over rare earth doped Ni based mesoporous catalysts with intensified low-temperature activity, *Int. J. Hydrogen Energy*, 42(23): 15523–15539.

[51] Zhang, Z., Zhang, X., Zhang, L., Gao, J., Shao, Y., Dong, D., Zhang, S., Liu, Q., Xue, L. and Hu, X. (2020). Impacts of alkali or alkaline earth metals addition on reaction intermediates formed in methanation of CO_2 over cobalt catalysts, *J. Energy Institute,* In Press.

[52] Zhi, G., Guo, X., Guo, X., Wang, Y. and Jin, G. (2011). Effect of La_2O_3 modification on the catalytic performance of Ni/SiC for methanation of carbon dioxide. *Catal. Comm.,* 16(1): 56–59.

Air Pollution Mitigation Using Artificial Intelligence: A Review

Lalita Josyula*, Shounak Chakraborty[1], Akanshu Mahajan[2], Varun Bharti[3] and Pradeep[4]

[1]Havells India Ltd., Noida, Uttar Pradesh, India

[2]KTH Royal Institute of Technology, Stockholm, Sweden

[3]iGloble Software Solutions Pvt. Ltd., New Delhi, India

[4]Luminous Power Technologies Pvt. Ltd., Gurgaon, Haryana, India

E-mail: [1]shounakcy@gmail.com; [2]akanshu@kth.se; [3]Varunbharti94@gmail.com; [4]Pra6081@gmail.com

ABSTRACT: When we mull over environmental problems, the nexus between green infrastructure, air pollution and human health and the struggles associated with them, have drawn worldwide attention to a great extent. Thus, predicting the same in advance will have great significance on the economy and controlling people's health. We would like to present a review of the current trends in mitigating air pollution using the latest technological advancements in the field of artificial intelligence. The non-linearity between the input and outputs in the prediction models to the optimization algorithms that study the air pollutant removal processes have been established for air pollution related clean-ups. Hybrid AI models are studied and applied to predict the pollutant concentrations and establish a warning system. AI also helps to identify and streamline measures which can effectively improve the air quality at any particular location. We provide an insight into the recent developments in alleviating air pollution and exploit the potential of artificial intelligence in order to assess air pollution objectively and further act on the processed information more specifically. Basically, technology can be an enabler to help address air pollution, it's prevention at source and helping different organizations optimize their operations and reduce their impact on worsening air quality.

Keywords: Air Pollution, Artificial Intelligence, AI Models, Air Pollutants, Hybrid, $PM_{2.5}$, CNN, LSTM.

**Corresponding author:* Department of Electronics, Sri Venkateswara College, University of Delhi, Delhi–110021, India. E-mail: lchamarthi@gmail.com

INTRODUCTION

Life on earth, from times immemorial, has always been dependent on environment. As civilization evolved, so has the awareness of the quality of life. Technology has influenced our lives in a revolutionary way during the past 25 years. One of the breakthrough technologies, worth mentioning, is the use of Artificial Intelligence (AI). AI helped us in many ways where our own intellect's perception was limited. Artificially intelligent tools and techniques have facilitated intelligent monitoring systems, analytics, predicting, decision and control systems, of many real-life applications. Management of environmental critical situations like air/water/soil pollution also can be effectively done using integrated intelligent systems as decision support tools. The reach of AI systems is such that they can be used to disseminate information to the population of certain regions such as the urban regions, metropolitan areas, or a whole county, continents, or intercontinental areas. To guide the development of strategies, an environmental decision support system (DSS) which is capable of handling environmental-related crisis requires an approach which is efficient in forecasting and analysis. Systems like these can be implemented by employing AI methods that are optimal and follows a hybrid manner. AI techniques that can be employed include fuzzy logic, machine learning techniques like artificial neural networks (ANNs), hybrid AI models, data mining algorithms, case-based reasoning and optimization methods like genetic algorithms. Due to the presence of a certain amount of uncertainty and dynamic environmental factors in the majority of the environmental-related processes and also as we do not always have complete datasets (with recorded time-series or real-time accurate measurements) the environmental forecasting and analysis systems have to deal with uncertain knowledge.

We review a number of approaches in AI and study their impact on various problems. It is followed by an evaluation of applying a similar concept to the problem of air pollution. The characteristic advantages and disadvantages of applying these novel algorithms are studied and an assessment is drawn. It is argued that using hybrid-AI models to the problem of air pollution produces better assessment results and therefore we can effectively extract the spatiotemporal features of air pollutant concentration, learn from such data and utilize that knowledge to prescriptively achieve specific goals and tasks. Also integrating auxiliary data which has a strong correlation with $PM_{2.5}$ (particulate matter with an aerodynamic diameter of ≤ 2.5 mm) has immensely contributed towards improved model prediction and therefore we can draw more insights from it. This information is then further utilized in devising a strategy in order to curb air pollution. Finally, future challenges and improvements are discussed and proposed.

By doing so, affected people, such as those with asthma and different allergies, can prepare for harsher conditions in advance. Basically, technology can be an enabler to help address air pollution, it's prevention at source and help different organizations optimize their operations and reduce their impact on worsening air quality.

BACKGROUND, SOURCES AND IMPACT

Despite dramatic progress in purifying the air since 1980, air pollution is one of the environmental entities that continues to harm the health of the people and the environment. It is well known that air pollution is associated with adverse health effects such as respiratory and cardiovascular diseases, cancer and even death (Zhiyong Wu *et al.*, 2019). The World Health Organization (WHO) report suggests that 91% of the world's population lives in places where the air quality exceeds the WHO standard limits while 4.2 million deaths were estimated as a result of exposure to ambient air pollution in the year 2016. Australia is the biggest greenhouse emitter during the period of wildfire. Household air pollution also leads to diseases and premature deaths (WHO, 2016). In under-developed countries, almost 98% of children under five years of age are gauged to breathe toxic air. Thus, being the main reason behind the demise of children below the age of 15, killing 600,000 annually (WHO, 2018).

Regionally, Southeast Asia, Middle-East, South Asian and Western Asian are worst-hit areas due to air pollution. To be particular cities of India, China and Pakistan have shown a significant rise in the AQI level in the year 2019. According to the 2019's World Air Report by IQAir, these regions tend to carry the highest burden of fine particulate matter ($PM_{2.5}$) pollution overall (IQAir, 2019). This particulate matter is a pollutant of concern. In addition to $PM_{2.5}$ there are other potential anthropogenic pollutants like CO, CO_2, HC, NO_x, Soot,

SO_2, chlorides and fluorides (Fly-Ashes) and natural occurring pollutants like H_2S, methane & sulphurous gases.

Sources of Air Pollution

Generally, the term "Pollutants" is given to substances which change the natural composition of air. These are only termed harmful or noxious if they have an adverse effect. The origin of air pollution has been classified into two categories natural and man-made (anthropogenic) sources. The high concentration of air pollutants emitted into the atmosphere due to some natural phenomena like volcanic eruptions, sand storms, forest fire, thunderstorms, plant pollens is considered as natural sources of air pollution. Anthropogenic pollutants primarily can be categorized in particulates, aerosols (micro size particles combined with fine droplets) and gases ex. Smoke, soot, oil mist (Gunter Baumbach, 1996).

Natural Sources of Air Pollution

Naturally formed particulate matter (PM) including dust from earth's surface, sea salt in coastal areas and gases from a volcanic eruption are inexorable quantities of air pollutants (Nurul Zakaria *et al.*, 2018).

- *Volcanic Eruptions:* When a volcano erupts it releases PM and gases such as SO_2, H_2S and methane, which remains in the atmosphere for a long period and harms the environment continuously. Every volcanic eruption on earth emits a huge amount of pollutants, which consists of not only ash and dust but also the strongest greenhouse gas on earth, known as CO_2 (Siegel and Ethan, 2017).
- *Forest Fires:* Whether forest fires are ignited by humans or lightning, they emit a huge amount of pollutants in the atmosphere in the form of smoke, hydrocarbons, CO, CO_2, NO_x and ash. Australia wildfire from September 2019 had affected 500 million animals, over 2000 become homeless and 25 people died (Neill and Pippa, 2020).
- *Dust Storms:* Violent meteorological phenomena like storms in many parts of the world particularly near-desert areas disperse huge concentrations of particulate matter in the atmosphere.
- *The Oceans:* Oceans continually release salt particles in the atmosphere which are corrosive for metal and paints. The large

wave hits rocks and reduces them to sand which can be airborne and affect in the same way as particulate matter.

- ***Hot Springs:*** Hot springs release sulphurous gases in nearby areas which can create obnoxious odours.
- ***Biogenic Emissions:*** Pollen grains, which contributes to biogenic emissions, are minute particles in the form of fine dust from catkins of some plant species. Pollens once airborne through wind can induce respiratory allergies.

Anthropogenic Emission Sources

The Discovery of fire by humans was the beginning of anthropogenic air pollutants. Smoke, CO, CO_2 and organic gases are the pollutants resulting from this. In early time coal was the main source of fuel for both industries and domestic use which contaminated both indoor and outdoor air with soot, sulphur dioxide and other pollutants. Emission sources are termed below according to their affected area (Gunter Baumbach, 1996).

- ***Point Sources:*** Pollutants released in high concentration at one location, e.g. Industrial stack.
- ***Small Sources:*** Chimneys, leakages in pipes and fittings in large scale industries, e.g. Refineries.
- ***Line Sources:*** High Traffic volume in urban areas.

The main sources of anthropogenic air contaminate are furnaces in industries, its processes, traffic and domestic furnaces which can be further divided into the following types:

- ***Pollution by Combustion:*** Combustion processes are major sources of anthropogenic air pollution. A constant increase in demand and consumption of energy leads to an exponential rise in the emission of pollutants. Industries, power generation in power plants, vehicles, aircraft, ship's engines, or domestic heaters generate energy through combustion processes. This process of combustion releases a high concentration of pollutants in the atmosphere.
- ***Vehicular Emissions:*** Traffic congestion and vehicular emission due to incomplete combustion of fuel lead to the generation of CO, hydrocarbons and soot types of pollutants.

- *Day to Day Gadgets/Utilities:* The utilities of the modern world are a major part of our life these days. We consume power from power plants not only to light our house but also for modern electronic gadgets which increase the load on power plants which further leads to more combustion.
- *Shipping Industry*: Shipping Industries are equally responsible for air pollution, thus contributing to 18% of the air pollutants. Large-sized diesel engines of shipping industries burn high sulphur content fuel also called bunker oil releases high concentrations of CO, CO_2 and hydrocarbons directly into the atmosphere.
- *Aviation*: Even fuel-efficient and less polluting turbofan/turboprop types engines of aviation industries can cause high-level emission of particulate matter and gases such as CO, hydrocarbons, CO_2, NO_x, SO_x and black carbon.

Meteorological Parameters Affecting the Dispersion of Air Pollutants

The impact of pollutants emitted into the atmosphere can be impelled by various factors which are dependent on atmospheric processes; these factors are called meteorological parameters (Gunter Baumbach, 1996). The dispersion and mass exchange of air pollutants in the atmosphere is majorly affected by wind direction, its velocity and its turbulence.

- *Wind:* Pollutant strewing occurs in the direction of the wind. The flue gas concentration is approximately proportional to the horizontal velocity of the wind. The exponential increase in wind velocity in upwards directions leads to the dispersions of flue gases from high stacks.
- *Turbulence:* Irregular movements of air flows in the atmosphere affect the dispersion rate of pollutants.
- *Inversion:* If temperature fall is less with an increase in elevation than corresponds with adiabatic lapse rate, then it creates sub adiabatic or stable layers which act as a barrier for air pollutants to mix with surroundings.
- *Mixing Layer and Barrier Layers:* The bottom part of the troposphere acts as a mixing layer for air pollutants and the surrounding air. Mixing is directly proportional to turbulence but this mixing of pollutants can be suppressed by a stable layer.

Type of Pollutants in Ambient Air

- *Particulate Matter (PM):* Mixture of fine solid particles and liquid droplets in the air are called particulate matter, namely dust, ash, sea-spray. Combustion of solid and liquid fuels in vehicle engines, power generation and domestic heating emits particulate matter which may vary in size. $PM_{2.5}$ and $PM_{10.5}$ are particulate matters with a diameter of less than 2.5 micrometres and 10 micrometres respectively (Daniel Vallero, 2008). $PM_{2.5}$ is generally emitted by moving vehicles, so eventually the concentration of $PM_{2.5}$ is higher in areas near roads. Oxidation of SO_2 and NO_2 results in the formation of PM_{10}. During Australia wildfire, Sydney in dec19 had reported worst air quality PM 2.5 level reached 400 (Neill and Pippa, 2020).

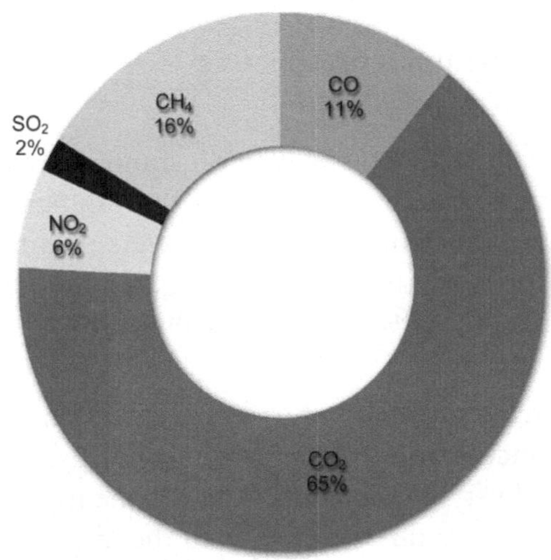

Figure 1: Percentage of Various Air Pollutants Level in Air

- *Sulphur Dioxide (SO_2):* Combustion engines that burn high sulphur content emits SO_2. Ex. Fossil fuel combustion at the thermal power plant. During the Taal lake, Philippines volcanic eruptions in January 2020, a huge amount of SO_2 was emitted, approximated to 5 tonnes per day (Ma and Cristina Arayata, 2020).
- *Nitrogen Oxides:* Advancement in combustion processes for automobiles engines to reduce fuel consumption and CO and hydrocarbons emission also leads to high temperature in the

combustion chamber which further leads to increase in emission of NO_x (NO_2 + NO).

- ***Benzene:*** The main component of crude oil and petrol is benzene so its main source of emission is vehicle exhaust and evaporation of oil at filling stations.
- ***Ozone (O_3):*** Another greenhouse gas, viz. O_3 is formed when organic compounds and nitrogen oxides react under sunlight. As it is dependent on sunlight, ozone has a higher concentration in summers. It's a major component of smog that reacts with tissue-reducing lung functions.
- ***Carbon Monoxide (CO):*** A toxic air pollutant CO is produced by incomplete combustion of carbon-based fuel. So, its main cause is the idling and deceleration of vehicles.
- ***Carbon Dioxide (CO_2):*** Any type of combustion whether it was done by humankind or naturally occurring disaster leads to the emission of CO_2 on a large scale. All different types of volcanoes on earth collectively emit around 645 million tons of CO_2 per year (Siegel and Ethan, 2017).

The comparative study between air pollutants is shown in Figure 1 (P. Karrupusamy *et al.*, 2020).

Impact

Air pollution, which has a strong diffusive capacity, is of mounting concern worldwide because of its negative impact on the environment, climate and public health. The exposure to such amounts of air pollutants has increased the mortality and morbidity over the years to a great extent. It has been examined and stated that air pollution consists of a heterogeneous mixture of gases and particles that include carbon monoxide, nitrates, sulphur dioxide, ozone, lead, toxic by-products of tobacco smoke and particulate matter. Therefore, the oxidative stress and inflammation caused by inhalation of these pollutants can lead to acute and chronic bronchitis disorders, heart attacks in the respiratory systems, as well as contribute to systemic inflammation and autoimmunity (Sylvia C.L. Farhat *et al.*, 2011). Indeed, the Global Burden of Disease (GBD) report shows ambient air pollution as the leading cause of death, accounting for more than three million premature deaths per year across the world (8 Lim SS *et al.*, 2010). Nevertheless, an alarming number of deaths is also because of indoor air pollution namely burnt coals, biomass. The magnitude of health problems due to air pollution and its impact on the eyes, cardiovascular issues, sleep apnea, chronic obstructive pulmonary diseases (COPD), asthma exacerbations are a matter of concern. Females have a higher risk of COPD (induced due to indoor pollutants namely PM, CO, NO_2, VOCs) due to frequent exposures due to cooking (Patrick Amoatey *et al.*, 2020). Studies reveal that black carbon particles from air pollution found in the placentae, viz, a temporary organ that works as in the nourishment barrier between the mother and the foetus during pregnancy, may not be as impermeable to air pollutants (AQI India, 2019). According to EIT Health (European Institute of Innovation and Technology-Health), sleep disorders, viz. Sleep apnea is affected by allergic rhinitis (AR) and is associated with asthma. Thus, AR and asthma cause social and health imbalances throughout the life cycle. Air pollution has a serious effect on AR severity and its repercussions (Jean Bousquet *et al.*, 2018). Over a period of time, studies have been recorded on the severity of ventricular and atrial arrhythmias linked with ambient air pollutant levels (David Q. Rich *et al.*, 2006). However, type 2 diabetes has also been associated with outdoor air pollution (Ikenna C. Eze *et al.*, 2015). It indicated that it is more common in cases of females than males. This evidence is also supported by reports of animal experiments. It indicated that systemic inflammation, immune responses in adipose tissue and peripheral insulin resistance can be caused by exposure to air pollutants (Sanjay Rajagopalan and Robert D. Brook, 2012).

Visibility is another known issue of air pollution that has impacted society and has become an important concern for both the society and the scientific community. The relationship between the pollutants present in the air (e.g. NO_2, CO, $PM_{2.5}$, PM_{10} and SO_2) and visibility is calculated for most of the polluted cities. The visibility had a negative correlation with the above pollutants (Souzana Achilleos *et al.*, 2019). There is also a higher risk of mortality during days of low visibility. A study shows that during a period from 2006–2016, there were more than 70,000 deaths from natural causes in Kuwait. The rate ratio comparing the mortality rate on low visibility days to high visibility days was 1.01 (95% CI: 0.99–1.03). Similar is the impact/estimated in case of dust storms too (1.02, 95% CI: 1.00–1.04) (Souzana Achilleos *et al.*, 2019).

Furthermore, environmental vicissitudes and increased human activities have induced alteration in biodiversity. Metagenomic and various other examinations of healthy and diseased individuals shows that reduced biodiversity and alterations in the composition of the gut and skin microbiota are linked with multiple inflammatory conditions, as well as asthma and allergic related diseases (Lucette Flandroy *et al.*, 2018). In accordance with the chemical composition, PM has the capability and is thus adversely affecting our surroundings and environment to a great extent. Along with depleting the nutrients in the soil, it's making the lakes acidic thus harming the habitat of aquatic animals. PM has also contributed to acid rains thus damaging the sensitive forests and farm crops (AQI India, 2019). Air pollution also plays an important role in affecting the life of materials used in buildings and infrastructure. Impacts of climate and air-pollutants derived chemical pathways and changing environmental conditions are resulting in corrosion and blackening of building materials. Such impacts are chronic and usually take place over a long period of time. The amount of material loss is usually carried out by the dose-response functions (DRFs) which relate the climatic parameters with the concentrations of air pollutants. The major pollutants used as a variable in DRFs are SO_2, O_3, NO_2 and particulate matter (Prashant Kumar and Boulent Imam, 2013).

AIR POLLUTION MEASUREMENT PARAMETER—AQI

In today's world, air pollution is ubiquitous and a certain amount of exposure cannot be avoided. The rapid increase rate of industrialization and urbanization in the modern world has contributed accordingly to the spread and exacerbation of health effects. This has brought a high level of research for air quality management and pollution control. The air quality index (AQI) is a dimensionless number used by the government to make the public aware of the levels of air pollution and its effects on humans in that region (Eder B. *et al.*, 2010). Earlier it was named as "Pollutant Standard Index" (PSI), it was developed and introduced by the United States Environmental Protection Agency. In 1999, the index was further revised and renamed as the "Air Quality Index" or AQI (USEPA, 2009), the most widely used index for air quality assessment and management (Mohammad

Hossein Sowlat *et al.*, 2011). But different countries still maintain their national air quality standards. Some of them are the Air Quality Health Index (Canada), the Air Pollution Index (Malaysia) and the Pollutant Standards Index (Singapore) (Wikipedia contributors, 2020).

This index (AQI) depends upon the concentration of multiple pollutants such as particular matter (PM), ozone (O_3), nitrogen dioxide (NO_2), sulphur dioxide (SO_2), lead (Pb) and carbon monoxide (CO) over a specified averaging period, obtained from an air monitor or model. The AQI level varies from 0 to 500 (Wikipedia contributors, 2020). There are six AQI categories, namely Good (0–50), Satisfactory (51–100), Moderately polluted (101–200), Poor (201–300), Very Poor (301–400) and Severe (401–500). The AQI level is directly related to air emissions. The Air Quality Index can be measured using the below relationship (N.A. Dahari *et al.*, 2016).

$$AQI_{pollutant} = \frac{\text{Pollutant Concentration}}{\text{Pollutant Standard Concentration}} \times 100 \qquad \dots (1)$$

While computing the AQI, the air quality index is measured as a piecewise linear function of the pollutant concentration. There is a discontinuous jump of one AQI unit at the boundary between AQI changes. To convert from concentration to AQI the following formula is used (Wikipedia contributors, 2020).

$$I = \frac{I_{High} - I_{Low}}{C_{High} - C_{Low}}(C - C_{Low}) + I_{Low} \qquad \dots (2)$$

where I = air quality index (AQI), C = pollutant concentration, C_{Low} = the concentration breakpoint i.e. $\leq$ C, C_{High} = the concentration breakpoint i.e. $\geq$ C, I_{Low} = the index breakpoint corresponding to C_{Low}, I_{High} = the index breakpoint corresponding to C_{High}.

The AQI system generally focuses on acute health impacts occurring over a short period – 24 hours or less – rather than chronic effects occurring over months or years. A notification is given to the public whenever an AQI value exceeds 100, a considerable amount of time is available for the public to react and take whatever steps they can to avoid exposure (George Kyrkilis *et al.*, 2007).

This approach is considered conservative, because:

- The standards have been created within a particular safety period and also been designed to save highly vulnerable people.

- As soon as the AQI level measured or recorded from a single sampling station exceeds 100, the public is made available to the information.
- AQI is a general index used to define the degree of pollution, thus it cannot take into account the damage that air pollutants cause to the flora, faunas and certain material (e.g. building surfaces, statues) (George Kyrkilis *et al.*, 2007).

ARTIFICIAL INTELLIGENCE (AI)

Background

One of the most disruptive technologies in the scientific community that has drawn the attention of almost all the researchers is the boom of AI and its remarkable usage in today's era. It has been explored and studied for years and thus remains the most elusive field in the Computer Science domain. In the field of AI, expectations have always seemed to have exceeded the reality. Put simply, AI is a system that can learn and make decisions similar to a human. With the boost in recent years, AI has emerged successfully and has helped businesses grow and achieve their utmost potential. Without improvements in the core programming languages, these advancements mentioned above wouldn't have been possible.

In general, AI models are implemented using software tools that are well designed and it's quite convenient for young researchers who are not directly working in the field of AI. This provides developers with an extensive set of libraries related to machine learning or deep learning and excellent community support to implement AI models. AI technology has seen a rapid increase over the last few years and plenty of AI methodologies are emerging and developing continuously.

The AI technology primarily refers to the artificial neural network (ANN), support vector machine (SVM), stats model, fuzzy logic (FL), etc., which have been applied to multiple disciplines like business sales, climate predictions, soil quality, engineering, healthcare, etc. It has been studied and examined that these approaches are more accurate, efficient and economical than conventional processes (Zhiping Ye *et al.*, 2020). It has been found that AI has a higher capability of tackling complex and dynamic issues more effectively than traditional mathematics.

The objective is to provide an up-to-date overview of AI in the process of air pollution control. Firstly, we review several approaches in AI technologies and study their fundamentals. And later, followed by an analysis of applying similar concepts to the problem of air pollution.

Categories and Fundamentals of AI

In past years, computer scientists have developed many algorithms to process and analyse biological data and likewise, biologists helped computer scientists to discover a new technique or ability through which machines can mimic the human brain to perform a task or solve a problem like a neural network. An artificial neural network is a nonlinear statistical computation model whose structure and function are based on biological neural networks (Techopedia, 2020).

ANN is broadly deployed for automation of industries to perform tasks like prediction, classification and optimization in various domains. ANN model gets trained through historical data to develop a non-linear relation between target variables and predictor variables, once a model is trained it can be used to predict, classify and forecasting the target values with high accuracy.

Multilayer Perceptron Network (MLP)

In the historical development of neural networks, a special place is occupied by a single perceptron. It is the simplest form of the neural network used for the classification of linearly separable patterns. Equation 4 defines a signal-flow in a perceptron (Simon S. Haykin, 2009). It consists of only single neurons with adjustable weights. A multilayer perceptron has one or more than one hidden layer and it has a remarkable impact on predictive capability (Zhiyong Wu *et al.*, 2019).

Figure 2 shows a fully connected network architectural graph of a multilayer perceptron with two hidden layers and an output layer. Fully connected means each neuron in each layer of the network is connected to all neurons or nodes of the previous layer. The direction of signal flow is in the forward direction from left to right and layer by layer basis.

In a multilayer perceptron, two types of signals flow through the network as shown in Figure 3.

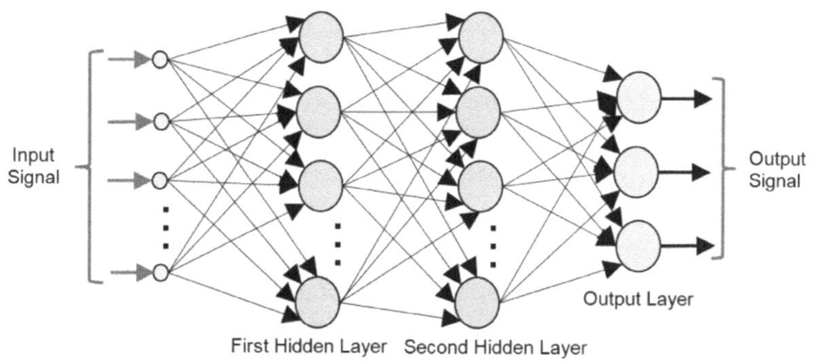

Figure 2: Architecture of Fully Connected Multi-layer Perceptron with 2 Hidden Layers

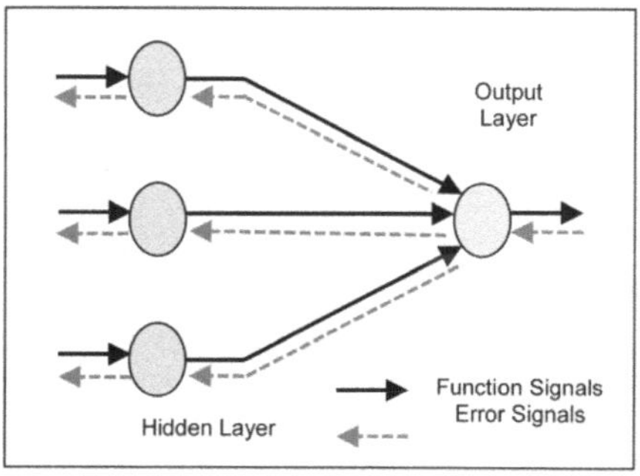

Figure 3: Direction of Flow of Two Signals in MLP

- ***Function Signal or Input Signal:*** A signal that propagates in forwarding direction neuron by neuron from input end and emerges out as an output signal at the output end of the network refers to a function signal. Each time a function signal passes through a neuron, a function with associated weight is calculated and applied to that neuron.
- ***Error Function:*** A signal originates at the output layer of the network and propagates in a backward direction layer by layer through the network termed as an error signal. An error dependent function is computed at each neuron while propagating back.

The input layer and output layer comprise of input neurons and output neuron respectively.

Neurons other than these constitute a hidden layer in the network, these are considered hidden because these are not included in the input and output layer.

Function of the Hidden Neurons: The hidden neurons perform a critical task in the operation of multi-layer perceptron they act as feature detectors. During the learning process, they characterize the training data based on their features through a nonlinear transformation on input data into feature space. An MLP-ANN is also called a back-propagation neural network (BP-ANN) because its model is trained by the back-propagation BP algorithm. This BP-ANN is widely employed in the field of environmental control and management.

Figure 3 shows that input signals are processed through the hidden layer and each neuron except input neurons computes a continuous nonlinear function of the input signal and associates a synaptic weight with it, to generate an output signal as shown in Figure 2. Each neuron in a layer can only affect the neuron in the next layers. The non-linear relation between input x_i and output y_i of a neuron in the hidden layer can be expressed as below:

$$y_j = f\left(\sum_{i=1}^{n} w_{ij}x_i + b\right) \quad \ldots (3)$$

where y_j = j^{th} output in hidden layer, $f(x)$ = transfer function, n = number if input variable,

w_{ij} = denotes weight from element *i* in the input layer to element *j* in the hidden layer, x_i = i^{th} output from the input layer, *b* = bias of the hidden layer (Zhiyong Wu *et al.*, 2019).

The signals are generated as a function of the inputs and associated weights applied to hidden layer neurons. In standard backpropagation, each neuron computes an estimate of the gradient of error surface to the weights connected to the input of the neuron (Simon S. Haykin, 2009). The learning algorithm employed for weights correction can be expressed as follows:

$$\Delta w_{ij}(s+1) = -\eta \frac{\partial E}{\partial w_{ij}} + \mu \Delta w_{ij}(s) \quad \ldots (4)$$

where, Δw_{ij}!= Correction of weight at s^{th} learning step, η = Training rate, E = Total sum square error of all data in the training set, μ = Momentum factor.

The weights and thresholds of all the neurons are being updated until all the errors set within the tolerance or the maximum number of iterations is achieved (Zhiyong Wu *et al.*, 2019).

Support Vector Machine (SVM)

Generally new machine learning technique is developed to minimize the upper bound of out-of-sample error/generalization error based on the convergence principle which gives it a remarkable capacity to regress the relationships between input and output values and to achieve good generalization results in both classification and regression (Zhiyong Wu *et al.*, 2019).

SVM after getting trained with training samples constructs an optimal hyperplane as the decision surface where the distance between the closest samples to the boundary (the margin) is maximized.

Particularly to deal with complex regression types problems and try to fit error within a certain threshold SVM is modified to support vector regression (SVR) or regression version of SVM. By introducing a selective loss function, it is considered as an efficient alternative technology to tackle regression problems in the field of environmental controls and management (Zhiyong Wu *et al.*, 2019).

Convolutional Neural Network (CNN)

A convolutional neural network (CNN) is one of the most popular deep learning frameworks. It comprises convolutional layers and sub-sampling layers which is then followed by fully connected (FC) networks and it's designed to adaptively learn the spatial features through backpropagation techniques It is mostly used to process data that generally has a grid pattern. Well-iterated weights and parameters are defined, followed by some form of loss functions (cross-entropy) and pooling (max pooling) which results in invariant features. As a deep learning method, CNN is widely used in time-series, video and image analytics and the main advantage of CNN is that it has the capability to extract important features without any human intervention.

The two characteristics of CNN are the local receptive field, where each neuron is connected to neurons of its corresponding neighbor-hood and weight sharing, which reduces the training parameters. The 2D-CNN model convolves two-dimensional matrices by sharing the weights followed by convolutional proposes. Since 2D-CNN does not consider the time dimension, 3D ConvNets are proposed to train the CNN layers and learn spatiotemporal features. After the convolu-tional layers and pooling layers, there may be numbers of fully connected (FC) layers to wrap up the CNN model. It is to be noted that the output of convolutional as well as pooling layers are 3D volumes, but for fully connected layers the output expects a 1D vector of numbers. So flattening is incorporated in this case. The training methodology is similar to that of ANN, backpropagation with gradient descent. The gradient descent is defined as an optimization algorithm that iteratively updates the parameters i.e. the filters and weights to minimize the loss. The 3D convolutional formula is provided below for reference (Congcong Wen *et al.*, 2019).

$$v^{xyz} = \sum_{p=0}^{P-1} \sum_{q=0}^{Q-1} \sum_{r=0}^{R-1} w^{pqr} u^{(x+p)(y+q)(z+r)} \quad \ldots (5)$$

where w is the convolutional kernel, u is input feature map, v is output feature map and p, q and r are the sizes of the three dimensions.

An overview of a convolutional network (CNN) architecture and training process is depicted in the below Figure 4 (Rikiya Yamashita *et al.*, 2018).

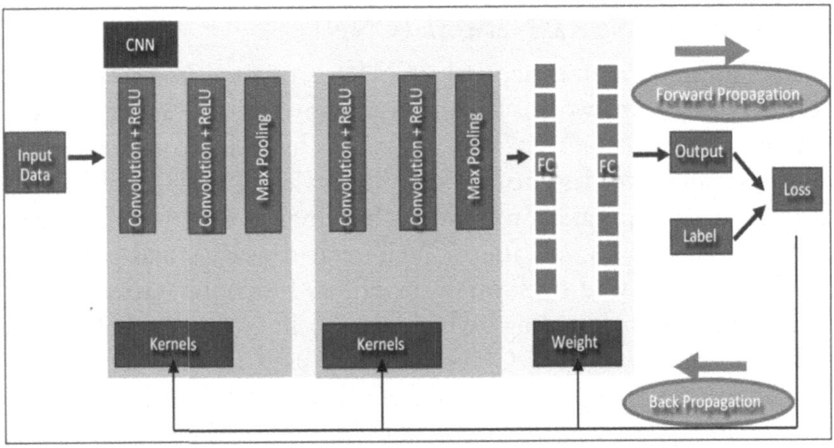

Figure 4: Convolutional Neural Network (CNN) Architecture and Training Process

Later in this chapter, we will analyse inputs using CNN as a deep learning method in conjunction with another neural network-based method named LSTM to determine the concentration of air pollutants. But before reviewing the role of AI techniques as a tool to manage air pollution, let's understand briefly about LSTM and hybrid AI models.

Long Short-Term Memory (LSTM)

LSTM neural network is a kind of recurrent neural network (RNN) represented by much intricate memory blocks than a simple or single neural network. RNNs are a class of neural networks wherein the result of the last step is fed into the present input step while having a hidden layer. This hidden layer maintains a memory about a sequence.

The memory blocks comprise of multiple gates to control the flow of information of the cells present in the internal memory. The network has 3 inputs. X_t is the input vector of the present time step, h_{t-1} is the output from the previous LSTM unit and the most important input, C_{t-1} is the "memory" of the previous unit. As for outputs, h_t is the output of the current network and C_t is the memory of the current unit. Thus, a single unit decides by considering the information of the present input, previous input and the last memory gathered, to generate a new output and change the memory. It has the capability to remember long term features and works well with time-series related datasets.

The LSTM also overcomes the problems of vanishing/exploding gradient associated with RNNs by introducing a memory unit called the cell in its network. Below shown is the diagram of LSTM building block Figure 5 (Qi Zhang *et al.*, 2020).

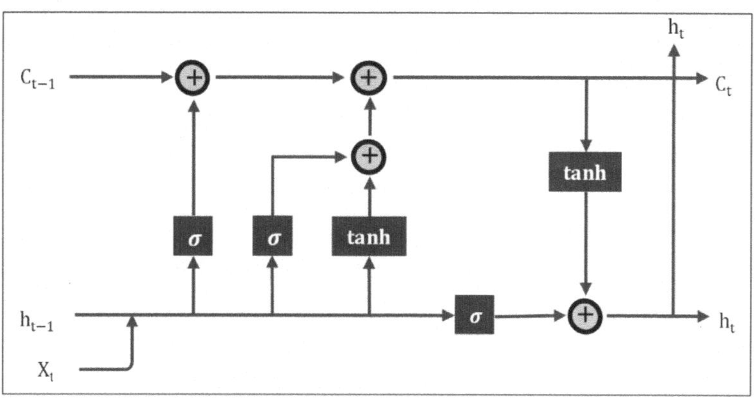

Figure 5: Network Structure of LSTM

where X_t = Input Vector, C_{t-1} = Memory from previous block, h_{t-1} = Output of previous block, C_t = Memory from current block, h_t = Output from current block, σ = Sigmoid, tanh = Hyperbolic tangent.

Hybrid AI Models

Hybrid AI models are systems that leverage the benefits of two or more AI technologies to overcome the shortcomings of using a single AI methodology. As we have discussed before AI technologies do provide an edge over traditional statistical or mathematical models, however using each AI method has its own limitations which in combination with another AI model can result in better results and predictions.

For instance, MLP has strong mapping abilities but lack in speed and overfitting issues. In such cases, hybrid intelligent systems come into play, where we can use another AI model in an effective way to solve such scenarios. SVM in conjunction with MLP helps in the optimization of the model thus eliminating the overfitting issue associated with this entire ANN framework (Zhiping Ye *et al.*, 2020). Similarly, novel convolutional neural networks (CNN) and long short-term memory (LSTM) type of hybrid AI systems are used together basically for better air-quality predictions at high-resolution than traditional ARIMA or pure LSTM models. The adaptive network-based fuzzy interference system (ANFIS), a neuro-fuzzy system, is a feedforward neural network that is coupled with fuzzy logic (FL) to represent non-linear behaviour in complex systems. This particular hybrid system has proved to be an effective tool in data mining, forecasting and also noise elimination. Thus, there are numerous hybrid systems such as MCSDE-CEEMD-ENN, combining complementary ensemble empirical mode decomposition (CEEMD), modified cuckoo search and differential evolution algorithm (MCSDE) and elman neural network (ENN), is proposed to improve the forecasting accuracy of air pollutant concentration making the learning process more robust and enhances the AI model to an extent (Zhongshan Yang and Jian Wang, 2017).

Currently, many machine learning techniques with statistical analysis methods are emerging as hybrid technologies. For instance, the response surface method (RSM) is used in conjunction with orthogonal experimental design to extract representative experimental data. The discussed hybrid technologies have ultimately made supervised learning much flexible and improved AI frameworks. However according to studies, hybrid AI systems can't be employed at all circumstances and hence it's not recommended to deploy a hybrid system in case a single AI model solves the problem with the required accuracy, as the hybrid systems are usually complicated and difficult to develop. Further information on hybrid models will be discussed in the chapter.

ARTIFICIAL INTELLIGENCE TO CURB AND MANAGE AIR POLLUTION

Air pollution is one of the major concerns of the hour these days across the world. The government is cautiously taking steps and remedies to control the same. AI technologies are extensively being employed by multiple organizations to curb air pollution. The AI methodologies are playing a leading role in the optimization, prediction and control of this environmental issue. Many studies related to environmental control especially in mitigating the issues related to air pollution have rapidly increased in the past decade. It's reported that a number of AI models have been used in order to estimate air pollutants such as $PM_{2.5}$, PM_{10}, SO_2, NO_2, CO and O_3, reduce emissions like NO_3^- or soot particles and establish model-based air-quality monitoring and early-warning systems. Numerous sources and large datasets have been required to develop these AI models for analysing as well as forecasting the concentration of these air pollutants. A period of one or more than one year of observational datasets was fundamental for the training and testing of the AI models. The concentration of air pollutants comprises characteristics of time-variability and uncertainty. The input data generally consists of information related to the meteorological factors and the pollutant's concentration.

$PM_{2.5}$ Prediction

Prediction of air pollution and pollutants such as $PM_{2.5}$ in advance has a great significance on the health control of people and the municipality's decision-making. The existing research methods to predict air pollution are deterministic methods and statistical methods. The deterministic methods require a very informative source of data which is difficult to obtain and has limited accuracy. It has been argued that statistical methods have shown satisfactory results. Commonly used statistical methods are multiple linear regression method (MLR), auto-regressive moving average method (ARMA), support vector regression (SVR), artificial neural network (ANN) and hybrid methods. But these models have failed to effectively integrate heterogeneous data, such as traffic flow and land use (Congcong Wen *et al.*, 2019). Deep learning (DL) models, which is a novel machine learning method, when fed with a significant amount of data can learn effective feature representation, thereby providing new possibilities to overcome the above stated problems.

A proposed DL model, long short-term memory neural network extended (LSTME), which is based on LSTM, integrates meteorological data as auxiliary data at the same time to fetch spatiotemporal correlations of air quality. However, the input data, in this case, consisted of data from multiple unrelated stations which resulted in negative impacts on the accuracy of the model. Thereafter, another model named spatial-temporal deep neural network (ST-DNN) was proposed, which employs multiple models like the LSTM model to fetch the temporal features, k-nearest neighbour (k-NN) and artificial neural network (ANN) to get the spatial features and integrates convolutional neural network (CNN) to extract the terrain features. However, since this model extracts these features separately, it undermines the inherent consistency of the data. Also, $PM_{2.5}$ is more relevant to the aerosol data but none of the above employed models considers the aerosol data.

In order to mitigate these restrictions, studies have proposed a novel spatiotemporal convolutional long short-term neural network extended (C-LSTME) for the $PM_{2.5}$ air pollution prediction. Along with the meteorological data, the aerosol data which has a strong correlation with $PM_{2.5}$, which contributes to the spatiality of the data, was included in the model (Simon S. Haykin, 2009). Here, the 3D-CNN was used to fetch the high-level spatiotemporal features and the stateful LSTM-NN was employed to maintain the state information/memory for a long time, achieving a more accurate and stable air-quality prediction. Hourly $PM_{2.5}$ concentration data were collected from multiple air quality monitoring stations. In the dataset 60% of the data were randomly selected as a training set, 20% as the validation set and the rest was left for the testing set. The main sources of the dataset are the Ministry of Environmental Protection, Health centres and even datasets issued by NASA (Congcong Wen *et al.*, 2019). In order to calculate the spatiotemporal distribution of $PM_{2.5}$ data, Euclidean distance and Pearson correlation coefficient between each of the two stations was measured. The formulas are represented as:

$$D(s_i, s_j) = \sqrt{(x_i - x_j)^2 + (y_i - y_j)^2} \quad \dots (6)$$

$$r(s_i, s_j) = \frac{\text{cov}(s_i, s_j)}{\sigma(s_i)\sigma(s_j)} \quad \dots (7)$$

where D and r denote the Euclidean distance and Pearson coefficient between two stations respectively, x and y denote the latitude and longitude of the corresponding stations, COV is the covariance and σ is the standard deviation.

It was observed that the correlation values were above 0.56 and with an increase in distance the overall trend in the correlation coefficient increases (Congcong Wen *et al.*, 2019). Thus, this showed that neighbouring stations could be used to improve the prediction of the present station. With the help of adaptive *k*-nearest, the spatiotemporal featured are extracted and the delayed $PM_{2.5}$ concentration is also introduced in the C-LSTME input model along with the auxiliary data.

Firstly, the adaptive *k*-nearest neighbouring stations for each station were selected and as a result, the spatiotemporal matrix $r*k*N$ is formed, where r is the time lag, k is the number of neighbouring stations and N is the total number of stations. Thereafter, the C-LSTME model extracts the spatiotemporal features of the air pollutant concentration from the spatiotemporal matrix mentioned above. The matrix is processed by the CNN and stateful LSTMs and as a result, the $N*1$ feature vector is obtained. Lastly, the auxiliary data is incorporated to improve prediction performance. A locally connected layer was used to connect the spatiotemporal eigenvectors of air pollution with the normalized auxiliary data. And finally, by integrating some fully connected layer, the $PM_{2.5}$ value of the station at time T was obtained (Congcong Wen *et al.*, 2019).

The role of CNN and LSTM individually has been mentioned before and it has been observed how these two different AI models complement each other to form an effective hybrid model. For CNN, a $k*1*N$ kernel-based 3D convolutional layer and ReLu as the activation layer is part of the two-layer structure. Along with these, two LSTM layers with 1000 nodes for each layer and 600 nodes for one fully connected layer were used. The model performance is then measured using statistical parameters such as root mean-squared error

(RMSE), mean absolute error (MAE) and mean absolute percentage error (MAPE).

$$\text{MAE} = \frac{1}{N}\sum_{i=1}^{N}|P_i - O_i| \quad \ldots (8)$$

$$\text{MAPE} = \frac{1}{N}\sum_{i=1}^{N}\frac{|P_i - Oi|}{O_i} \times 100\% \quad \ldots (9)$$

$$\text{RMSE} = \sqrt{\frac{1}{N}\sum_{i=1}^{N}(P_i - O_i)^2} \quad \ldots (10)$$

where O_i is the observed air pollutant concentration, P_i is the predicted air pollutant concentration and N is the number of test samples. The employed C-LSTME model showed higher prediction accuracy than classical LSTME, ARMA, SVR, or ST-DNN model. The comparison Table 1 is shown below.

Table 1: Comparison of the Performance of Different Methods Using Three Indicators: the Root Mean-Squared Error (RMSE), the Mean Absolute Error (MAE) and the Mean Absolute Percentage Error (MAPE). The Bold Ones are the Best Predicted Results with Minimum Error.

Method	*RSME (µg/m³)*	*MAE (µg/m³)*	*MAPE (%)*
C-LSTME	12.08	5.82	17.09
ST-DNN	17.76	8.02	21.93
LSTME	18.25	8.48	22.82
LSTME (meteorological)	22.22	11.09	29.93
ARMA	34.40	28.05	37.54
SVR	39.92	29.98	37.23

SO_2 and Total Suspended Particle Prediction

In recent years, powerful AI methods such as fuzzy logic and neural networks have been adopted to a large extent in tackling environmental problems. Since neural networks have the ability of learning, adapting, fault-tolerance and generalization, in some cases the concept of fuzzy logic, which performs an inference mechanism, are incorporated into the neural networks in order to enable a system that can deal with the cognitive uncertainties of fuzzy logic in a way that is more like a human. The resulting hybrid AI model is termed as the adaptive neuro-fuzzy inference system (ANFIS) which has already been briefed about in the hybrid AI model section. This study aims to estimate the

SO_2 and total suspended particle (TSP) pollution levels in the west black sea region of Turkey depending on various meteorological parameters.

Theoretically, the fuzzy inference interprets the values in the input vector and by some defined fuzzy logic it sets values to the output. A fuzzy inference system consists of a rule-base, database and a reasoning mechanism. The first-order Sugeno model is usually preferred (Yilmaz Yildirim and Mahmut Bayramoglu, 2006). Conversely, a neural-network consists of nodes that connect through directional links. In order to minimize the error, the basic learning rule is the back-propagation mechanism, which is a sum of squared differences between the network's outputs and desired outputs (Yilmaz Yildirim and Mahmut Bayramoglu, 2006). The ANFIS system is faster than the traditional back-propagation methodology, by involving the gradient method and least-squares estimate together to recognize the consequent and antecedent parameters.

The materials and required meteorological data are usually obtained from the Department of Meteorology or the respective health centres. Arithmetic averages were used to show the daily average of the meteorological data. In the study relative humidity, temperature, precipitation, wind speed, atmospheric pressure and solar radiation were employed as the input data to be fed into the model. In addition to this, the previous day's SO_2 and TSP concentrations were also taken into account. So, a total of seven parameters are used as a set of inputs. These input variables are then characterised by four gaussian membership functions as mentioned below. The rule base then consists of four rules of first-order Sugeno type and the connections and networks are then trained by the hybrid AI method. The model building, training and testing part are performed using MATLAB's fuzzy toolbox. The subtractive clustering method is used to generate the ANFIS structure as shown in Figure 6. Taken into account the statistical aspect of air pollution modelling, the Gaussian type membership function is employed in the study as described in the below equation.

$$\mu_{i_i}(x) = \exp\left\{-\left(\frac{x - c_i}{a_i}\right)^2\right\} \quad \ldots (11)$$

where a_i and c_i are membership function parameters.

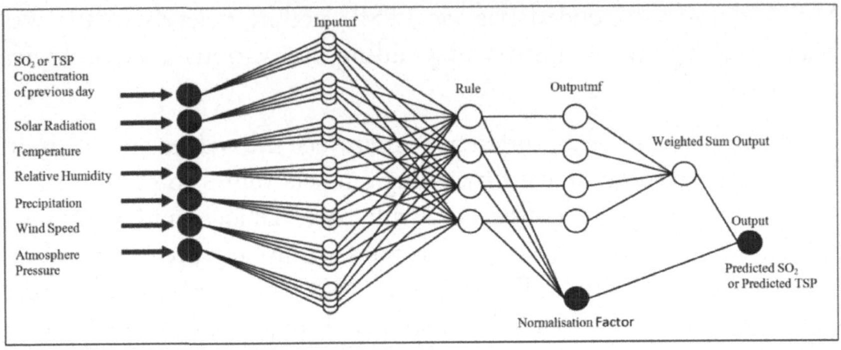

Figure 6: The ANFIS Model Structure Used in Training and Testing of the Model. Seven Input Parameters, Four Gaussian Membership Functions and Four Rules of First-Order Sugeno Type.

The model performance is then measured using statistical parameters such as root mean-squared error (RMSE) from equation 9 and index of agreement (IA) defined as mentioned below:

$$IA = 1 - \frac{\sum_{i=1}^{N}(O_i - P_i)^2}{\sum_{i=1}^{N}[|O_i'| + |P_i'|]^2} \quad \ldots (12)$$

where O_i and P_i are observed and predicted SO_2 and TSP concentrations value on a day *i*, *N* is the number of days in the test set, $P_i' = P_i - O_m$ and $O_i' = O_i - O_m$, wherein O_m is the average observed of SO_2 and TSP concentrations.

It was found that the measured and estimated concentrations of SO_2 and TSP pollutants showed peak points together. The RMSE (μg/m^3) of SO_2 and TSP was found to be 19 and 30 respectively and IA for the same was 0.82 and 0.78 respectively. When forecasted over a period of time the model indicating acceptable forecast limits between 75–90% and 69–80% for SO_2 and TSP respectively (Yilmaz Yildirim and Mahmut Bayramoglu, 2006). For better results and air quality management, precise daily air quality forecasts are required for an individual region when proper health advice and strategies are to be issued to the public using monitoring or warning systems.

Exhaust Emission (NO_3^-, soot particles) Prediction

Diesel engines are the most popular types of the engine which is known for its efficiency and power. But apart from its known capabilities, it is also known for its toxic and carcinogenic exhaust emissions into nature. ANN is an effective technique to predict the diesel engine exhaust emission. It is a powerful tool for estimating nitrate and soot emissions from diesel engines by modelling the systems.

Exhaust emission from diesel engines is one of the leading sources of soot particles that are carcinogenic. There are many methods to reduce these emissions, ANN is one of the possible solutions which can accurately predict engine performance and exhaust emission characteristics. As mentioned earlier architecture of ANN consists of three layers namely the input layer, hidden layer and the output layer. For estimating nitrate levels and soot levels at the output layer, the ANN network is trained by considering parameters like speed, fuel injection pressure and valve position in the input layer.

Once the optimum values of weights associated with hidden layer neurons are determined, it can be used to predict nitrate and soot level accurately for different values inputs.

Before feeding to nodes/neurons of the input layer, absolute values of input parameter are normalized to the range of [v_l = 0.001, v_h = 1.0] by using equation 13. To evaluate the accuracy and performance of the model, Mean square error (MSE) of output values are calculated using equation 14.

$$\text{v}_{\text{normalised}} = \frac{\text{v}_a - \text{v}_{\text{min}}}{\text{v}_{\text{max}} - \text{v}_{\text{min}}} \times (\text{v}_h - \text{v}_\iota) \qquad \dots (13)$$

$$\text{MSE} = \frac{1}{N}\sum_{i=1}^{N}(P_i - Q_i)^2 \qquad \dots (14)$$

where O_i and P_i denote the observed and predicted values and N is the number of data sample, respectively.

ANN predicted nitrate level more accurately with regression coefficient 0.99 and less than 5% relative error but the model has low accuracy in predicting soot level as the formation of soot is complex and cannot be defined by limited parameters (Caneon Kurien and Ajay Kumar, 2018).

PM_{10}, SO_2, NO_2, Solid Waste Emissions and Airborne Particle-Bound Metal Prediction

In order to predict the daily concentrations of air pollutants such as PM_{10}, NO_2 and SO_2, a chinese researcher suggested an MLPNN based AI model which is trained by back-propagation (BP) algorithms, also called back-propagation neural network (BP-ANN) (Jianjun He *et al.*, 2016). The input dataset consisted of daily average PM_{10}, NO_2 and SO_2 concentration data along with other local meteorological parameters such as wind speed at 200 m, temperature lapse rate at 50 m, wind direction index at 800 m, boundary layer height and stable energy. It was found that the meteorological data have significant contributions to the daily variation in the pollutant concentration. The model performance is then measured using statistical parameters such as R^2 and root mean-square error (RMSE). R^2 is the degree of correlation between experimental and estimated values, which is expressed as the equation below:

$$R^2 = 1 - \sum_{i=1}^{N} (P_i - O_i)^2 \Big/ \sum_{i=1}^{N} (P_i - \bar{O}_i)^2 \qquad \text{... (15)}$$

where N = number of model output, P_i = predictive output value, O_i = experimental value, $\bar{O}_i$ = average values of the experiments.

While in the case of RMSE as mentioned in equation 9 the errors between the experimental values and model output are calculated and it is to be noted that a higher value of R^2 (0–1) and lower value of RMSE indicates a better prediction performance.

In the study discussed a relatively low correlation coefficient value, the R^2 was observed of 0.71 to 0.83 for all the daily averages of PM_{10}, NO_2 and SO_2. And the RMSE of each of PM_{10}, NO_2 and SO_2 were 64.5, 20.1 and 15.7 respectively (Jianjun He *et al.*, 2016). Later in another study, the data was gathered for a longer-term using the meteorological and PM concentrations as the input to predict the size-fractioned airborne particle-bound metals, wherein the concentrations of fourteen different elements were estimated as the output. It was found that amongst BP-ANN, MLR and SVM, SVM models were the superior ones. In this case, SVM models were then proposed to predict the concentration of airborne particle-bound metals namely As, Ba, Cd, Cr, Cu, Fe, Mn, Ni, Pb, Sr, Ti, V and Zn. The results revealed that higher concentrations of almost all the pollutants were mostly observed during the winter season in the industrial areas. The R^2 and RMSE of Nickel (Ni) were observed as 0.82 and 1.179 respectively (Zhiping Ye *et al.*, 2020).

BP-ANN model has also been used to predict the emissions released from a municipal solid waste incinerator. Polychlorinated dibenzo-p-dioxins (PCDDs) is one of those emissions which have been affecting the air quality to a large extent. In the studies, five factors have been reported, which include injection of activated carbon, injection frequency, HCl concentration in the flue gas at the stack emission, the temperature at the mixing chamber and effluents. These factors were extracted using a data mining process called principal component analysis (PCA), wherein the percentage of variance was 75.78% of the total variance in the 4-year monitoring dataset of an incinerator in that particular region. The results illustrated that the proposed BP-ANN model's result provided high performance in the prediction of PCDDs emission with a higher R^2 value of 0.998 (Zhiping Ye *et al.*, 2020).

PM_{10}, $PM_{2.5}$, SO_2, NO_2, CO and O_3 Prediction

It has been seen that the air quality index (AQI) of China is the second worst amongst all 180 countries according to the Global Environmental Performance Index. Therefore, to assist people in keeping fit and improve the AQI level a hybrid model MCSDE-CEEMD-ENN was proposed, combining complementary ensemble empirical mode decomposition (CEEMD), modified cuckoo search and differential evolution algorithm (MCSDE) and elman neural network (ENN) to improve the forecasting accuracy of air pollutant concentration making the learning process more robust, accurate and make the system highly desirable. Detailed information on each of these models has been presented separately (Zhongshan Yang and Jian Wang, 2017).

Based on the historical concentrations of PM_{10}, $PM_{2.5}$, SO_2, NO_2, CO and O_3 the model can forecast the concentrations of these six main air pollutants. The proposed improved optimization algorithm MCSDE combines cuckoo search (CS) and differential evolution (DE). This improves the forecasting capability, given better initial weights and thresholds to ENN. The model can also avoid getting trapped into local optima, resulting in faster convergence rate and global searching capability. The CEEMD model is introduced to extract the original concentrations of the air pollutants into several intrinsic mode functions (IMFs). The new series employed is then reconstructed with the highest frequency IMF removed (Zhongshan Yang and Jian Wang, 2017).

The model performance is then measured using multiple perspectives, namely the mean absolute error (MAE), mean squared error (MSE) and mean absolute percentage error (MAPE). The MAE, MSE and MAPE equations are 10, 14 and 11 respectively.

The experiment was conducted by gathering the pollution data of a city of China, named Xi'an. Multiple models were considered like BP, ENN, MCSDE-ENN, but the proposed model showed the least MSE, MAE and MAPE error values as shown in the below Table 2 (Zhongshan Yang and Jian Wang, 2017). The proposed model also indicated that PM_{10} and $PM_{2.5}$ are the main pollutants. Thus, the prediction of air-pollutant concentration is a major part of the air quality early warning system.

Table 2: Forecasting Results Obtained Using Air Pollution Data. The Bold Ones are the Best Predicted Results with Minimum Error.

Pollutants	*Performance*	*BP*	*ENN*	*MSCDE-ENN*	*MCSDE-CEEMD-ENN*
SO_2	MSE	17.69	16.37	15.80	**3.49**
	MAE	2.70	2.64	2.56	**1.32**
	MAPE (%)	11.78	11.46	10.74	**6.12**
NO_2	MSE	26.51	19.78	18.96	**3.64**
	MAE	3.81	3.40	3.30	**1.45**
	MAPE (%)	13.50	11.58	10.94	**4.88**
CO	MSE	0.01	0.01	0.01	**0.00**
	MAE	0.08	0.08	0.07	**0.04**
	MAPE (%)	5.13	4.59	4.34	**2.67**
O_3	MSE	3.18	3.47	2.46	**0.60**
	MAE	1.25	1.38	1.16	**0.59**
	MAPE (%)	3.86	4.16	3.64	**2.05**

(Contd...)

(Table 2 contd...)

Pollutants	*Performance*	*BP*	*ENN*	*MSCDE-ENN*	*MCSDE-CEEMD-ENN*
PM_{10}	MSE	147.83	131.38	125.69	**36.56**
	MAE	9.17	8.52	8.35	**4.57**
	MAPE (%)	9.45	8.61	8.23	**4.76**
$PM_{2.5}$	MSE	27.86	25.28	24.44	**6.73**
	MAE	3.47	3.53	3.41	**1.97**
	MAPE (%)	9.04	9.29	8.46	**5.82**
Average	MSE	37.18	32.72	31.21	**8.50**
	MAE	3.47	3.26	3.14	**1.66**
	MAPE (%)	9.04	8.28	7.73	**4.38**

Case Study of Hashemi Nejad Gas Refinery of Khorasan

Many industries such as refineries, petrochemical plants and natural gas plants use an amine gas treatment process to remove H_2S and CO_2 from gases. This process is also known as gas sweetening (Mahdi *et al.*, 2011).

ANN which is a powerful tool for estimating output is similar to biological neural network/structures. It takes a certain dataset as the input which it trains to predict output and does not require any formulation. It consists of a series of multiple layers with the number of nodes including the input layer, hidden layers and output layer. In this case study, we will see the MLP model, one of the most widely used neural network model and it's generally trained with back-propagation of error algorithm. In MLP neural network the hidden layer neurons do the maximum compute and input/output layer neurons distribute and collect signals respectively. During the training of the model, the network adapts synaptic weights by comparing network output signals with desired signals (Mahdi *et al.*, 2011).

The rate of adaption may vary according to the learning rate (n). If the learning rate is high enough, the network will adapt its weights quickly but the network may become unstable and if the rate of adaption is zero weights will remain constant. In this study to accelerate the network convergence, a linear weight is also added (α_s) as shown in Figure 7. Backpropagation method is used to update the parameters linearly after each iteration using the following equation:

$$(\Phi^T \Phi)\,\alpha = \Phi^T y \qquad \text{... (16)}$$

where $\Phi i, j = \varphi(u\ (zi, j), i = 1,..., N; j = 1,..., M; y = N \times 1$ vector of measured values, N = Number of training data, M = Number of Neurons.

Mean square error (MSE) and coefficient of determination (R^2), normalized mean-squared error (NMSE), mean absolute error (MAE), root mean square error (RMSE), for each output, were calculated by using equations (14, 15, 17, 18, 19) for training and testing data.

$$\text{NMSE} = \frac{1}{\sigma^2}\frac{1}{N}\sum_{i=1}^{N}(O_i - P_i)^2 \qquad \text{... (17)}$$

$$\text{MAE} = \frac{1}{\sigma^2}\frac{1}{N}\sum_{i=1}^{N}|O_i - P_i| \qquad \text{... (18)}$$

$$\text{RMSE} = \sqrt{MSE} \qquad \text{... (19)}$$

where O_i is the ith experimental value, P_i is the ith predicted value, N is the number of data, R^2 is the variance of experimental data.

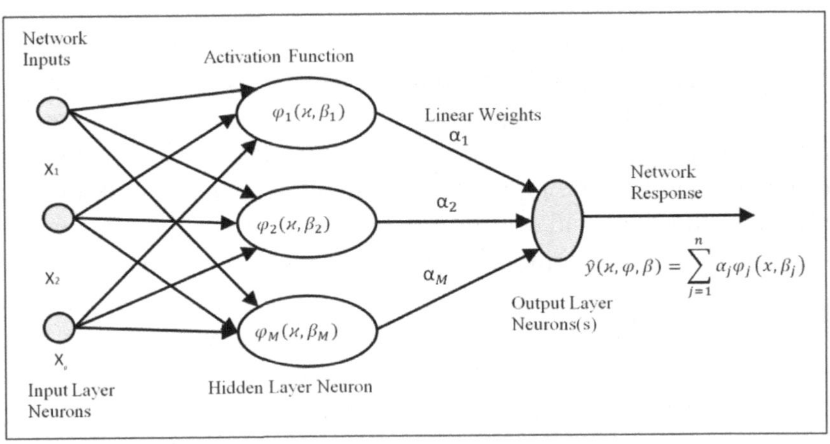

Figure 7: Architecture Single-Layer Perceptron with 1 Hidden Layer

In this study, the work was done on a prototype refinery located in the SARAKHS gas field Iranian onshore. The data obtained from gas refinery regeneration columns with min and max range are inlet reflux temperature (36.25–59.75), inlet/outlet condenser temperature difference (44.5–64.58), amount of H_2O(7.87–13.43) and inlet amine temperature (99.67–104.33), outlet down the temperature (116.5–121) and amount of reflux (21.25–25.42). The dataset was divided into two groups: training & cross-validation (80%) set and verification set (20%). Training set to train the model and its output was used to adapts weights in the network in such a way to minimize the difference between actual and predicted output. The cross-validation was set to ensure that the model would not overfit the data.

In the network, input layers consist of inlet temperatures of reflux, the difference between inlet and outlet temperatures of condenser, amount of H_2O and inlet amine temperatures and the output layer consists of outlet down temperatures of the tower and the amount of reflux. The correlation coefficient of desired and predicted values obtained down temperature of the tower and amount of reflux was calculated as 0.94498 and 0.92312 respectively. The ANN performance matrix for the outputs has been mentioned in the below Table 3. Values that are closest to 1, indicates that the model was trained has a high accuracy of prediction. The higher value of the correlation coefficient (equation 15) and a smaller value of RMSE (equation 9) mean better performance of the model. This study showed the capability and applicability of MLP-ANN in the prediction of gases emitted during gas sweetening in the regeneration column.

Table 3: ANN Performance Matrix for both Outputs

Performance	*Down Temperature of the Tower*	*Amount of Reflux*
MSE	0.0429392	0.0356498
EMSE	0.2072177	0.1888111
MAE	0.2061	0.186314
NMSE	0.654214	0.435612
R^2	0.9498	0.9231

DISCUSSION

Distribution of both the quantitative and qualitative analysis of the data using AI/ML techniques back to all the intelligent systems using the network infrastructure is a crucial process. For e.g., if it is found that a particular area is contributing more towards the air pollution, as specified by the authorities, there should be intelligent decisions taken by the robust AI/ML systems in that area to curb the increase. One classical case of that is the inflow of traffic during the peak hours of the day leading to congestion and a steep rise in air pollution in those parts of the city. It is evident from other studies that AI/ML techniques can accurately find and predict the causes in this steep rise in air pollution but then these causes are not being used anywhere. If it is communicated back to connected infrastructure and air quality monitoring system using Internet of Things (IoT) can be employed that can make informed decisions such as requesting the authorities for increasing the frequency of public transport, recommending people to detour/take alternative transportation solutions and the effects of being exposed to this high levels of pollution.

The development of IoT has the potential to mitigate air pollution by real-time environmental monitoring and forecasting of the level of air pollutants in an area using AI methodologies in the backend. In the long run, monitoring provides information on pollution sources and their levels, which can be analysed to establish patterns that help in the development of strategies and policies to control pollution. Further, the authorities can recommend scheduling for the organizations in order to curb such high inflow.

PPM (Part Per Million) is a mathematical term used to measure the gas level using sensors. For accurate calculations of AQI, an IoT system integrated with different gas sensors unfolds many techniques to enhance the methods to reduce air pollution. Level of pollutants and other meteorological parameters such as relative humidity, carbon monoxide (CO), carbon dioxide (CO_2) and luminosity can be recognized by using SHT10, MQ7, T6615, LDR respectively. MQ-135 and DHT-11 sensors to detect the temperature, humidity, CO, CO_2, smoke and alcohol (Nurul Zakaria *et al.*, 2018).

Along with sensors, real-time data of the environment images of a particular region are also used as an input for monitoring of air-pollution using IoT. The gathered input data is then streamed wirelessly to the cloud where certain rules and AI models are formulated to get the readings of the above-mentioned pollutants. The calculated average data of each pollutant can be compared to the average AQI level of each of the pollutants. Accordingly, the presence of pollutants can be categorised as high, medium, or low which can be further published through multiple interfaces. This will help the citizens to stay aware of the current pollution level and moreover, people can avoid visiting such polluted areas.

It can also be beneficial that such corridors are pushed for infrastructural developments for e-vehicles and then adoption of e-vehicles is made rapid in that particular corridor such as allowing only e-vehicles to pass through during peak hours. Further, in terms of energy, if everyone can be made aware of the benefits of renewable resources of energy and its long-term impact, then the adoption can be rapid. It's evident that affordance would be subjective and the policymakers would need to step in alongside financing agencies. Such strategies and technologies need to be developed to achieve the necessary reduction in air pollution and to record or show the rate of progress towards attaining ambient air quality standards.

CONCLUSION

Over the past few decades, one remarkable development was the adaptability of AI technologies in the area of air pollution control and the prediction of the air pollutants concentration. The AI methods implemented have been considered as efficient and effective alternative methods to confront the uncertainties and complexities of the air pollution problems. However, since the performance of single AI methods is restricted to relatively poor reproducibility and limited global search capability, many researchers thereafter emphasized on hybrid AI methods wherein the limitations of a single AI model can be overcome by the usability of the other. The same has been discussed while selecting hybrid models like ANFIS, C-LSTME. It is therefore clear from the studies that hybrid models are becoming more popular in the last few years and it seems to overtake the single AI methods to become the most popular AI models for environmental or air pollution control methods. These novel hybrid ANN methods have been effective for risk mapping and early warning system to a great extent, by outperforming the existing statistical and classical methods in accuracy and stability. Overall, this review summarizes and gives a brief overview of the causes, sources and the impact in the environment and human health due to air pollution initially, followed by how AI is used as a tool to predict the harmful air pollutants so that required precautions can be adapted accordingly. The effectiveness of AI methods related to air pollution has been extensively discussed. The non-linearity of the artificial neural network has been able to precisely predict air pollutant removal. As a summary of the multiple AI applications reported in this review, the AI models are validated by testing the historical data from the real system. The error percentage of AI models are calculated to understand the accuracy of the predicted values compared to other models.

Though AI models are computationally expensive, consumes more time in processing, training and validation and also requires large datasets for deep learning models to achieve considerable accuracy, the sample of AI methods discussed shows that its capabilities and accuracies are enough to establish the potential of AI techniques to tackle the challenges caused due to air pollution. The emergence of AI technologies to replace and overcome the limitation of traditional statistical methods due to its reliability and rapid response has made this technology successful in handling these environmental related crises.

ACKNOWLEDGEMENTS

The authors are grateful to the Organizers of the workshop, Sri Venkateswara College and Centre for Science & Technology of the Non-Aligned and Other Developing Countries (NAM S&T Centre), New Delhi for the invitation extended to us to contribute our work and views on the most challenging environmental issue.

REFERENCES

[1] Air Pollution. World Health Organization, www.who.int/health-topics/air-pollution (accessed 14 Apr. 2020).

[2] Air Quality Index. Wikipedia, Wikimedia Foundation, (27 Mar. 2020). en.wikipedia.org/wiki/Air_quality_index. (accessed 26 Apr 2020)

[3] Caneon Kurien and Ajay Kumar (2018). Scope of artificial intelligence techniques for exhaust emission prediction of ci engines and renewable energy applications.

[4] Congcong, Wen; Shufu, Liu; Xiaojing, Yao; Ling, Peng, Xiang, Li; YuanHu and Tianhe, Chi (2019). A novel spatiotemporal convolutional long short-term neural network for air pollution prediction. *Science of The Total Environment,* 654:1091–1099.

[5] Dahari, N.A., Wahid, H., Md. Yunus, M.A., Abdul Rahim, H. and Sulaiman, M.D. (Jul. 2016). "A review on air pollutants measurement techniques and its future advancement," *Jurnal Teknologi,* Vol. 78, No. 7–4.

[6] Daniel Vallero (2008). Fundamentals of Air Pollution. Elsevier.

[7] David, Q. Rich; Murray, A. Mittleman; Mark, S. Link; Joel, Schwartz; Heike, Luttmann-Gibson; Paul, J. Catalano; Frank, E. Speizer; Diane, R. Gold and Douglas, W. Dockery (2006). Increased risk of paroxysmal atrial fibrillation episodes associated with acute increases in ambient air pollution. *Environmental Health Perspectives*, 114(1):120–123.

[8] Eder, B., Kang, D., Rao, S.T., Mathur, R., Yu, S., Otte, T., Schere, K., Wayland, R., Jackson, S. and Davidson, P., Using national air quality forecast guidance to develop local air quality index forecasts. *Bull. Am. Meteorol. Soc.*, 2010; 91: 313–326.

[9] George, Kyrkilis; Arhontoula, Chaloulakou and Pavlos, A. Kassomenos (2007). Development of an aggregate air quality index for an urban Mediterranean agglomeration: Relation to potential health effects. *Environment International*, 33(5):670–676.

[10] Gunter, Baumbach (1996). Air Quality Control. Springer Berlin Heidelberg.

[11] Ikenna, C. Eze; Lars, G. Hemkens; Heiner, C. Bucher; Barbara, Hoffmann; Christian, Schindler; Nino, K Àunzli; Tamara, Schikowski and Nicole, M. Probst-Hensch (2015). Association between ambient air pollution and diabetes mellitus in Europe and North America: Systematic review and meta-analysis. *Environmental Health Perspectives,* 123(5):381–389.

[12] Jean, Bousquet; Josep, M. Anto; Isabella, Annesi-Maesano; Toni, Dedeu; Eve, Dupas; Jean-Louis, P´epin; Landry, Stephane Zeng Eyindanga; Sylvie, Arnavielhe; Julia, Ayache; Xavier, Basagana; Samuel, Benveniste; Nuria, Calves Venturos; Hing, Kin Chan; Mehdi, Cheraitia; Yve,s Dauvilliers; Judith, Garcia-Aymerich; Ingrid, Jullian-Desayes; Chitra, Dinesh; Daniel, Laune; Jade, Lu Dac; Ismael, Nujurally; Giovanni, Pau; Robert, Picard; Xavier, Rodo; Renaud, Tamisier; Michael, Bewick; Nils, E. Billo; Wienczyslawa, Czarlewski; Joao, Fonseca; Ludger, Klimek; Oliver, Pfaar and Jean-Marc, Bourez. Pollar (Sep. 2018). Impact of air pollution on asthma and rhinitis; a European institute of innovation and technology health (eit health) project. *Clinical and Translational Allergy*, 8(1):36.

[13] Jianjun, He; Ye, Yu; Yaochen, Xie; Hongjun, Mao; Lin, Wu; Na, Liu and Suping, Zhao (Jun. 2016). Numerical model-based artificial neural network model and its application for quantifying impact factors of urban air quality. *Water, Air, & Soil Pollution*, 227(7):235.

[14] Karrupusamy, P.; Chen, Joy and Shi, Yong, editors (2020). Sustainable Communication Networks and Application. Springer, Switzerland.

[15] Know How Particulate Matter 2.5 and 10 Are Endangering Your Life. AQI India (28 Mar. 2019). www.aqi.in/blog/what-is-particulate-matter-and-its-health-effects (accessed 27 Apr. 2020).

[16] Lim, S.S., Vos, T., Flaxman, A.D., *et al.,* A comparative risk assessment of burden of disease and injury attributable to 67 risk factors and risk factor

clusters in 21 regions, 1990–2010: A systematic analysis for the Global Burden of Disease Study 2010. Lancet. 2012; 380:2224–2260.

[17] Lucette, Flandroy; Theofilos, Poutahidis; Gabriele, Berg; Gerard, Clarke; Maria-Carlota, Dao; Ellen, Decaestecker; Eeva, Furman; Tari, Haahtela; Śebastien, Massart; Hubert, Plovier; Yolanda, Sanz and Graham, Rook (2018). The impact of human activities and lifestyles on the interlinked micro-biota and health of humans and of ecosystems. *Science of the Total Environment,* 627:1018–1038.

[18] Ma and Cristina, Arayata. "SO_2 Emission Continues as Taal Remains under Alert Level 3." Philippine News Agency RSS, www.pna.gov.ph/articles/1092536 (accessed 27 Apr. 2020).

[19] Mahdi, Koliva Salooki; Reza, Abedini; Hooman, Adib and Hadis, Koolivand (2011). Design of neural network for manipulating gas refinery sweetening regenerator column, ELSEVIER.

[20] Mohammad, Hossein Sowlat; Hamed, Gharibi; Masud, Yunesian; Maryam, Tayefeh Mahmoudi and Saeedeh, Lotfi (2011). A novel, fuzzy-based air quality index (faqi) for air quality assessment. *Atmospheric Environment,* 45(12):2050–2059.

[21] Neill and Pippa. "The Horror of the Australian Bushfires and Air Pollution." Air Quality News, 7 Jan. (2020), airqualitynews.com/2020/01/06/the-horror-of-the-australian-bush-fires-and-air-pollution (accessed 24 Apr. 2020).

[22] Nurul, Zakaria; Zaheera, Zainal Abidin; Norharyati, Harum; Low, Hau; Nabeel, Ali and Fairul, Jafar (2018). Wireless internet of things-based air quality device for smart pollution monitoring. *International Journal of Advanced Computer Science and Applications,* 9:65–69.

[23] Patrick, Amoatey; Hamid, Omidvarborna; Mahad, Said Baawain; Abdullah, Al-Mamun; Aynul, Bari and Warren, B. Kindzierski, (2020). Association between human health and indoor air pollution in the gulf cooperation council (GCC) countries: a review. *Reviews on Environmental Health,* (0):20190065.

[24] Prashant, Kumar and Boulent, Imam (2013). Footprints of air pollution and changing environment on the sustainability of built infrastructure. *Science of The Total Environment,* 444:85–101.

[25] Qi, Zhang; Jacqueline, CK Lam; Victor, OK Li and Yang, Han. Deep-air (2020). A hybrid CNN-LSTM framework for fine-grained air pollution forecast.

[26] Rikiya, Yamashita; Mizuho, Nishio; Richard, Kinh Gian Do and Kaori, Togashi (Aug. 2018). Convolutional neural networks: An overview and application in radiology. *Insights into Imaging,* 9(4):611–629.

[27] Sanjay, Rajagopalan and Robert, D. Brook (2012). Air pollution and type 2 diabetes. Diabetes, 61(12):3037–3045.

[28] Siegel and Ethan. "How Much CO_2 Does A Single Volcano Emit?" Forbes, Forbes Magazine, 6 June 2017, www.forbes.com/sites/startswithabang/

2017/06/06/how-much-co2-does-a-single-volcano-emit (accessed 24 Apr. 2020).

[29] Simon, S. Haykin (2009). Neural networks and learning machines. *Pearson Education, Upper Saddle River,* NJ, third edition.

[30] Souzana, Achilleos; Ebaa, Al-Ozairi; Barrak, Alahmad; Eric, Garshick; Andreas, M. Neophytou; Walid, Bouhamra; Mohamed, F. Yassin and Petros, Koutrakis (2019). Acute effects of air pollution on mortality: A 17-year analysis in Kuwait. *Environment International,* 126:476–483.

[31] Study Says Black Carbon Particles from Air Pollution Found on Fetal Side of Placentas. AQI India, 23 Sept. 2019, www.aqi.in/blog/black-carbon-particles-from-air-pollution (accessed 27 Apr. 2020).

[32] Sylvia, C.L. Farhat; Clovis, A. Silva; Maria, Angelica M. Orion; Lucia, M.A. Campos; Adriana, M.E. Sallum and Alf,´esio L.F. Braga (2011). Air pollution in autoimmune rheumatic diseases: A review. *Autoimmunity Reviews,* 11(1):14–21.

[33] What is an Artificial Neural Network (ANN)? - Definition from Techopedia. Techopedia.com, www.techopedia.com/definition/5967/artificial-neural-network-ann (accessed 26 Apr. 2020).

[34] World's Most Polluted Cities in 2019 - $PM_{2.5}$ Ranking: AirVisual. www.iqair.com/world-most-polluted-cities (accessed 19 Apr. 2020).

[35] Yilmaz, Yildirim and Mahmut, Bayramoglu (June 2006). Adaptive neuro-fuzzy based modelling for prediction of air pollution daily levels in city of zonguldak. *Chemosphere,* 63(9):1575–1582.

[36] Zakaria, N.A., Zainal, Z., Harum, N., Chen, L., Saleh, N. and Azni, F. (2018). "Wireless Internet of Things-Based Air Quality Device for Smart Pollution Monitoring," *International Journal of Advanced Computer Science and Applications,* Vol. 9, No. 11.

[37] Zhiping, Ye; Jiaqian, Yang; Na, Zhong; Xin, Tu; Jining, Jia and Jiade, Wang (2020). Tackling environmental challenges in pollution controls using artificial intelligence: A review. *Science of the Total Environment,* 699:134279.

[38] Zhiyong, Wu; Yuqiang, Z.; Leiming, Zhang; Minjuan, Huang; Liuju, Zhong; Duohong, Chen and Xuemei, Wang (2019). Trends of outdoor air pollution and the impact on premature mortality in the pearl river delta region of southern china during 2006–2015. *Science of the Total Environment,* 690:248–260.

[39] Zhongshan, Yang and Jian, Wang (2017). A new air quality monitoring and early warning system: Air quality assessment and air pollutant concentration prediction. *Environmental Research,* 158:105–117.

ABBREVIATIONS

AI	:	Artificial Intelligence
ANFIS	:	Adaptive Network-Based Fuzzy Interference System
ANN	:	Artificial Neural Networks
AQI	:	Air Quality Index
AR	:	Allergic Rhinitis
ARIMA	:	Auto Regressive Integrated Moving Average
ARMA	:	Autoregressive Moving Average
BP-ANN	:	Back-Propagation Artificial Neural Network
CEEMD	:	Complementary Ensemble Empirical Mode Decomposition
CI	:	Confidence Interval/Clarity Index
C-LSTME	:	Convolution Long Short-Term Memory Neural Network Extended
CNN	:	Convolution Neural Network
COPD	:	Chronic Obstructive Pulmonary Diseases
DL	:	Deep Learning
DRF	:	Dose Response Functions
DSS	:	Decision Support System
ENN	:	Elman Neural Network
FC	:	Fully Connected
FL	:	Fuzzy Logic
GBD	:	Global Burden Disease
IMF	:	Intrinsic Mode Functions
IoT	:	Internet of Things
K-NN	:	K-Nearest Neighbour
LSTM	:	Long Short-Term Memory
LSTME	:	Long Short-Term Memory Neural Network Extended
MAE	:	Mean Absolute Error
MAPE	:	Mean Absolute Percentage Error
MCSDE	:	Modified Cuckoo Search and Differential Evolution Algorithm
ML	:	Machine Learning

MLP : Multilayer Perceptron
MLPANN : Multilayer Perceptron Artificial Neural Networks
MLR : Multiple Linear Regression
MSE : Mean Square Error
NMSE : Normalised Mean-Squared Error
PCA : Principal Component Analysis
PCDDs : Polychlorinated Dibenzo-p-dioxins
PM : Particulate Matter
PPM : Parts Per Million
PSI : Pollutant Standard Index
RMSE : Root Mean-Squared Error
RNN : Recurrent Neural Network
RSM : Response Surface Method
SPM : Suspended Particulate Matter
ST-DNN : Spatial Temporal Deep Neural Network
SVM : Support Vector Machine
TSP : Total Suspended Particle
VOC : Volatile Organic Compounds
WHO : World Health Organization

Impact of Air Pollution on Public Health: Current and Future Challenges

Surya Prakash Bhatt and Randeep Guleria

Department of Pulmonary, Critical Care and Sleep Medicine; All India Institute of Medical Sciences, Ansari Nagar, New Delhi–110029, India

ABSTRACT: Air pollution consists of highly variable and complex mixture of gas and particulate matter that is recognized as major contributors to morbidity and mortality. According to the World Health Organization (WHO), more than seven million deaths each year can be attributed to the effects of air pollution. The report further states that the most health damaging particles are those with a diameter of 10 microns or less. This is due to the ability of the smaller particles to travel deeper into the lungs and pass into the bloodstream. Over the last 10 years, there has been a substantial increase in research which suggests that particulate matter is not only exerting a greater impact on established health endpoints, but is also associated with other health issues such as chronic obstructive pulmonary disease, asthma, bronchiolitis, lung cancer, cardiovascular events, central nervous system dysfunction, and cutaneous diseases. This effect is more in developing countries with the rise in the use of fossil fuels in industrialization, urbanization, transport, power generation, building heating systems, agriculture/waste incineration, heating and household fuels. The level of air pollution in India has reached alarming proportions affecting the health of millions of citizens in both urban and rural India. According to the WHO survey in 2014, 13 of the most polluted 20 cities in the world are in India. However, there is still insufficient data that directly correlates the poor quality of ambient air to poor health indicators especially respiratory and cardiovascular illnesses. India is currently exposed to high levels of ambient and household air pollutants. Respiratory adverse effects of air pollution are significant contributors to morbidity and premature mortality in India. The only way to tackle this problem is through public awareness coupled with a multidisciplinary approach by scientific experts. National and international organizations must address the emergence of this threat and propose sustainable solutions.

Keywords: Air Pollution, Particulate Matter, Respiratory Disease, Public Health, India.

INTRODUCTION

Air pollution is a big problem in India and other part of the world. The major components of air pollution is an ambient particulate matter (PM) pollution, household pollution, ozone in the troposphere and the lowest layer of atmosphere. Air pollution contributes extensively to premature mortality and disease burden. Globally it has greater impact in low-income and middle-income countries as compared to high-income countries (Cohen *et al.,* 2017; Landrigan *et al.,* 2018, Guleria R. and Collaborators, 2019). It has been observed that India has one of the maximum exposure levels to air pollution in the world (Cohen *et al.,* 2017).

According to the World Health Organization (WHO), air pollution exposure contributes more than seven million premature deaths each year (Kuehn *et al.,* 2014) and 13 of the highest polluted twenty cities in the world are in India. The level of air pollution in India is estimated to kill 1.8 million people every year; it is the fifth largest killer in India. In India, particularly Delhi, has reached alarming pollution levels affecting the health of millions of citizens in both urban and rural India. In 7th November 2017, the $PM_{2.5}$ levels in Delhi was 710 $\mu g/m^3$ and the smog reached its peak on November 8, 2017, which was a red day in the history of Delhi as the air quality index (AQI) was 999, above the upper limit of worst category (Chaitanya *et al.,* 2018). In 2018, the Central Pollution Control Board (CPCB) data reported the air quality index (AQI) of Delhi at 412, which drops in the 'severe category', while according to the Centre-run System of Air Quality and Weather Forecasting (SAFAR) recorded an AQI of 388 occurs the 'very poor' category. The overall $PM_{2.5}$ level (fine particulate matter in the air with a diameter of less than 2.5 micrometer) (Figure 1) was recorded at 339 and the PM_{10} level at 506 in Delhi.

In 2015, the annual ambient $PM_{2.5}$ exposure in India has been 55 $\mu g/m^3$. In Delhi, the largest exposure (> 100 $\mu g/m^3$) was observed (Chowdhury *et al.,* 2019). Figure 2 shows annual ambient $PM_{2.5}$ exposure across India in 2015. In Northwest India, all districts have annual exposures higher than the National Ambient Air Quality Standard (NAAQS). In Central and West India, 82 and 50% of districts increased the NAAQS, while in North, Northeast, and South India, only 22, 19 and 2% districts disturb the NAAQS.

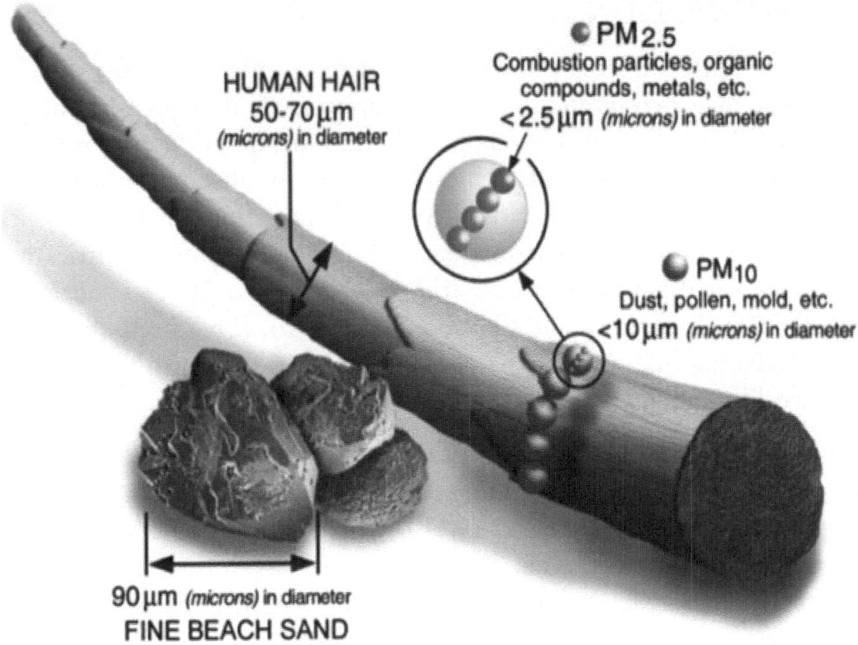

Figure 1: Source from US Environmental Protection Agency

The poor quality of ambient air has been linked directly with increased incidences of chronic bronchitis, asthma, lung cancer and heart disease (Williams *et al.,* 2012; Pope *et al.,* 2015; Nhung *et al.,* 2018). Current research studies show that air pollution from the burning of fossil fuels produces different health effects from allergy to death (Robert *et al.,* 2012). Recent investigations suggest that the public health impacts may be considerable (Howard *et al.,* 2008). In addition, air pollution is also associated with spectrum of acute and chronic health effects, the nature of which may vary, depending on type of the pollutants present in that area as well as the group of the population.

Poor air quality leads to poor health outcomes for both adults and children and also one of the most important environmental problems in the world. Materials occurring in increased levels in the air and causing air pollution include: carbon dioxide (CO_2), nitrogen (NOx), sulfur dioxides (SO_2, fluorine (F), ozone (O_3), hydrocarbons and PM. Depending on the regular aerodynamic particle size, PM10, $PM_{2.5}$ and ultrafine particulate matter (UFPM) of < 0.1 µm are distinguished.

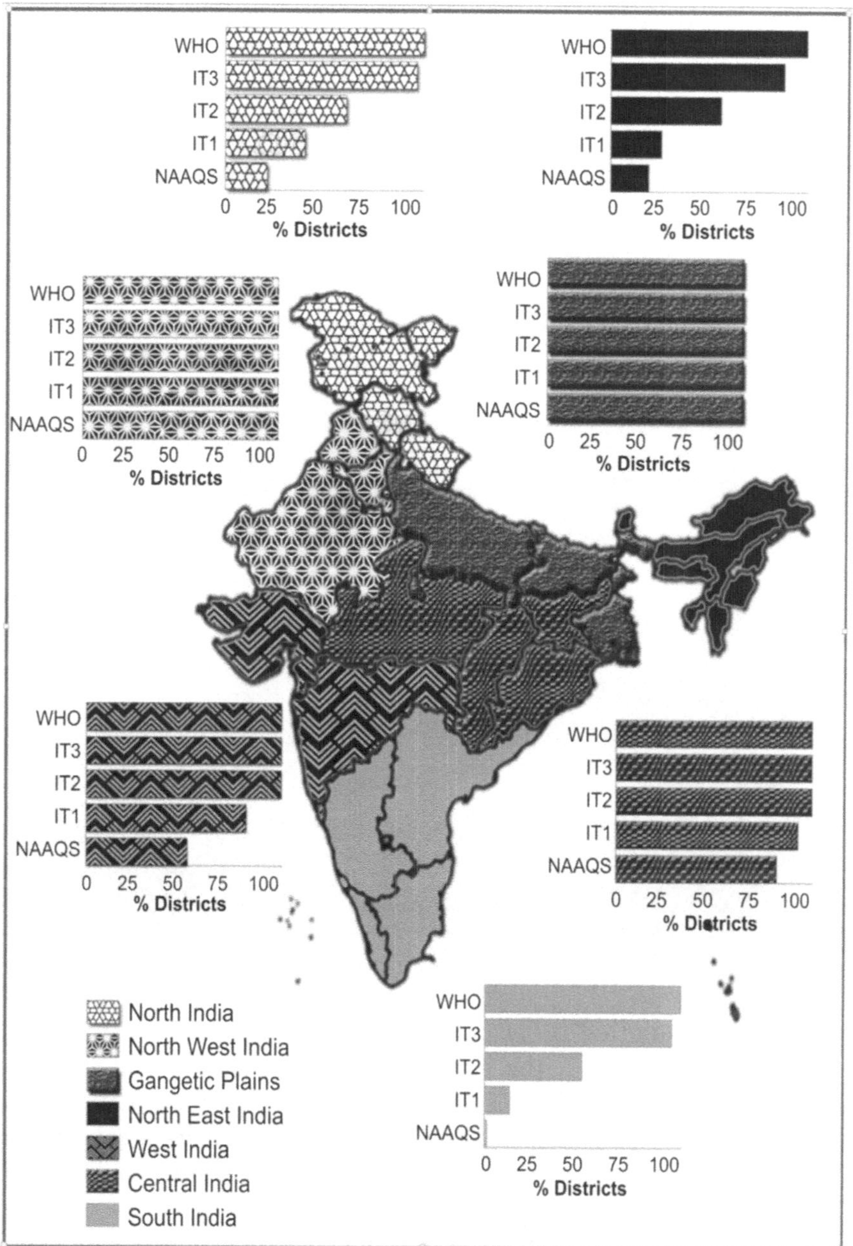

Figure 2: Annual ambient PM2.5 exposure in India with reference to Indian NAAQS and WHO air-quality guidelines. Indian NAAQS: 40 μg m^{-3}; WHO-IT1: 35 μg m^{-3}; WHO-IT2: 25 μg m^{-3}; WHO-IT3: 15 μg m^{-3}; WHO-AQG: 10 μg m^{-3}. Statistics are shown as percentage of districts in each region exceeding the standard/guidelines (Chowdhury *et al.*, 2019).

SOURCE OF AIR POLLUTANTS

The common sources of ambient particulate matter pollution are coal burning for thermal power production, burning of fossil fuels, construction activity and brick kilns, industrialization, urbanization, transport, power generation, building heating systems, agriculture/ waste incineration, heating and household fuels. There are four main types of air pollution sources: *mobile* (cars, buses, planes, trucks, and trains), *stationary* (power plants, oil refineries, industrial facilities, and factories), *area* (agricultural, cities, and wood burning fireplaces) and *natural sources (*wind-blown dust, wildfires, and volcanoes) (Figure 3).

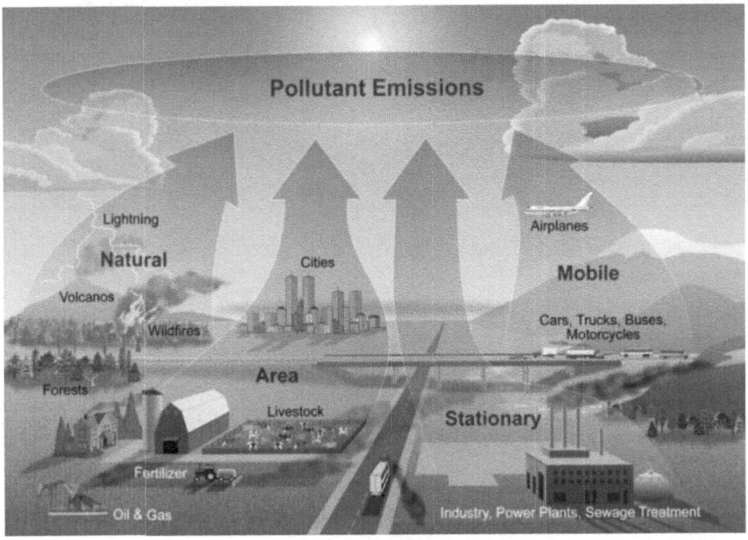

Figure 3: Sources of Air Pollution

Outdoor air pollution is one of the biggest environmental and health and problem. In 2012, the global burden of diseases estimated 3.4 million premature deaths due to outdoor air pollution (GDB, 2017, Figure 4). This means that outdoor air pollution was estimated for 6% of global deaths. In some other countries, it accounts for 10% or higher incidence of deaths. According to GDB (2017), the outdoor pollution in Africa, North America and Oceania, Europe and Latin America and Oceania and many countries in Asia, Egypt, Turkey and China and India was 2%, 2–3%, 4–6%, 6%, 12%, 10% and 8% respectively. Outdoor air pollution is a risk factor for many causes of death, including stroke, cardiovascular diseases, lung cancer, and respiratory diseases, such as asthma (WHO 2018).

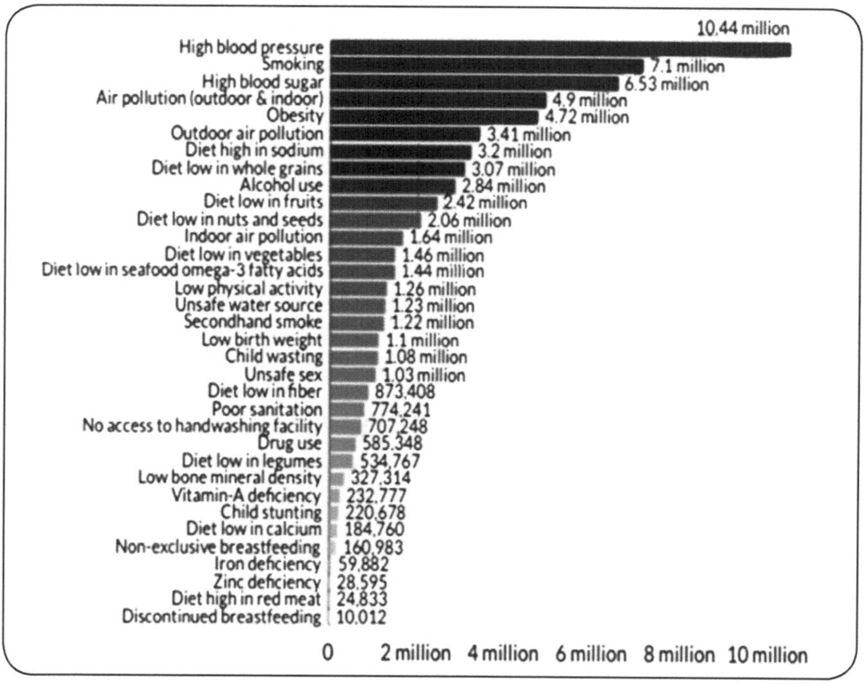

Figure 4: Number of Death by Risk Factor. Total Annual Number of Deaths by Risk Factor, Measured across All Age Groups and Both Sexes (GBD 2017).

Outdoor air pollutants consist of NOx, SO_2, O_3, CO, HC, and PM of different particle sizes (Table 1). NOx, SO_2, O_3, PM pollutants are commonly found in both indoor and outdoor environments. These pollutants can be inorganic, organic, biological and radioactive. The effect of these pollutants on humans depends on their toxicity, concentration and exposure time. In urban areas, these pollutants are mainly released from road vehicles, but there are also contributions from power and petrochemical plants, industrial boilers, aircrafts, ships, depending on the locations and prevailing winds. Comparatively, the contribution from cross border sources is lesser in urban areas due to their increased distance from the pollution sources. However, urban air quality is highly variable from city to city. Densely distributed and deep street canyons can block and weaken the approaching wind, thus reducing its air dispersion capability (Cheng *et al.,* 2009; Li *et al.,* 2009). A master plan urban design with focus on air pollution control may help in dispersing air pollutants and alleviating the problems of air pollutant contributing in one location (Santamouris *et al.,* 2013).

Table 1: Source of Air Pollution

Pollutant	*Sources*	*Health Effects*
Nitrogen dioxide (NO_2)	• Motor vehicles. • Power and Industrial plants.	• Reduction in lung functions and increases symptoms of chronic lung disease. • Lung inflammation, irritation and damage. • Symptoms of bronchitis, wheezing and exacerbation of pneumonia, asthma, bronchial symptoms. • High risk factor of emphysema and premature death. • Aggravate existing hearth disease exacerbation of lung.
Particulate matter	• Burning coal and solid fuel in power and industrial plants. • Combustion—including emission from vehicles, ships, power generation and households. • Natural sources, such as sea salt, wind-blown soil and sand. • Road dust, Sea spray. • Construction.	• Wheezing and exacerbation of asthma. • Respiratory infections. • Chronic bronchitis and obstructive pulmonary disease. • Exacerbation of chronic obstructive pulmonary disease. • Decreased lung function. • Eye, nose, and throat irritation. • Irregular heartbeat and premature death • Mutations, reproductive problems or even cancer.
Sulfur dioxide (SO_2)	• Produced by volcanoes, power plants, and other industrial facilities. • Mainly generated from combustion of sulfur-containing fossil fuels, such as coal and petroleum.	• Wheezing and aggravation of asthma and chronic bronchitis/Breathing problems/Respiratory illness. • Exacerbation of chronic obstructive pulmonary disease. • Cardiovascular disease. • Eye, nose and throat irritation. • Increases chronic lung disease.

(Contd...)

(Table 1 contd...)

Pollutant	*Sources*	*Health Effects*
Ammonia (NH_3)	• Decaying organic matters. • The excreta of humans and animals. • Agricultural processes • Industrial processes. • Vehicular emissions. • Volatilization from soils and oceans.	• Eye, nose, and throat irritation. • Burning the skin eyes, throat, or lungs might be cause permanent blindness, lung disease, or death. • Can cause life-threatening accumulation of fluid in the lungs (pulmonary edema). • Long-term exposure may harm the respiratory system.
Volatile organic compounds (VOCs)	• Fossil fuel combustion. • Industrial activities. • Solvents, paints, glues, and other products • Oil and gas fields. • Road vehicles. • Household heating. • Power generation. • Natural emissions from vegetation and fires.	• Eye, nose, and throat irritation. • Headaches, Dizziness, Fatigue, Nausea. • Emesis, Epistaxis, Loss of co-ordination. • Damage to liver, Damage to kidney. • Damage to central nervous system. • Allergic skin reaction dyspnea. • Cancer
Ozone (O_3)	• Is not emitted directly from any source but is formed in sunlight when certain chemicals react. These chemicals from: • Motor vehicles. • Electric utilities. • Refineries and Factories. • Petrochemicals. • Vegetation and Landfills. • Miscellaneous small sources such as gas stations.	• Coughing, chest tightness, chest pain, and wheezing. • Throat irritation and congestion. • Worsen bronchitis, emphysema, and asthma. • Asthma attacks. • Decreased lung function. • Permanent lung damage. • Damage to lung tissue. • Aggravates chronic lung disease. • Irritates respiratory system. • Headaches and weariness. • Increase premature death.

Indoor air pollution is one of the major human health problems in modern culture as many people spend more than 90% of their time indoors, particularly at their own homes. It has been estimated that 3 billion people worldwide are exposed daily to aero contaminants of indoor air pollution owing to the use of solid fuels such as coal/ biomass fuels for combustion (Sussan *et al.,* 2014). The most common syndrome is called sick building syndrome (SBS), in which patients suffer acute health problems such as irritation of nose, eyes and throat, skin ailments and allergies (Wargocki *et al.,* 2000). These symptoms may not be identified, but may disappear after an affected person leaves the office or building. It has been also observed that indoor air quality can be improved and SBS can be reduced when the ventilation rate of the room is improved (Wargocki *et al.,* 2000). The indoor air quality also affects the performance of workers and office staff. Wyon *et al.,* 2004 showed that the performance of office work was significantly affected by changes in indoor environmental air quality and that the work performance could be significantly enhanced by removing common indoor sources of air pollution.

EFFECTS OF OUTDOOR AIR POLLUTION ON HEALTH

Previous studies have indicated that higher levels of outdoor air pollution are linked to a number of health problems including; abnormal lung function, cough, mucous production, shortness of breath, wheeze, asthma attacks, pneumonia in infants and elderly, chronic COPD, heart attacks, stroke and other respiratory problems (Gordon & Guleria *et al.,* 2019).

The burden of ambient air pollution exposure is borne by developing countries. In Asia, the research studies conducted in these settings has not been significant. A review of the literature of air pollution studies conducted in 13 countries in Asia, identified around 43 studies on the health impacts of ambient air pollution. Most of the studies, conducted in the metropolitan cities of Delhi and Mumbai (Health Effects Institute, 2010), reported association between ambient air pollution and the prevalence of acute and chronic respiratory symptoms. In addition, studies showed significant increases in respiratory illness (Bladen *et al.,* 1983), all-cause mortality (Cropper *et al.,* 1997) and cardio-respiratory problems (Pandey *et al.,* 2005).

Most metropolitan cities in Asia are facing problems due to an increase in the ambient PM and NO_2 concentrations as an outcome of increasing urbanization. Worldwide, in most of the metro cities, motorized road transport is one of the largest pollution sources. Road transport is responsible for 70% of environmental pollution and 40% of greenhouse gas emissions in European cities (Martin *et al.,* 2010). In the UK, automobile pollution frequently violates the national ambient air quality standards (Gulia *et al.,* 2015). Likewise, in the Asian sub-continent, some fast-developed countries, such as Singapore, Japan, and Hong Kong, are facing problems due to an increase in the number of motorized vehicles (Gulia *et al.,* 2015). Thus, traffic-related air pollution has been associated with respiratory diseases and cardiovascular complications. From a public health angle, the effects of air pollution are both chronic and acute. Breathing in high capacities of exhaust fumes can produce short-term irritation to the respiratory tract within a few minutes of exposure. Short-term inhalation of air pollutants may worsen ongoing irritation such as cough, mucous buildup, and inflamed airways. Although the acute, short-term effect is of the least concern to the general public, over time, long-term exposure eventually merges with the chronic effect.

Published research studies have reported higher rates of mortality as a result of short-term exposure to PM and other pollutants. Some of them have estimated premature mortality as a result of exposure to ambient air pollution in individual cities (Balakrishnan *et al.,* 2011; Kumar *et al.,* 2010). The research shows similar results in other countries such as China, South Korea, Japan, etc. Table 2 shows the estimated premature mortality due to outdoor air pollution in individual cities in India.

Table 2: Estimated Premature Mortality Due to Ambient Air Pollution in India

Reference	*Indian City*	*Study Year*	*Pollutant*	*Premature Mortality*
IHME (2013)	All India	1990	PM10	438,000
	All India	2010	PM2.5 + ozone	695,000
Guttikunda *et al.,* 2013	Delhi	2010	PM2.5	7350 to 16,200
Dholakia *et al.,* 2013	Delhi	2030	PM2.5	22,000

(Contd...)

(Table 2 contd...)

Reference	*Indian City*	*Study Year*	*Pollutant*	*Premature Mortality*
Guttikunda *et al.*, 2012	Pune	2010	PM10	3600
	Chennai	2010	PM10	3950
	Indore	2010	PM10	1800
	Ahmedabad	2010	PM10	4950
	Surat	2010	PM10	1250
	Rajkot	2010	PM10	300
Nema *et al.*, 2010	Delhi	2001	PM10	5000
	Kolkata	2001	PM10	4300
	Mumbai	2001	PM10	2000
	Chennai	2001	PM10	1300
	Ahmedabad	2001	PM10	4300
	Kanpur	2001	PM10	3200
	Surat	2001	PM10	1900
	Pune	2001	PM10	1400
	Bhopal	2001	PM10	1800
Kandlikar *et al.*, 2000	Delhi	1993	PM10	3800–6200
	Mumbai	1993	PM10	5000–8000

EFFECTS OF INDOOR AIR POLLUTION ON HEALTH

The health effect of indoor air pollution result in about two million premature deaths per year, where in 44% deaths are due to pneumonia, 54% from chronic obstructive pulmonary disease (COPD), and 2% from lung cancer [World Health Organization (WHO)], 2014, 2013; (Table 3). In addition, the most affected group is women and children because they spend maximum time at home (WHO, 2013). The morbidities associated with indoor air pollution are respiratory problems (Dherani *et al.,* 2008) and COPD (Kurmi *et al.,* 2010). Few studies have been reported to show the effect of wood smoke on cardiovascular health (Unosson *et al.,* 2013). McCracken *et al.,* 2007 reported that reduction in wood smoke exposure by use of improved chimney stove resulted in lowering of systolic and diastolic blood pressure. Another study has reported the reduction in ST-segment depression on electrocardiogram after stove intervention (McCracken *et al.,* 2011). Studies showed sizeable

Table 3: Health Impacts of Indoor Air Pollution, WHO (2012)

Health Outcome	*Evidence*	*Population*	*Relative Risk*	*Risk (95% CI)*
Infections of the lower respiratory tract	Strong	Children aged 0–4 years	2.3	1.9–2.7
Chronic obstructive pulmonary disease	Strong	Women aged ≥ 30 years	3.2	2.3–4.8
	Moderate I	Men aged ≥ 30 years	1.8	1.0–3.2
Lung cancer (coal)	Strong	Women aged ≥ 30 years	1.9	1.1–3.5
	Moderate I	Men aged ≥ 30 years	1.5	1.0–2.5
Lung cancer (biomass)	Moderate II	Women aged ≥ 30 years	1.5	1.0–2.1
Asthma	Moderate II	Children aged 5–14 years	1.6	1.0–2.5
	Moderate II	Adults aged ≥ 15 years	1.2	1.0–1.5
Cataracts	Moderate II	Adults aged ≥ 15 years	1.3	1.0–1.7
Tuberculosis	Moderate II	Adults aged ≥ 15 years	1.5	1.0–2.4

differences in diurnal averaged PM levels between households using cleaner cooking fuels such as LPG (75 $\mu g/m^3$) and biomass (≈ 360 $\mu g/m^3$) 13 in indoor living areas. Table 4 shows typical indoor pollutant levels reported in India under usage of various domestic fuels.

In India, previous studies have reported harmful effects of indoor air pollution. In a case-control study, showed solid-fuel use significantly increased child deaths at ages 1–4 years and more girls died from exposure to solid fuels compared to boys. In addition, solid fuel use was also associated with nonfatal pneumonia (Bassani *et al.,* 2010). Ramesh *et al.,* 2012 showed that liquefied petroleum gas (LPG) was significantly associated with acute lower respiratory tract infection (24.8% had pneumonia, 45.5% had severe pneumonia, and 29.7% had very severe disease). The use of biomass fuel was associated with significantly prolonged nasal mucociliary clearance time in comparison

Table 4: Cooking Time Indoor Total Suspended Particles, PM10, PM2.5 Levels with the Usage of Biomass, Dung Cakes, and LPG as Domestic Fuels

Author	*State*	*Type of Fuel*	*Area of Sampling*	*Duration of Sampling*	*Concentration (µg_/m^3)*		
Menon *et al.*, 1988	Madhya Pradesh, Pondicherry	Wood	Kitchen	2 hours	TSP		2000–5000
Parikh *et al.*, 2001	Tamil Nadu	Biomass and LPG	Kitchen, Living room	2 hours	PM10		
						Kitchen	Living room
					Biomass	2000	1800
					LPG	80	70
Balakrishnan *et al.*, 2013	Madhya Pradesh, Tamil Nadu, Uttaranchal, West Bengal	Biomass, Dung cake, and LPG	Kitchen, Living room	24 hours	PM2.5		
						Kitchen	Living room
					Biomass	590	157
					Dung cake	741	190
					LPG	179	95

to clean fuel users, and reduced peak expiratory flow rate (Priscilla *et al.*, 2010). Another research has shown that use of biomass as a cooking fuel was found to be significantly associated with a high prevalence of tuberculosis. The prevalence remained large and significant even after analyzing separately for men and women and for urban and rural areas. The prevalence of active tuberculosis was 51% who has cooking smoke in the age group 20 years and above (Mishra *et al.*, 1999).

ENVIRONMENTAL IMPACT OF AIR POLLUTION

Air pollution is damaging not only human health but also the environment (Ashfaq *et al.*, 2012). Acid rain containing toxic amounts of nitric and sulfuric acids and able to acidify the water and soil environments and damage trees and plantations. *Haze* is created when fine particles are dispersed in the air and reduce the transparency of the atmosphere and also produced by gas emissions in the air coming from industries, power plants and automobile. *Ozone* occurs both at ground level and in the upper level of the Earth's atmosphere. Stratospheric ozone is protecting us from the Sun's harmful ultraviolet

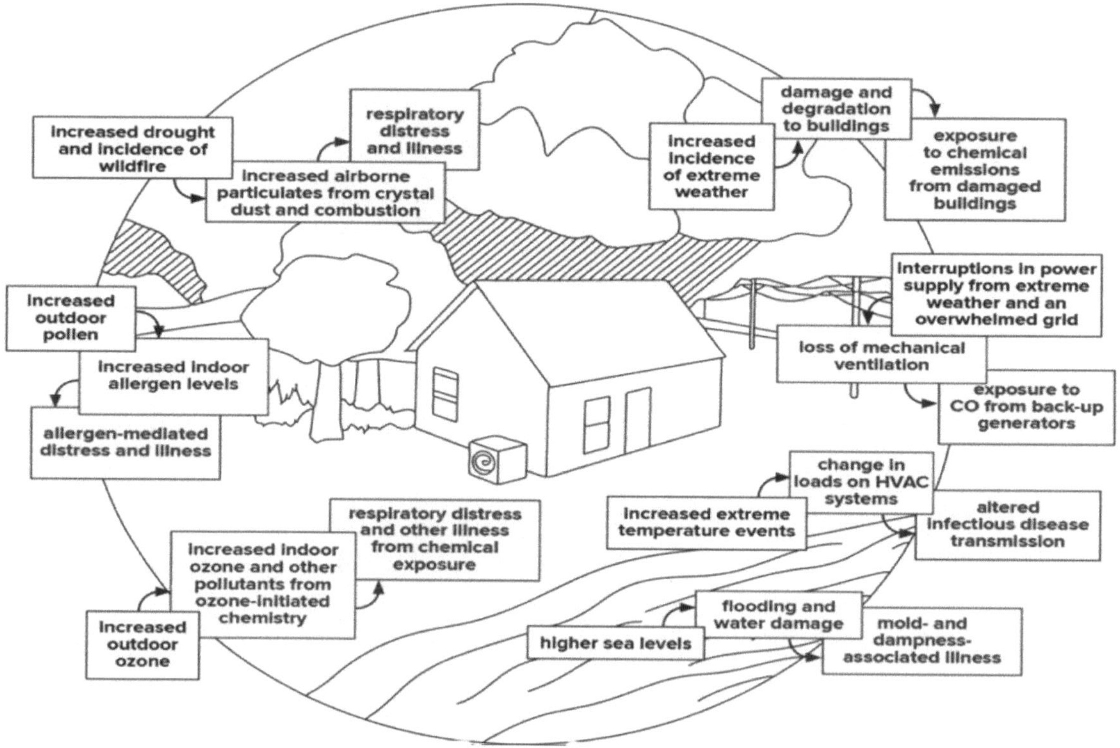

Figure 5: Global warming, intensified weather events, and other effects of climate change may affect buildings and the health of their occupants through many different routes.

Source: Climate change, indoor environments, and health, Indoor Air, 2012.

rays and damaged by ozone-depleting substances can reach our Earth, with harmful effects for human life (skin cancer) (Madronich *et al.*, 1993) and crops (Teramura *et al.*, 2006). In plants, ozone penetrates through the stomata, inducing them to close, which blocks CO_2 transfer and induces a reduction in photosynthesis (Singh *et al.*, 2009). Moreover, plant processes and plan microbial interactions are known to be adversely affected due to pollution (Sharma *et al.*, 2020). Khilnani *et al.*, (2018) reported that air pollution is most important environmental contributor to morbidity and mortality. The respiratory adverse effects of air pollution are acute asthma, COPD, exacerbation of preexisting respiratory diseases, tuberculosis, and lung cancer (Khilnani *et al.*, 2018). Climate change is an important problem that concerns mankind. A report showed that global warming is adding to the health risks of poor people (Manderson *et al.*, 2019) (Figure 5). In addition, people living in poorly constructed buildings in warm climate are at high risk for health problems (Manderson *et al.*, 2019). Wildlife is troubled by toxic pollutants and animals can develop health problems when exposed to high levels of pollutants. Eutrophication is occurring when increased levels of nutrients including nitrogen, often results from air pollution.

CONTROL MEASURES

Air pollution damages human health and the environment. In particular, heavy metals and gaseous air pollutants have harmful effects on human health. In India, research studies provide us evidence that air pollution is a cause of increasing morbidities and mortalities, and there is a need for an urgent intervention. Following is a list of suggested measures, which should be adopted to curb the menace of air pollution:

- ***Public Awareness:*** One of the most significant steps in prevention of air pollution is educating among people about the issue and the serious threat it poses. The proper tutoring should benefit public in finding different ways of decreasing air pollution exposures. Public should also be educated about the use of alternate source of energy to replace combustion of biomass fuel. This education should be extended to politicians and administrators to ensure their assurance and awareness about health effects of air pollution. Poor air quality can have a significant impact on human health, and the previous sections have drawn upon research conducted over recent years that supports the notion that risks are increasing. The general consensus is that society would benefit from being better engaged and educated about the complex relationship between air quality and health problems (Kelly *et al.,* 2012). In turn, such educational awareness has the potential to create a cleaner environment and a healthier population. This can be implemented by engaging organizations working on mobilization and strategic communications to carry out a large-scale public awareness campaign. Creating broad based awareness of the impact of air pollution on health, and identifying pathways for action at an individual and community level can both empower citizens to take action as well as demand their right to clean air.

- ***Localized Evidence Generation:*** The health evidence shows that many of the studies conducted on ambient air pollution exposure are limited to the metro cities such as Delhi and Mumbai. There is a need to develop a confirmation base in tier 2 and 3 towns as well as in industrial hotspots. The Indian Government's smart cities program lists many tier 2 and 3 cities, which currently have no monitoring capacity whatsoever. Developing short-term studies on air pollution health in rapidly industrializing tier 2 and 3 towns will aid in air quality becoming a serious consideration in their master plans and future development. More studies need to be done in large in scale or complex in design. Even simple, time-series or cross-sectional studies could have tremendous impact, with the data for these already available to some extent.
- ***Change in Pattern of Fuel Use:*** Fuel use is dependent on its availability, and affordability. At present, most of low income families trust exclusively on direct combustion of biomass fuels for their cooking needs as this is the cheapest and easiest option available to them. However, this could be rectified by promoting the use of cleaner energy sources such as gobar gas, which utilizes cow dung to produce gas for cooking.
- ***Modification of Design of Cooking Stove:*** The stoves should be modified from traditional smoky and leaky cooking stoves to those which are fuel efficient, smokeless and have an exit (e.g., chimney) for indoor pollutants. Improvement in ventilation: During construction of a house, importance should run to adequate ventilation; for poorly ventilated houses, measures sort of a window

above the kitchen stove and cross ventilation though doors should be instituted.

- ***Intersectoral Coordination and Global Initiative:*** Indoor pollution can only be controlled with coordinated and committed efforts between different sectors concerned with health, energy, environment, housing, and rural development.
- ***Future Directions:*** Increased prospective studies and randomizes intervention trials are needed in several environments, including indoor and outdoor. Additional epidemiological studies using gene-environment interaction and natural phenomenon and biomarkers of health are linked to age, sex, allergy, personality traits, stress and dietary habits. There is a need to continuously estimate the health implications of latest technologies and to execute small-scale testing of latest building materials before commonly utilized in large-scale production. Moreover, there is requirement to use modern statistical methods in order to estimate and predict model for cleaner air and pollution free environment.

CONCLUSION

Though evidence exists for increase in air pollution in India, and its association with both increased morbidity and mortality, there is still a need of further studies to assess the exposure levels of indoor and outdoor pollutants and to further strengthen the evidence for their association with outcomes like tuberculosis, asthma, COPD, cardiovascular health, and lung cancers. At the same time, through public awareness is wanted coupled with a multidisciplinary approach by scientific experts. National and international organizations must address the emergence of this threat and propose sustainable solutions.

REFERENCES

[1] Ashfaq, A. and Sharma, P. Environmental effects of air pollution and application of engineered methods to combat the problem. *J. Indust Pollut Control* (2012) 29.

[2] Balakrishnan, K., Ghosh, S., Ganguli, B., Sambandam, S., Bruce, N., Barnes, D.F., *et al.,* State and national household concentrations of PM2.5 from solid cookfuel use: Results from measurements and modeling in India for estimation of the global burden of disease. *Environ,* 2013; 52–64.

[3] Balakrishnan, K., Ramaswamy, P., Sambandam, S., Thangavel, G., Ghosh, S., Johnson, P., *et al.,* Air pollution from household solid fuel combustion

in India: an overview of exposure and health related information to inform health research priorities. *Glob Health Action,* 2011; 4:1–9.

[4] Bassani, D.G., Jha, P., Dhingra, N. and Kumar, R., Child mortality from solid-fuel use in India: A nationally-representative case-control study. *BMC Public Health,* 2010; 10:491.

[5] Bladen, W.A., Relationship between Acute Respiratory Illness and Air Pollution in an Indian *Industrial City. J. Air Pollut Control Assoc.,* 1983; 33(3):226–7.

[6] Chaitanya, V.S., Air Pollution in Delhi Essay—Case Study. [Last accessed on 2018 May 03]. Available from: https://www.kitchenarena.in/delhi-air-pollution.

[7] Cheng, W.C., Liu, C.H. and Leung, D.Y.C., On the correlation of air and pollutant exchange for street Canyons in combined wind-buoyancy-driven flow. *Atmos. Environ.,* 2009; 43, 3682–3690.

[8] Cohen, A.J., Brauer, M., Burnett, R., *et al.,* Estimates and 25-year trends of the global burden of disease attributable to ambient air pollution: An analysis of data from the Global Burden of Diseases Study 2015. *Lancet,* 2017; 389: 1907–18.

[9] Cropper, M.L., Simon, N.B., Alberini, A., Arora, S. and Sharma, P.K., The health benefits of air pollution control in Delhi. *Am J Agric Econ.,* 1997; 79(5): 1625–9.

[10] Dherani, M., Pope, D., Mascarenhas, M., Smith, K.R., Weber, M. and Bruce, N., Indoor air pollution from unprocessed solid fuel use and pneumonia risk in children aged under five years: A systematic review and meta-analysis. *Bull World Health Organ.,* 2008; 86:390–8C.

[11] Dholakia, H.H., Purohit, P., Rao, S. and Garg, A., Impact of current policies on future air quality and health outcomes in Delhi, India. Atmos. *Environ,* 2013. 75, 241e248.

[12] Garima, Sharma; Rahul, Randeep Guleria and Vartika, Mathur. Differences in Plant Metabolites and Microbes Associated with Azadirachta Indica with Variation in Air Pollution, 2020; 257:113595.

[13] GBD (2017). Risk Factor Collaborators—Global, regional and national comparative risk assessment of 84 behavioural, environmental and occupational and metabolic risks or clusters of risks for 195 countries and territories, 1990–2017: A systematic analysis for the Global Burden of Disease Study, 2017.

[14] Gulia, S., Nagendra, S.S., Khare, M. and Khanna, I., Urban air quality management—A Health Effects Institute. Public Health and Air Pollution in Asia (PAPA): Coordinated Studies of Short-Term Exposure to Air Pollution and Daily Mortality in Four Cities. *Public Health,* 2010 (154).

[15] Guttikunda Sarath, K., Goel, Ra and Pant, P., Nature of air pollution, emission sources and management in the Indian cities. *Atmospheric Environment,* 95 (2014), 501e510.

[16] Guttikunda, S.K. and Jawahar, P. (2012). Application of SIM-air modeling tools to assess air quality in Indian cities. *Atmos. Environ,* 62, 551e561.

[17] Howard, Frumkin; Jeremy, Hess; George, Luber; PJosephine, Malilay, M.P.H. and Michael, McGeehin, Climate Change: The Public Health Response. *American Journal of Public Health (AJPH),* March 2008.

[18] IHME (2013). The Global Burden of Disease 2010: Generating Evidence and Guiding Policy. Institute for Health Metrics and Evaluation, Seattle, USA.

[19] India State-Level Disease Burden Initiative Air Pollution Collaborators. The Impact of Air Pollution on Deaths, Disease Burden and Life Expectancy Across the States of India: The Global Burden of Disease Study, 2017.

[20] Jon Unosson, Anders Blomberg, Thomas Sandström, Ala Muala, Christoffer Boman, Robin Nyström, Roger Westerholm, Nicholas L. Mills, David E. Newby, Jeremy P. Langrish and Jenny A. Bosson. Exposure to wood smoke increases arterial stiffness and decreases heart rate variability in humans *Part Fibre Toxicol.*, 2013; 10: 20.

[21] Kandlikar, M. and Ramachandran, G. (2000). The causes and consequences of particulate air pollution in urban India: A synthesis of the science. *Annu. Rev. Energy Environ.,* 25, 629e684.

[22] Kelly, F.J., Fuller, G.W., Walton, H.A. and Fussell, J.C. Monitoring air pollution: Use of early warning systems for public health. *Respirology,* 2012; 17(1): 7–19.

[23] Khilnani, Gopi C. and Tiwari, Pawan. Air pollution in India and related adverse respiratory health effects. *Current Opinion in Pulmonary Medicine,* 2018: 24 (2); 108–116.

[24] Kuehn, B.M., WHO: More Than 7 Million Air Pollution Deaths Each Year. *JAMA.* 2014; 311(15): 1486.

[25] Kumar, R., Sharma, S.K., Thakur, J.S., Lakshmi, P.V. and Sharma, M.K. ST. Association of air pollution and mortality in the Ludhiana city of India: A time-series study. *Indian J. Public Health,* 2010 (54): 98–103.

[26] Kurmi, O.P., Semple, S., Simkhada, P., Smith, W.C. and Ayres, J.G., COPD and chronic bronchitis risk of indoor air pollution from solid fuel: A systematic review and meta-analysis. *Thorax,* 2010; 65:221–8.

[27] Li, X.X., Liu, C.H. and Leung, D.Y.C., Numerical investigation of pollutant transport characteristics inside deep urban street canyons. *Atmos. Environ.* 2009; 43, 2410–2418.

[28] Madronich, S., de Gruijl, F., Skin cancer and UV radiation. Nature (1993). 366:23–9. doi: 10.1038/366023a0.

[29] Manderson, L., How global Warming is Adding to the Health Risks of Poor People. The Conversation. University of the Witwatersrand. Available online at: http://theconversation.com/how-global-warming-is-adding-to-the-health-risks-of-poor-people-109520 (accessed October 5, 2019).

[30] Martin, J., The European environment: State and outlook 2010: Synthesis. *European Environment Agency,* 2010:31–33.

[31] McCracken, J.P., Smith, K.R., Díaz, A., Mittleman, M.A. and Schwartz J. Chimney stove intervention to reduce long-term wood smoke exposure lowers blood pressure among Guatemalan women. *Environ Health Perspect,* 2007; 115: 996–1001.

[32] McCracken, J.P., Smith, K.R., Stone, P., Díaz, A., Arana, B. and Schwartz, J., Intervention to lower household wood smoke exposure in Guatemala reduces ST-segment depression on electrocardiograms. *Environ Health Perspect,* 2011; 119:1562–8.

[33] Menon, P., Indoor spatial monitoring of combustion generated pollutants (TSP, CO and BaP) by Indian cookstoves. *DTIC Document*; 1988.

[34] Mishra, V.K., Retherford, R. and Smith, K.R., Biomass cooking fuels and prevalence of blindness in Indian. *J Environ Med.,* 1999; 1:189–99.

[35] Nema, P. and Goyal, S.K. (2010). Estimation of health impacts due to PM10 in major Indian cities (Chapter 11). Air Pollution: Health and Environmental Impacts. CRC Press—Taylor & Francis Group, pp. 297e310.

[36] Nguyen Thi, Trang Nhung, Christian Schindler, Tran Minh Dien, Nicole Probst-Hensch, Laura Perez and Nino Künzli, Acute Effects of Ambient Air Pollution on Lower Respiratory Infections in Hanoi Children: An Eight-Year. *Environ Int.,* 2018; 110:139–148.

[37] Pandey, J.S., Kumar, R. and Devotta, S., Health risks of NO2, SPM and SO_2 in Delhi (India). *Atmos Environ [Internet].* 2005; 39(36):6868–74.

[38] Parikh, J., Balakrishnan, K., Laxmi, V. and Biswas, H. (2001). Exposure from Cooking with bio-fuels: Pollution Monitoring and Analysis for Rural, Tamil Nadu. *Energy—The International Journal,* 26 (10), 949–962.

[39] Pope, C.A. III. Epidemiology of Lung Cancer Associated with Outdoor Air Pollution. *Article in Environmental and Molecular Mutagenesis,* 56:S41–S41 · August 2015

[40] Priscilla, J., Padmavathi, R., Ghosh, S., Paul, P., Ramadoss, S., Balakrishnan, K., *et al.,* Evaluation of mucociliary clearance among women using biomass and clean fuel in a periurban area of Chennai: A preliminary study. Lung India, 2011; 28:30–3.

[41] Ramesh Bhat, Y., Manjunath, N., Sanjay, D. and Dhanya, Y., Association of indoor air pollution with acute lower respiratory tract infections in children under 5 years of age. *Paediatr Int Child Health,* 2012; 32:132–5.

[42] Robert, J. Laumbach and Howard, M. Kipen, Respiratory Health Effects of Air Pollution: Update on Biomass Smoke and Traffic Pollution. *J Allergy Clin Immunol,* 2012. 129(1): 3–11; quiz 12–3.

[43] Ron, Williams; Robert, Brook; Robert, Bard; Teri, Conner; Hwashin, Shin and Richard, Burnett Impact of Personal and Ambient-Level Exposures to Nitrogen Dioxide and Particulate *Matter on Cardiovascular Function Int J. Environ Health Res.,* 2012; 22(1):71–91.

[44] Santamouris, M. (2013). "On the built environment—the urban influence," in Energy and Climate in the Urban Environment, ed M. Santamouris, 3–15. Available online at: http://www.google.com.hk/books?hl=zh-TW&lr=&id=_r_9lPbjxX8C&oi=fnd&pg=PP1&dq=Santamouris,+M.+On+the+built+environment&ots=4BegLwd1b1&sig=pYVwpjQwPDAMGbRqyN_Sms0Tsbw&redir_esc=y#v=onepage&q=Santamouris%2C%20M.%20On%20the%20built%20environment&f=false

[45] Singh, E., Tiwari, S. and Agrawal, M. Effects of elevated ozone on photosynthesis and stomatal conductance of two soybean varieties: a case study to assess impacts of one component of predicted global climate change. *Plant Biol Stuttg Ger.* (2009), 11(Suppl. 1): 101–8.

[46] Sourangsu, Chowdhury; Sagnik, Dey; Sarath, Guttikunda; Ajay, Pillarisetti; Kirk, R. Smith and Larry, Di Girolamo, Indian annual ambient air quality standard is achievable by completely mitigating emissions from household sources. *Proceedings of the National Academy of Sciences,* May 2019, 116 (22), 10711–10716.

[47] Sussan, T.E., Ingole, V., Kim, J.H., McCormick, S., Negherbon, J., Fallica, J., *et al.,* Source of biomass cooking fuel determines pulmonary response to household air pollution. *Am J Respir Cell Mol Biol.*, 2014; 50(3): 538–48.

[48] Teramura, A., Effects of UV-B radiation on the growth and yield of crop plants. *Physiol Plant.* (2006), 58:415–27.

[49] Terry, Gordon; Kalpana, Balakrishnan; Sagnik, Dey; Sanjay, Rajagopalan; Jonathan, Thornburg; George, Thurston; Anurag, Agrawal; Gwen, Collman; Randeep, Guleria; Sneha, Limaye; Sundeep, Salvi, Vasu Kilaru and Srikanth, Nadadur. Air pollution health research priorities for India: Perspectives of the Indo-U.S. *Communities of Researchers. Environ Int.,* 2018; 119: 100–108.

[50] The impact of air pollution on deaths, disease burden, and life expectancy across the states of India: The Global Burden of Disease Study 2017. India State-Level Disease Burden Initiative Air Pollution Collaborators. *Lancet Planet Health,* 2019; 3(1): e26–e39.

[51] Wargocki, P., Wyon, D.P., Sundell, J., Clausen, G. and Ole Fanger, P. (2000). The effects of outdoor air supply rate in an office on perceived air quality, *Sick Building Syndrome (SBS).* Indoor Air 10, 222–236.

[52] WHO (2018). Fact sheet—Ambient air quality and health. Updated May 2018.

[53] WHO (2014). Outdoor Air Pollution in the World Cities. World Health Organization, Geneva, Switzerland. http://www.who.int/phe/health_topics/outdoorair/ databases/en.

[54] WHO (2014). Household (Indoor) Air Pollution. Available online at: http://www.who.int/indoorair/en/ (Accessed 2 September, 2014).

[55] Wyon, D.P. (2004). The effects of indoor air quality on performance and productivity. Indoor Air 14, 92–101.

Effects of Rising Air Pollution on Public Health and Measures Adopted Globally to Improve Air Quality

Keerti Mishra

Science and Engineering Research Board,
Department of Science & Technology, Government of India
E-mail: keertim033@gmail.com

ABSTRACT: Air pollution is one of the biggest threats for the environment and affects everyone including humans, animals, crops, cities, forests, aquatic ecosystems etc. It is the fifth leading risk factor for mortality throughout the world. It is among the highest 5 risk factors for population health globally, shortening life on average by 20 months. Major studies reveal that air pollution is one of the leading risk factors for death. It is responsible for more deaths than many better-known death factors such as road traffic injuries, malaria, malnutrition and alcohol use. Air pollution exposure is linked with increased hospitalizations, disability and early death from respiratory diseases, heart disease, stroke, lung cancer, and diabetes as well as communicable diseases like pneumonia. As per the recent estimation of WHO there are 7 million deaths every year around the globe, the bulk of which i.e. 70 percent occurs in Asia Pacific due to air pollution. Major research indicates that by improving air quality globally the life expectancy can be increased by an average of 3.2 years. The key air pollutants are broadly classified into Outdoor and Indoor. Outdoor Air Pollution has alone caused some 4.2 million deaths in 2016, while 1.6 billion people died due to exposure to Indoor Air Pollution from the use of solid fuels for cooking. More than 90% of air pollution related deaths occur in low- and middle-income countries, mainly in Asia and Africa. As per the UNEP "Emission Gap Report" the top four emitters are China, USA, EU and India. Considering the seriousness of the problem, various countries are taking appropriate measures to tackle and reduce air pollution. While addressing air pollution, there is also a need to address critical and easy-to-implement solution to climate change.

Keywords: Air Pollution; Green House Gases; Emission Trading; Air Quality; Particulate Matter; Climate Change.

INTRODUCTION

Air Pollution is one of the leading factors of mortality and disability in the world. Keeping in view of the degrading air quality time bound strategy is must to tackle rising air pollution around the globe. The rise in air pollution with the advent of time has pushed everyone globally to take crucial steps in order to improve air quality. Our health is deteriorating day by day by intake of polluted air. People inhaling polluted air are prone to various chronic diseases including respiratory and cardiovascular diseases. As per IQAirAirVisual 2018 World Air Quality Report and interactive World's Most Polluted Cities ranking, India has 15 out of 20 most polluted cities in the world with Gurgaon, Ghaziabad, Faridabad, Noida and Bhiwadi in the top six positions. Breathing in polluted air can give rise to various chronic diseases in human i.e. chronic respiratory diseases, heart disease, cancer, lung infections etc.

Eminent international agencies working in this direction like the World Health Organisation (WHO) consider only few set of air pollution indicators in order to get closer to quantify and monitor the air pollution as it is not feasible yet to quantify all air pollutants. It specifically focuses on four air pollutants, i.e., particulate matter (PM), particles with an aerodynamic diameter lesser than 10 μm (PM10) and lesser than 2.5 μm (PM2.5), Sulphur dioxide (SO_2), Nitrogen Dioxide (NO_2), and Ozone. PM10 are inhalable coarse particles, which are particles with a diameter between 2.5 and 10 micrometers (μm) and PM2.5 are fine particles with a diameter of 2.5 urn or less. The particulate matters are the deadliest form of air pollutant as they have ability to penetrate deep into the lungs and blood streams unfiltered. The smaller PM2.5 are more harmful as due to their small size it can penetrate deeper into the lungs.

UNEP EMISSION GAP REPORT

United Nations Environment Programme (UNEP) was created in 1972 as a specialized body of United Nation. Its headquarters is in Nairobi, Kenya. It aims to protect the global environment and sets the global environmental agenda for sustainable development within the United Nations system.

UNEP "Emission Gap Report" 2019

The top four emitters are China, USA, EU and India.[1] These countries contributed to over 55% of the total emissions over the last decade, excluding emissions from land-use change such as deforestation. (Figure 1)

The Emissions Gap Report measures and projects three key trend lines:

1. The amount of greenhouse gas emissions every year up to 2030.
2. The commitments countries are making to reduce their emissions and the impact these commitments are likely to have on overall emission reduction.
3. The pace at which emissions must be reduced to reach an emission low that would limit temperature increase to 1.5°C, affordably.

The report also identifies key opportunities for each country to increase the pace of emission reduction necessary to close the gap.

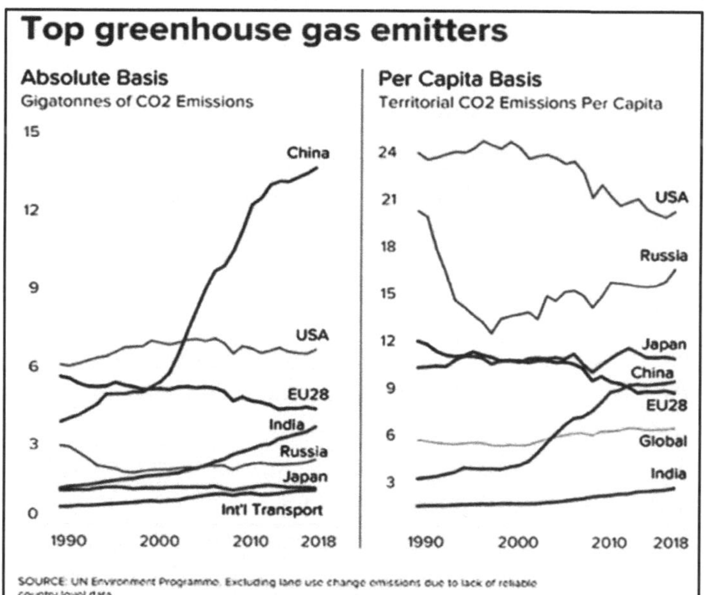

Figure 1: Top Green House Gas Emitters

(*Source:* UN Environment Programme-land use change emissions due to lack of reliable country level data excluded)

Key Findings of the Report

1. The world will fail to meet the 1.5°C temperature goal of the Paris Agreement unless global greenhouse gas emissions fall by 7.6 per cent each year.
2. Global temperatures are set to rise about 3.2°C by 2100, the report says, bringing catastrophic weather including hotter, deadlier heat waves and more frequent floods and drought.
3. The top four emitters (China, USA, EU and India) contributed to over 55% of the total emissions over the last decade, excluding emissions from land-use change such as deforestation.
4. The rankings would change if land-use change emissions were included, with Brazil likely to be the largest emitter.
5. Status of India: India is the fourth-largest emitter of Green House Gases (GHGs). It is among a small group of countries that are on their way.

Carbon Disclosure Project (CDP) India Annual Report

This *report* has been released by *CDP (Carbon Disclosure Project) India.* It analyses *carbon reduction activities of companies.* The CDP is a global disclosure system that enables companies, cities, states and regions to measure and manage their environmental impacts. It collects and collates self-reported environmental data in the world.

It is aimed at measuring the carbon reduction activities undertaken by different companies and firms operating in various countries across the globe.[2]

Key Findings

1. The *boards of 98% of the firms directly monitor climate change risks* with top management integrating these concerns in performance evaluation.
2. 2/3rd of the 59 firms that were surveyed use *climate analysis tools* to formulate their business strategies.
3. Improvement in disclosure rate has primarily been driven by investors who are actively pushing companies to reveal climate risks and take steps to reduce their carbon footprint.
4. Among the key focus areas of Indian firms is renewable energy. According to the report, 23 companies reported renewable energy

targets in 2019, a 44% rise over 2018. Of these, *Infosys, Dalmia Cement and Tata Motors have reported 100% RE consumption.*

5. United States of America secured the top spot followed by Japan, United Kingdom and France.

Overall Performance of India

1. India is now among *the top five countries globally when it comes to adopting science-based target initiatives (SBT)* with as many as 38 Indian companies in 2019 committing to going beyond policy requirements to plan urgent climate action, a significant rise from 25 firms in 2018.
2. In 2019, up to 57 of the 59 responding companies stated that they have a process for risks assessment; 51 declared that their process of identifying, assessing and managing climate related risks is integrated into the multi-disciplinary, company-wide risk identification, assessment, and management process which is considered a best practice.

Sustainable Development Goals Dealing with Air Pollution

Achieving a sustainable and healthy future for all requires action on air pollution, which is a major cause of morbidity and mortality globally.[3] The list of targets with special emphasis on air pollution includes:

- **SDG 3.9:** Calls for a substantial reduction in deaths and illnesses from air pollution.
- **SDG 7.1:** Aims to ensure access to clean energy in homes.
- **SDG 11.2:** Aims to provide access to safe, affordable, accessible and sustainable transport systems for all.
- **SDG 11.6:** Aims to reduce the environmental impact of cities by improving air quality.

Types of Air Pollution

Air Pollution is broadly classified into two subheads i.e. Indoor Air Pollution and Outdoor or Ambient Air Pollution. Harmful pollutants when degrade indoor air quality generally occurs by burning of solid cooking fuels like coal, cow dung, firewood etc. which leads to Household or Indoor Air Pollution (Figures 2 & 3). Outdoor air pollution[4, 5] results from combustion of fossil fuels that are used in industries, transportation, power generation and other economic activities [Figure 4]. The Indoor Air Pollution is more concentrate than Outdoor Air Pollution and its effect can be seen more frequently on Human health.

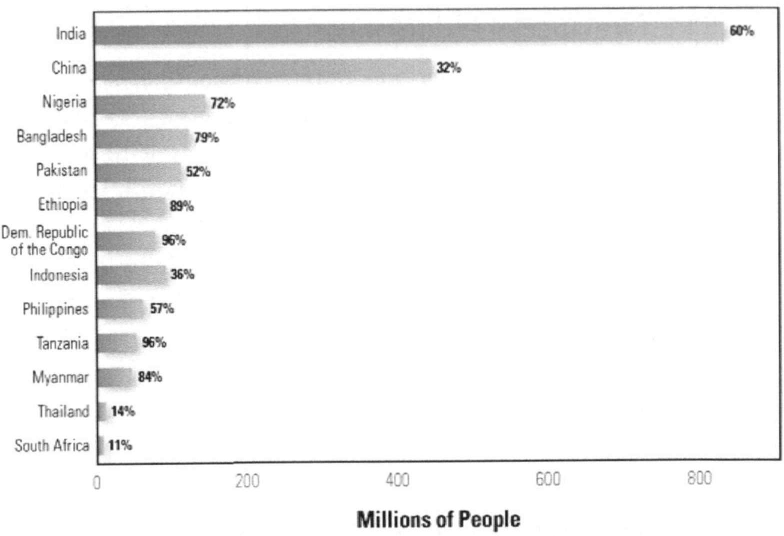

Figure 2: Numbers of People and Percentage of Population Exposed to Household Air Pollution in 13 Countries with Populations over 50 Million

(*Source:* State of Global Air Report, 2019)

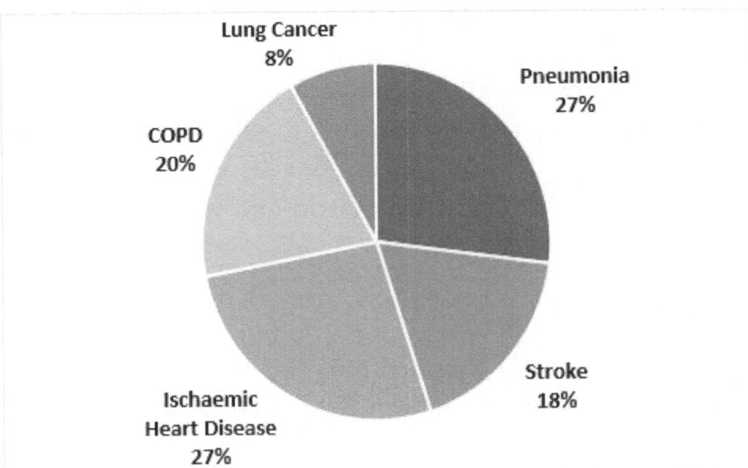

Figure 3: Mortality Rate Due to Indoor Air Pollution

(*Source:* World Health Organisation Report, 2019)

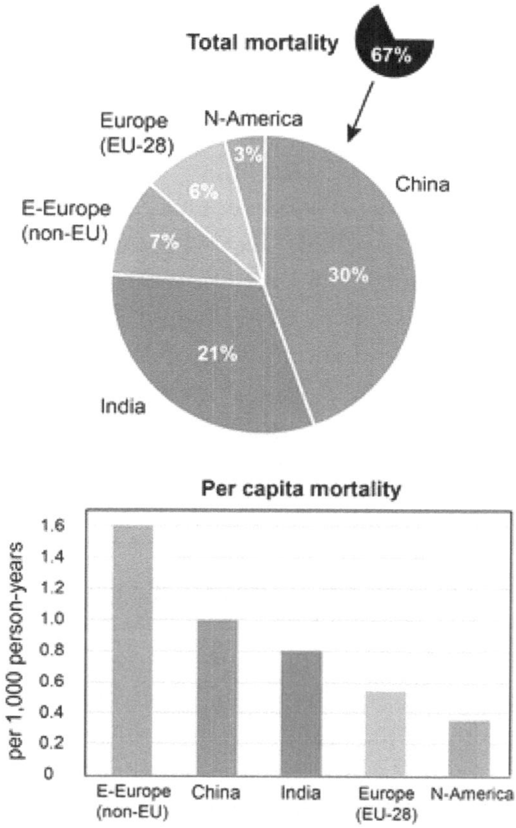

Figure 4: Mortality Due to Outdoor Air Pollution
(*Source:* Royal Society of Chemistry)

Measures Adopted both at National as well as International Level to Combat Air Pollution

Various Countries including international organisations have taken bold measures to combat air pollution. The efforts have shown its results up-to some extent but keeping in view of the growing air pollution more efforts are needed to curb its rising level. Some global and national efforts are discussed for a broader view on this direction.

GLOBAL EFFORTS

Kyoto Protocol

The Kyoto Protocol focuses on rising level of Green House Gas Emission. It binds developed countries and places a heavier burden on them under Common but Differentiated Responsibility and Respective Capabilities. It also sets Emission Reduction Target for 37 Industrialized and European Union Countries.[6] It recognizes that the developed countries are largely responsible for the current rise in the levels of GHG emissions into the atmosphere. Currently it is working over a set of Commitment Period i.e. First Set (From 2008–2012) and Second Set (From 2013–2020). During the first commitment period, 37 industrialized countries and the European Community committed to reduce GHG emissions to an average of five percent against 1990 levels. During the second commitment period, Parties committed to reduce GHG emissions by at least 18 percent below 1990 levels in the eight-year period from 2013 to 2020; however, the composition of Parties in the second commitment period is different from the first.

Emission Trading Scheme

It is a market toll which aims to reduce the global emission of Green House Gases. The Article 17 of Kyoto Protocol sets out Emission Trading norms globally. The countries committed under Kyoto Protocol obtain emission units from other countries which is committed under the Protocol and use them towards meeting a part of their targets. It uses a software-based accounting system in order to ensure secure transfer of emission reduction units between countries.

Cap and Trade System

Under this system, the total mass of pollution is first defined by the regulator that can be put into the air over a defined period by all factories put together. Thereafter, a set of permits is generated, each of which allows a certain amount of pollution, and the total is equal to the cap. These permits are the quantity that is bought and sold. Each factory is allocated a share of these permits (this could be equal or based on size or some other rule). After this, plants can trade permits with each other, just like any other commodity on the National Commodity and Derivatives Exchange Limited (NCDEX).

European Union Emissions Trading System (EU ETS)

It aims to reduce the Greenhouse Gas Emission in Europe. It is the largest Emission Trading Scheme in the world. The efforts of all members i.e. 28 EU member countries along with Iceland, Norway, and Liechtenstein has almost halved the anthropogenic carbon di-oxide emission as well as there is 40% reduction in total greenhouse gas emission by these countries. This scheme works under four phases (Figure 5). This scheme works on 'cap and trade' principle, i.e. a maximum (cap) is set on overall amount of greenhouse gases that can be emitted by all participating installations.[7] "Allowances" for emissions are then auctioned off or allocated for free, and can subsequently be traded.

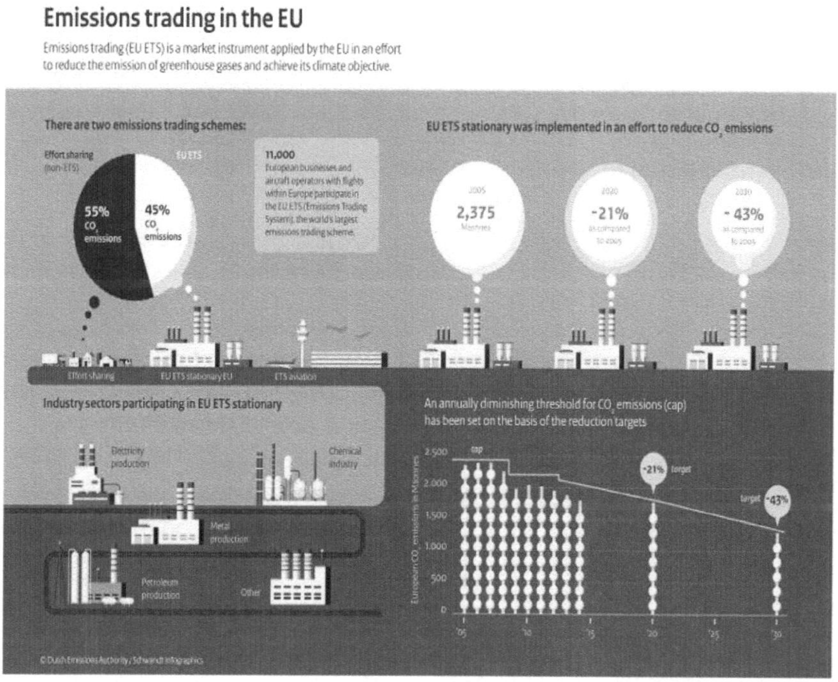

Figure 5: Overview of European Union Emission Trading Scheme (*Source:* Dutch Emission Authority)

Reducing Emissions from Deforestation and Forest Degradation Plus (REDD+) Efforts

It is estimated that globally, deforestation and forest degradation accounts for around 11 percent of CO_2 emissions. Halting deforestation is a cost-effective action that has a clear impact in reducing global GHG emissions. This is a global effort to reduce the emission of greenhouse gas to reduce air pollution thereby mitigating Climate Change (Figure 6). It is designed by the parties of United Nation

Framework for Combating Climate Change (UNFCCC). This programme provides incentives to the developing countries for reducing carbon emission by keeping their forest standing.

UN REDD Programme facilitates the developing countries in developing capacities for meeting the REDD+ targets. It promotes the involvement of all the stakeholders including the indigenous people of forest area, tribal communities residing nearby the forested land and various other communities dependent on forest for their livelihood.[8]

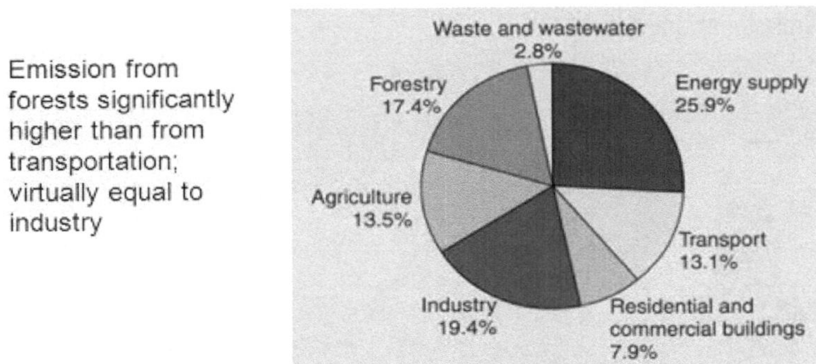

Figure 6: Comparative Analysis of Green House Gas Emission from Various Sources Source: UN Reducing Emissions from Deforestation and Forest Degradation (REDD) Programme

Food and Agriculture Organization of the United Nations (FAO) supports the countries in meeting the implementation challenges of the goals of REDD+ programme. [Figure 7]

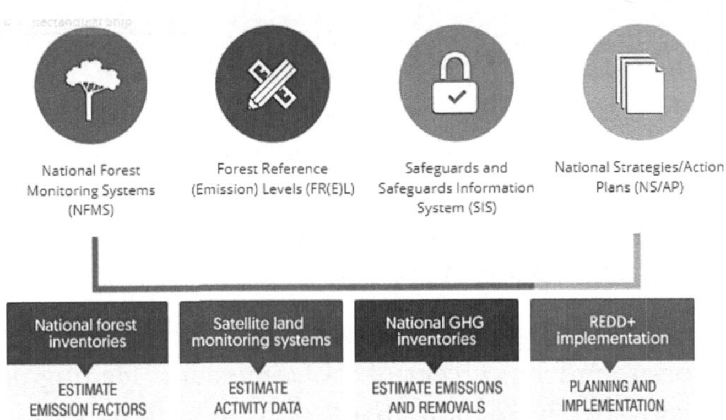

Figure 7: Areas of Food and Agriculture Organization Support to Parties of REDD+ Programme

(*Source:* United Nation, Food and Agriculture Organization www.fao.org)

UNITED NATION CLIMATE ACTION SUMMIT 2019

The summit recently held in New York, USA. It reinforced that 1.5°C[9] is the safe limit of global warming by the end of this century and in order to achieve this target, the world needs to work efficiently to curb the rising level of Air Pollution leading to Climate Change. This summit primarily aims at reducing Green House Gas emission by 45 per cent over the next decade, and to net zero emissions by 2050. The worsening global scenario can be seen with the rising level of natural hazards. It can be evident from the rising sea level, forest fires, droughts, flooding, melting of ice and various other sequence of events.

Key Commitments

- Overall, 77 countries have pledged to reduce their emission to the net zero level by 2050.
- Many Small Island Nations have agreed to achieve 100% renewable energy use targets by 2030.
- Two countries i.e. Hungary and Greece have pledged to shut their coal powered plant by 2030 and 2028 respectively.
- India pledge to increase their renewable energy capacity by more than 175 GW heading towards the target of 450 GW. Eighty countries have joined International Solar Alliance.
- Total number of signatories increased to 187 with Russia ratifying the Paris Agreement.
- The Central American country i.e. El Salvador pledged to reduce its Green House Gas Emission up-to 40% by 2030.
- China pledged to reduce 12 billion tons of Green House Gas Emission.
- The Central African Forest Initiative pledged to save forest. This will reduce the emission of 70 Gigatons of CO_2 into the atmosphere.

NATIONAL EFFORTS

Pradhan Mantri Ujjwala Yojana (PMUY)

The Government of India aims to safeguard the health of women & children by providing them with a clean cooking fuel—LPG, so that they don't have to compromise their health in smoky kitchens or wander in unsafe areas collecting firewood reducing indoor air pollution up-to a remarkable level. Under this scheme, 50 million LPG connections will be provided to BPL families with a support of US$ 21.92 (₹ 1600) per connection.[10] Ensuring women's empowerment, especially in rural India, the connections will be issued in the name of women of the households. Approximately US$ 1096 (₹ 8000 Cr). has been allocated towards the implementation of the scheme. Identification of the BPL families will be done through Socio Economic Caste Census Data. As per latest data; in just two years the scheme has benefitted 37 million women living below the poverty line with free LPG connections to support them to switch to clean cooking fuels (Figure 8). This scheme has been partly funded by 'Give It Up' initiative.

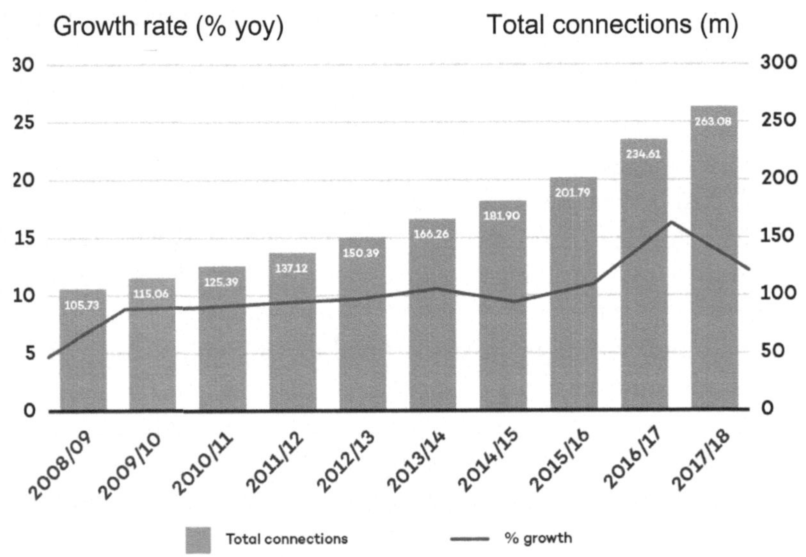

Figure 8: Total registered LPG Connection (FY 2008–09 to FY 2017–18)
(*Source:* Ministry of Petroleum and Natural Gas, 2018)

National Clean Air Programme (NCAP)

This Scheme was introduced in 2019 to tackle rising Air Pollution in a time bound manner. It is a five-year action plan and 2019[11, 12] will be considered as first year. The tentative national level target of 20%–30% reduction of $PM_{2.5}$ and PM_{10} concentration by 2024 is proposed under the NCAP taking 2017 as the base year for the comparison of concentration [Figure 9].

The following are the broad objectives of NCAP:

(a) To augment and evolve effective and proficient ambient air quality monitoring network across the country for ensuring comprehensive and reliable database

(b) To have efficient data dissemination and public outreach mechanism for timely measures for prevention and mitigation of air pollution and for inclusive public participation in both planning and implementation of the programmes and policies of government on air pollution.

(c) To have feasible management plan for prevention, control and abatement of air pollution.

Figure 9: Cities Covered under Nation Clean Air Programme
Source: Public Information Bureau (PIB India)

Graded Response Action Plan (GRAP)

It is an initiative of Ministry of Environment, Forest and Climate Change, Government of India. It is an emergency measure adopted by Government to fight air pollution when the air quality shifts from poor to very poor. Based on the air quality the grades have been classified as Emergency, Severe, Very Poor and Moderate poor.[13] It will be enforced by Environment Pollution Control Authority (EPCA).

Recently this measure has been implemented to tackle worsening Air Quality in Delhi NCR region.

The major steps announced are:

Severe+ or Emergency—(PM 2.5 over 300 µg/cubic metre or PM10 over 500 µg/cu. m. for 48+ hours):

1. Stop entry of trucks into Delhi (except essential commodities).
2. Stop construction work.
3. Introduce odd/even scheme for private vehicles and minimise exemptions.
4. Task Force to decide any additional steps including shutting of schools.

Severe—(PM 2.5 over 250 µg/cu. m. or PM10 over 430 µg/cu. m.):

1. Close brick kilns, hot mix plants, stone crushers.
2. Maximise power generation from natural gas to reduce generation from coal.
3. Encourage public transport, with differential rates.
4. More frequent mechanised cleaning of road and sprinkling of water.

Very Poor—(PM 2.5 121–250 µg/cu. m. or PM10 351–430 µg/cu. m.):

1. Stop use of diesel generator sets.
2. Enhance parking fee by 3–4 times.
3. Increase bus and Metro services.
4. Apartment owners to discourage burning fires in winter by providing electric heaters during winter.
5. Advisories to people with respiratory and cardiac conditions to restrict outdoor movement.

Moderate to poor—(PM 2.5 61–120 µg/cu. m. or PM 10 101–350 µg/cu. m.):

1. Heavy fines for garbage burning.
2. Close/enforce pollution control regulations in brick kilns and industries.
3. Mechanised sweeping on roads with heavy traffic and water sprinkling.
4. Strictly enforce ban on firecrackers.

Bharat Stage VI (BS VI) Emission Norms

Government of India launched Bharat Stage standards to regulate emission of air pollutants from motor vehicles. The norms were introduced in 2000. With appropriate fuel and technology, they limit the release of air pollutants such as Particulate Matters (PM), Nitrogen Oxides, Carbon Monoxide, Hydrocarbons, and Sulphur Oxides from vehicles using internal combustion engines. With the progress of stages, the control norms become stricter. [Figure 10] Thus, under Bharat Stage VI Emission norms[14] Indian Government has decided that only BS VI compliant vehicles to be sold from April 2020. Currently the vehicles sold are BS stage IV compliant.

Difference between BS-IV & BS-VI:

- The major difference is presence of sulphur.
- Reduce the amount of sulphur by 80%, from 50 ppm to 10 ppm.
- The emission of NOx (Nitrogen Oxides) from diesel cars is also expected to reduce by nearly 70% and 25% from cars with petrol engines.

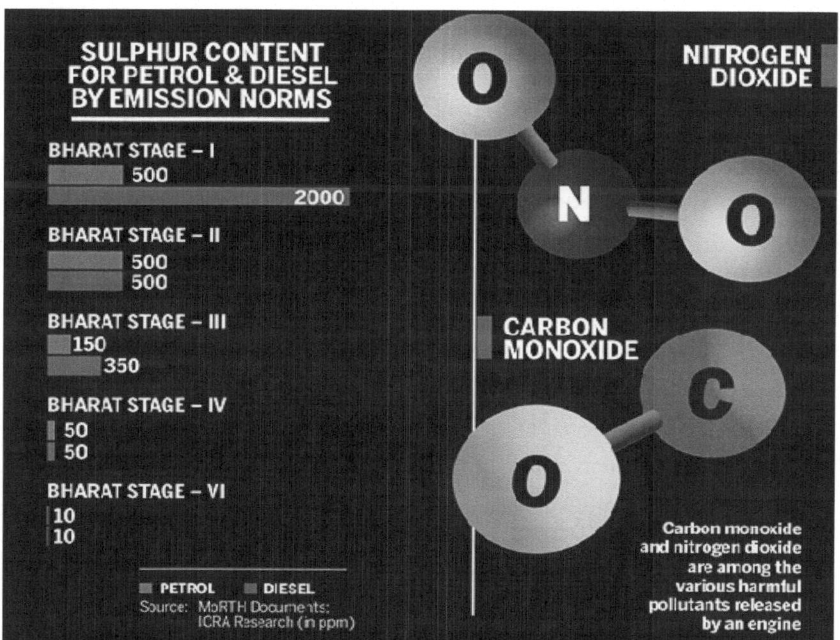

Figure 10: Comparison between Bharat Stages Emission Norms

Source: Ministry of Road Transport and Highways (MoRTH).

Impact and Cost Analysis

The shifting away from BS IV to BS VI technology will prove costly not only to Automobile manufacturers but also to general public who will purchase BS VI compliant vehicle. BS VI compliant fuel will also become costly. As the government will issue more guidelines in this direction in the near future because as per the orders, April 2020 is the deadline after which only BS VI compliant vehicles will be sold in 13 metropolitan cities of India. Also, as these vehicles will help in controlling the air pollution level there is a need for making this technology cost effective so that its broader objective to curb air pollution can be served.

CONCLUSION

It is very crucial that the air quality should be regularly monitored and reported. While there are number of initiatives throughout the world but more active measure should be taken by the Government as well as by the individual to tackle rising Air Pollution as it is taking a toll on our health. Some measures that can be adopted are building a greater number of Air Quality Monitoring Stations, Installing Smog Towers, Complete ban on Stubble Burning of any kind, Ban on Plastics, Check on polluting Vehicles etc. There are many dimensions that effect Air Pollution level i.e. Population Density including average age of population, Economic Status of Country, Human Development Status of its population, Previous Health Burden. To tackle the problem of polluted air measures should be taken by keeping in view of various dimensions which is not only country but also region specific.

REFERENCES

[1] United Nation Environment Programme, (2019), Emissions Gap Report 2019, Nairobi Kenya. Date of access: 28.11.2019. https://www.unenvironment.org/resources/emissions-gap-report-2019

[2] Carbon Disclosure Project India (2019). New Delhi, India. Date of Access: 28.01.2020 https://www.cdp.net/en/india

[3] United Nation, Sustainable Development Goals (2015). Date of access: 03.12.2019 https://www.un.org/sustainabledevelopment/sustainable-development-goals

[4] Royal Society of Chemistry (2019). Tackling Global Challenges, Cambridge, UK. Date of access: 12.12.2020. https://www.rsc.org/campaigning-outreach/global-challenges/environment

[5] WHO (2015). Database: Outdoor Air Pollution in Cities; 2007–2015, Geneva, Switzerland. Date of access: 30.01.2020.
http://www.who.int/phe/health_topics/outdoorair/databases/cities-2011/en

[6] United Nation Framework Convention on Climate Change, (2020), Kyoto Protocol, Bonn, Germany. Date of access: 03.01.2020.
https://unfccc.int/kyoto_protocol

[7] European Commission (2019). European Union Emission Trading Scheme (EU-ETS). Date of access: 15.12.2019.
https://ec.europa.eu/clima/policies/ets_en

[8] United Nation, Reducing Emissions from Deforestation and Forest Degradation, Food and Agriculture Organization Database. Date of access: 23.12.2019. http://www.fao.org/redd/initiatives/un-redd/en/; www.fao.org

[9] United Nation Framework Convention on Climate Change, United Nation Climate Action Summit, 2019. Date of access: 01.01.2020.
https://climateaction.unfccc.int/views/events-summit.html

[10] Ministry of Petroleum and Natural Gas (2019). Prime Minister Ujjwala Yojana, New Delhi, India. Date of access: 30.01.2020.
https://pmuy.gov.in/about.html

[11] Ministry of Environment, Forest & Climate Change (Govt. of India), (2019). National Climate Action Plan, 2019 (Delhi NCR), Delhi, Date of access: 01.12.2019.
http://moef.gov.in/wp-content/uploads/2019/05/NCAP_Report.pdf

[12] Public Information Bureau (PIB), (2019). Ministry of Environment Forest and Climate Change, National Climate Action Plan, New Delhi, India. Date of access: 25.01.2020.
https://pib.gov..in/newsite/PrinRelease.aspx?relid=187400

[13] Ministry of Environment, Forest & Climate Change, (Govt. of India), (2019). Central Pollution Control Board, Graded Response Action Plan Report (Delhi NCR), Delhi, India. Date of access: 03.01.2020.
http://moef.gov.in/environment/pollution

[14] International Council on Clean Transportation (2016). Bharat Stage VI Emission Policy, New Delhi, India. Date of access: 29.11.2019.
https://theicct.org/sites/default/files/publications/India%20BS%20VI%20 Policy%20Update%20vF.pdf

Air Pollution Abatement Technologies: A Brief Introduction on Filters, Membranes and Coatings

C.P. Leo

School of Chemical Engineering, Engineering Campus,
Universiti Sains Malaysia, 14300 Nibong Tebal, Penang, Malaysia
E-mail: chcpleo@usm.my

ABSTRACT: According to the New Malaysia Ambient Air Quality Guideline, a total of six air pollutants (PM_{10}, $PM_{2.5}$, SO_2, NO_2, SO_2, O_3 and CO must be controlled below the standard in 2020. Particulate matters can be removed using sedimentation, filtration, centrifugation, washing and electrostatic precipitation. Settling chambers are only effective to remove particulate matters larger than 10 microns. Fabric filter such as high efficiency particulate air (HEPA) filter can collect particles as small as 2.5 micron. Cyclone facilitates centrifugal and inertial forces to separate submicron particles as the gas spiral through the cyclone. Water droplets can be sprayed to collect particulate matters before a cyclone separator. In electrostatic precipitator, the gas flows between the vertical wires and a stack of metal plates. An electric corona discharge ionizes the air and the particulate matters are collected at plates which are positive electrode. Other air pollutants such as nitrogen oxides (NOx), sulphur oxides (SOx), and CO can be reduced by adsorption and absorption. Adsorption involves the capture of air pollutants on the surface of porous materials (adsorbent). Unlike adsorption, absorption involves the use of solvent such as water, mineral oil and other aqueous solution to remove air pollutants such as CO_2, HCl, H_2SO_4, H_2S, NH_3, Cl_2, formaldehyde, ethylene and benzene. Although gas separation membranes can capture carbon from different gas sources, the polymeric membranes are suceptible to the plasticization caused by NOx and SOx. Pervaporation membrane can remove volatile organic compouns (VOCs) from wastewater. Polydimethylsiloxane (PDMS) membranes are used to remove VOC from air but they are not commonly reported. Since ground level ozone is not emitted but is created from chemical reaction *between NOx and VOCs in the presence of sunlight and heat. It can be only controlled by reducing NOx and VOC. Recently, paint containing photocatalyst is a new option to degrade NOx and VOC.*

Keywords: Air Pollution, Filter, Membrane, Coating.

AIR POLLUTANTS AND AIR QUALITY GUIDELINE IN MALAYSIA

Air Pollution Index (API) is the indicator of ambient air quality in Malaysia, but it does not signify the actual concentration of air pollutants. API indicates the effects of air quality on health as shown in Table 1, ranging from good to emergency as specified in the National Haze Action Plan.[1] API is determined from the maximum sub-index of six pollutants, namely carbon monoxide (CO), nitrogen dioxide (NO_2), sulphur dioxide (SO_2), ozone, particulate matter with particle size between 2.5 and 10 µm (PM_{10}) as well as particulate matter with particle size smaller than 2.5 µm ($PM_{2.5}$) (Figure 1). The average

Table 1: Air Quality Indicated by API

API	*Air Quality*
0–50	Good
51–100	Moderate
101–200	Unhealthy
201–300	Very Unhealthy
>301	Hazardous
>500	Emergency

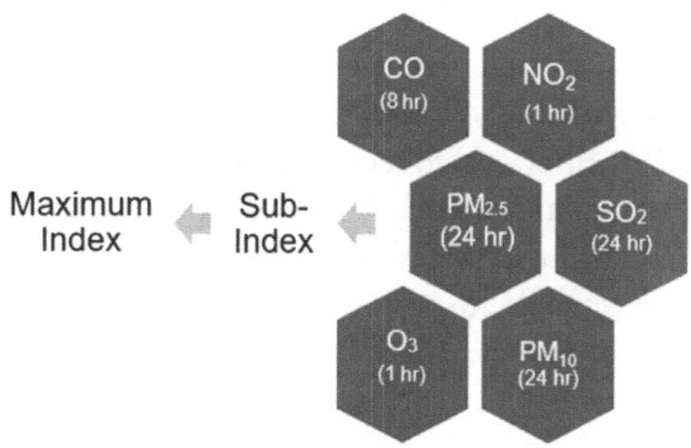

Figure 1: The Calculation of API

concentration of these pollutants within certain duration is measured according to World Health Organization (WHO) standard. Pollutant Standard Index (PSI) developed by the United States Environmental Protection Agency (US-EPA) is the major reference for the Malaysian API system.

A total of six air pollutants must be controlled below the standard in 2020 according to the New Malaysia Ambient Air Quality Guidelines.[2] The standard for PM_{10} is 40 μg/m^3 per year or 100 μg/m^3 per day, while the standard of $PM_{2.5}$ is 15 μg/m^3 per year or 35 μg/m^3 per day. The concentration of SO_2 and NO_2 is limited below 250 μg/m^3 per year (80 μg/m^3 per day) and 280 μg/m^3 per year (80 μg/m^3 per day), respectively. The standard of ground level O_3 is set at 180 μg/m^3 per hr (100 μg/m^3 per 8 hr), but CO concentration is set at 30 mg/m^3 per hr (10 mg/m^3). The motorised vehicles have contributed about 70.4% of these air pollutants in Malaysia, followed by power plants (24.5%), and industrial activities (2.9%) in 2018.[3] Transboundary haze episodes in Malaysia occurred more frequently in the recent years as well. Particulate matter (PM) remains to be the dominant pollutant during haze.[4]

ADVANCED FILTER FOR PARTICULATE MATTERS

PM in the air is a mixture of organic matters (carbon and organic carbon), inorganic matters (NO_3, SO_4 and SiO_2), liquid droplet and small particles. Vehicles and combustion in power plants are the major sources of PM. Compared to other air pollutants; $PM_{2.5}$ has attracted more attention due to its frequent occurrence. Furthermore, $PM_{2.5}$ can permeate through lungs and bronchi.[5] PM is conventionally removed using sedimentation, filtration, centrifugation, washing and electrostatic precipitation in industrial practice.[6] Settling chambers are only effective to remove PM_{10}. Fabric filter such as high efficiency particulate air (HEPA) filter can collect particles as small as $PM_{2.5}$. Cyclone facilitates centrifugal and inertial forces to separate submicron particles as the gas spiral through the cyclone as shown in Figure 2(a). Water droplets can be sprayed to collect particulate matters (Figure 2(b) before a cyclone separator. In electrostatic precipitator, the gas flows between the vertical wires and a stack of metal plates. An electric corona discharge ionizes the air and the particulate matters are collected at plates which are positive electrode as shown in Figure 2(c).

To reduce PM concentration in the indoor air, air filters made from fibrous filters are commonly installed in commercial and residential buildings. The fibrous filters dominate the air filter market more than 70%.[7] The common air filters include fiberglass filter, polyester and pleated filters as well as high-efficiency particulate air (HEPA) filters. HEPA filters are specifically used to capture $PM_{2.5}$. They can filter 99.97% of particles larger than 0.3 μm according to ASME standard. The filtration performance of fibrous filters depends on the capture mechanism such as interception, inertial collection, Brownian diffusion, electrostatic interaction and gravity.[8] PM can be trapped in the filter matrix via interception.

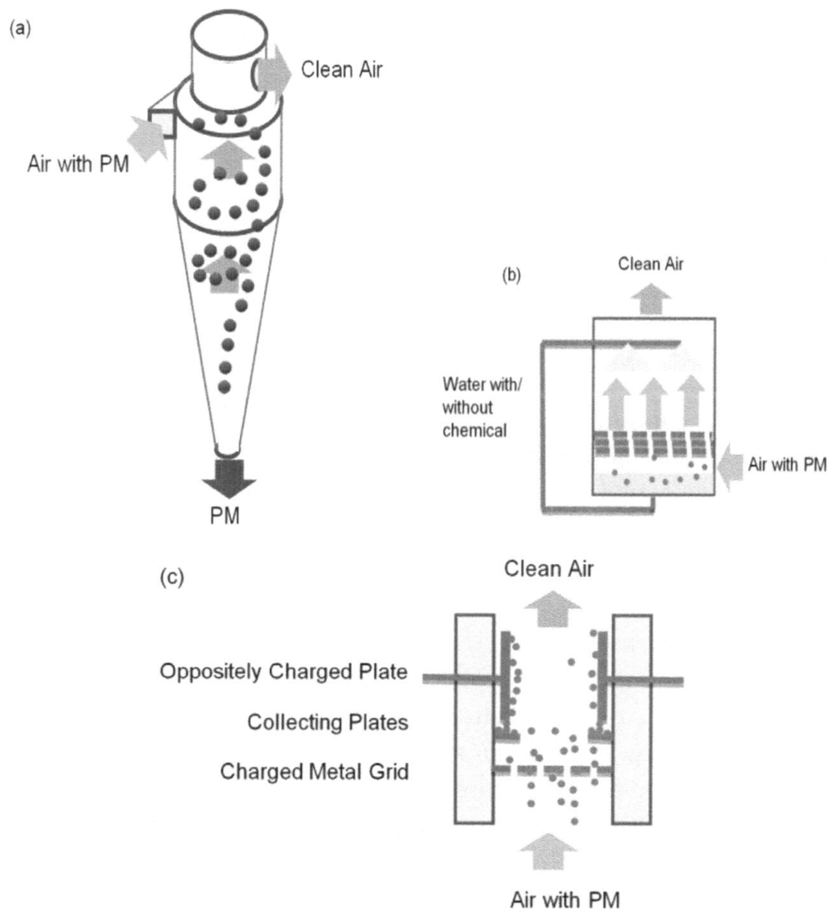

Figure 2: PM Collection Unit Operations: (a) Cyclone, (b) Wet Scrubber Chamber and (c) Electrostatic Precipitator

The large PM is collected by inertial effects at high velocity while the small PM is collected on filter surface by Brownian diffusion. Furthermore, the electrostatic interaction between PM and the filter helps to collect more PM as shown in Figure 3. Besides the removal efficiency, air permeability of air filter is another important indicator of filter performance. This parameter is usually associated with pressure drop. The quality factor is used when both removal efficiency and pressure drop are considered in the filter selection.[9] Modern filters are equipped with more features by incorporating nanoparticles. Working like the lotus leaf, super-hydrophobic filter with nano-roughness exhibits self-cleaning and antibacterial properties.[10] Antibacterial agent such as; TiO_2 nanoparticles are incorporated to reduce the bacterial growth on air filters.[11]

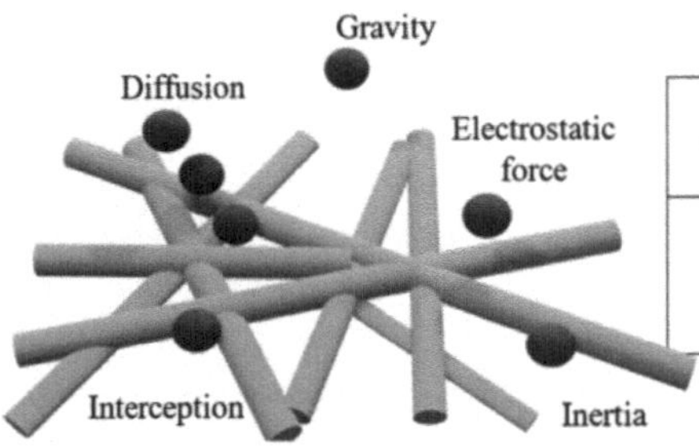

Particle diameter (μm)	Filtration mechanism
<0.3	Diffusion and electrostatic
0.3-1.0	Interception
1-10	Inertial impaction and sieving
>10	Gravity, clogging and sieving

Figure 3: Filtration Mechanisms According to Particle Size

Respiratory equipment such as face mask and cartridge respirator provide personal protection from air pollutant when outdoor activities must be conducted under unhealthy air quality. Similar to the air filters, respiratory equipment contains non-woven media such as glass fibers, split film fibres, melt blown fibres and expended PTFE (polytetrafluoroethylene) membranes.[9] The right respiratory equipment not only remove PM from inhale air, but also can provide personal protection from harmful gases, toxic vapours, aerosol hazards, and transmissible diseases including influenza, diphtheria, SARS (sever acute respiratory syndrome) and swine flu.[12] Adsorbent such as activated carbon in the cartridge can even capture volatile organic compounds (VOCs) but it requires replacement once the breakthrough is achieved. National Institute of Occupational Safety and Health (NIOSH) has classified the respirator equipment according to oil resistance with labels of N, R and P as listed in Table 2.[13] It is important to note that surgical face masks can only prevent the spread of mouth generated particles from the wearer or the blood splatter on the wearer's mouth and nose. The surgical mask does not protect the wearer from air pollution, unless surgical respirator is chosen.

Table 2: NIOSH Approved Respiratory Equipment

Respiratory	*Function*	*Oil Resistance*
N95	Remove 95% of airborne particles.	No
Surgical N95	Prevent the spread of mouth generated particles from the wearer or the blood splatter on the wearer's mouth and nose.	No
N99	Remove 99% of airborne particles.	No
N100	Remove more than 99.97% of airborne particles.	No
R95	Remove more than 95% of airborne particles.	Moderate
P95	Remove more than 95% of airborne particles.	Strong
P99	Remove more than 99% of airborne particles.	Strong
P100	Remove more than 99.97% of airborne particles.	Strong

3. MEMBRANE TECHNOLOGY FOR GASEOUS POLLUTANT

Other air pollutants such as NOx, SOx, and CO can be reduced by adsorption and absorption.[14] Adsorption involves the capture of air pollutants on the surface of porous materials (adsorbent) as shown in Figure 4(a). The polluted gas is fed into a bed of adsorbent to facilitate the formation of physical or chemical bonding between the pollutants and the surface of adsorbent. The common adsorbents are activated carbon, silica gel, zeolite and clay. These materials offer more than few hundred m^2 of area per g of adsorbent weight for the formation of physical and chemical bonding. Adsorption is effective at high concentration and low temperature. After adsorption, the air pollutants on the adsorbent can be desorbed by increasing the temperature and reducing the pressure. Unlike adsorption, absorption involves the use of solvent such as water, mineral oil and other aqueous solution to remove air pollutants such as CO_2, HCl, H_2SO_4, H_2S, NH_3, Cl_2, formaldehyde, ethylene and benzene. The difference between adsorption and absorption is shown in Figure 4. A packed absorption tower containing packing materials with large surface area is commonly used to improve the gas-liquid mass transfer. Since ground level ozone is not emitted but is created from chemical reaction between NOx and VOCs in the presence of sunlight and heat. It can be only controlled by reducing NOx and VOC.

Compared to adsorption and absorption, membrane technology is more attractive since it offers high energy efficiency, small footprint, easy operation and low-cost operation.[15] Different from filter, membrane is a thin layer material which works as a selective barrier that allow a specific component in air to permeate based on solubility and diffusivity. Figure 4(c) shows the illustration of typical membrane separation. The component that can pass through a membrane is known as permeate while the component which remained in another side is known as retentate. In gas separation membrane; concentration and pressure gradient across the membrane are the driving forces that allow the gas components to be separated.[16] In general, the performance of gas separation membrane is evaluated in terms of permeation and selectivity. The permeability can be interpreted as the ability of a gas component to permeate across a membrane and it also indicates how fast the transport of a gas component in that membrane. Meanwhile, the selectivity is defined as the permeability ratio between two components. The gas separation membranes are commonly used to purify natural gas and biogas as they are capable to separate methane from carbon dioxide (CO_2).[17] Some membranes can even capture CO_2 from flue gas, yet they still require improvement under pilot studies up to date.[18] NOx and SOx are the major plasticizers of polymeric membranes. Hence, they must be removed prior to membrane separation. The volatile organic compound (VOC) can be separated from industrial wastewater using pervaporation membrane.[19] In pervaporation, separation is achieved by the partial vaporization through membrane. Polydimethyl siloxane (PDMS) membranes are often used to remove volatile organic compound (VOC) from industrial gas. They can separate halogen-derived hydrocarbon and aromatic compound in the pressure-driven separation process.[20] Due to the high-pressure requirement in membrane gas separation, membrane is also integrated with liquid absorbent to achieve more practical removal of acid gases.[21] The absorption of acid gases including CO_2 and hydrogen sulphide (H_2S) is promoted by the microporous membrane in membrane gas absorption. The microporous membrane with large surface area enhances gas-liquid interfacial contact for absorption.

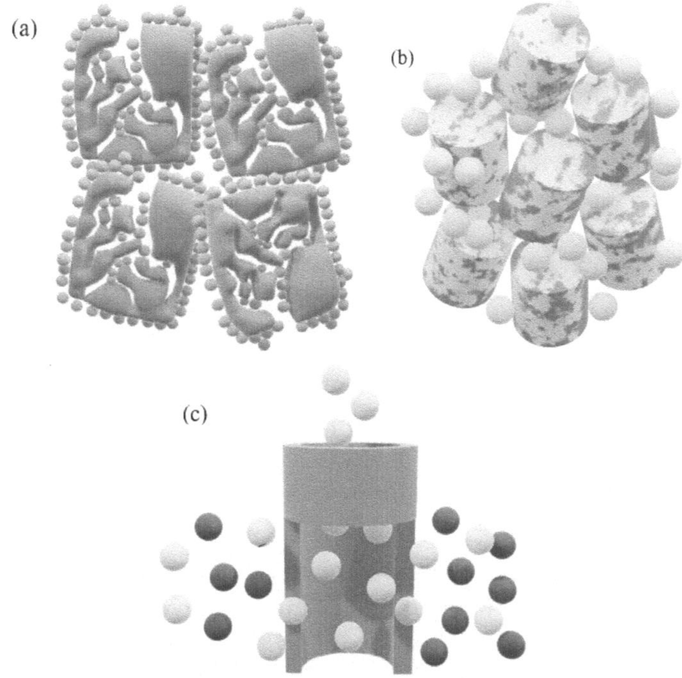

Fig. 4: Air Pollutant Removal via (a) Adsorption on Porous Adsorbent (b) Absorption on Liquid Absorbent Wetted on Packing and (c) Permeation Through Selective Membrane

PHOTOCATALYTIC COATING FOR INDOOR AIR QUALITY

In the recent years, the oxidation of air pollutant by photocatalyst in paint and coating has been extensively studied.[22] Photocatalyst especially TiO_2 has been long added into paint and coating because it is a well-known white pigment. TiO_2 also absorbs ultraviolet (UV) light from the sun to generate radicals that can convert NOx, VOC and other organic pollutants in the air as illustrated in Figure 5. Crystalline TiO_2 in anatase phase with a band gap of 3.2 eV absorbs less UV than rutile TiO_2 with a band gap of 3.0 eV, respectively. However, the radicals also shorten the life span of organic paint. Hence, pigment manufacturer applied silica layer on TiO_2 to reduce the radical formation.

$$NO, NO_2 \text{ (NOx)} \rightarrow NO_3 \text{ (Nitrate)}$$
$$VOC, Organics \rightarrow CO_2 + H_2O$$

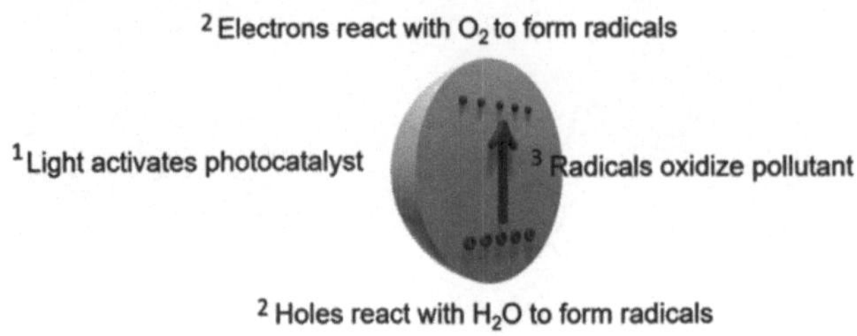

Figure 5: Photocatalytic Oxidation of Air Pollutants

CONCLUSIONS

Fibrous filters remove PM10 and PM2.5 in the air effectively, but not other air pollutants such as NOx, SOx, CO and ozone. Air filter and respiratory equipment must be equipped with adsorbent to remove other air pollutants. In common industrial practice, NOx and SOx are removed from waste gas using adsorbent and adsorbent in packed columns. Although gas separation membranes can capture carbon from different gas sources, the membranes made from low-cost polymer are suceptible to the plasticization caused by NOx and SOx. On the other hand, pervaporation membrane can remove VOC from wastewater. PDMS membranes are used to remove VOC from air but they are not commonly reported. Photocatalysts can be further incorporated in paint and coating to generate radicals to oxidize NOx and VOC under UV irradiation. By reducing NOx and VOC in the air, the occurance of ozone can be reduced as it is generated from the chemical reaction between NOx and VOCs in the presence of sunlight and heat.

ACKNOWLEDGMENT

The author would like to acknowledge Centre for Science & Technology of the Non-Aligned and Other Developing Countries (NAM S&T Center) as well as Sri Venkateswara College, University of Delhi for providing the financial support to participate International Workshop on Air Pollution and Public Health: Challenges and Interventions at New Delhi, 5th–7th February 2020.

REFERENCES

[1] Department of Environment, Air Pollutant Index of Malaysia, 2020. http://apims.doe.gov.my/public_v2/aboutapi.html (accessed 20 Jan. 2020).

[2] Department of Environment, New Malaysia Ambient Air Quality Standard, 2020. http://www.doe.gov.my/portalv1/wp-content/uploads/2013/01/Air-Quality-Standard-BI.pdf. (accessed 20 Jan. 2020).

[3] Department of Statistics. Malaysia, Compedium of Environment Statistics, Malaysia 2018. https://www.dosm.gov.my/v1/index.php?r=column/cthemeByCat&cat=162&bul_id=U3p3RVY0aGtGS08yTDY2cEpraDFlUT09&menu_id=NWVEZGhEVlNMeitaMHNzK2htRU05dz09. (accessed 20 Jan. 2020).

[4] Department of Environment, Chronology of Haze Episodes in Malaysia, 2020. https://www.doe.gov.my/portalv1/en/info-umum/info-kualiti-udara/kronologi-episod-jerebu-di-malaysia/319123. (accessed 20 Jan. 2020).

[5] Anderson, J.O., Thundiyil, J.G. and Stolbach, A., (2012). Clearing the air: A review of the effects of particulate matter air pollution on human health, *Journal of Medical Toxicology,* 8(2), pp. 166–175.

[6] Harrison, R.M., Chester, R. and Slater, D.H. (2001). Pollution: Causes, Effects and Control, Royal Society of Chemistry, Piccadilly, London.

[7] Hutten, I.M. (2016). Chapter 8 - Air Filter Applications, in: Hutten I.M. (Ed.), Handbook of Nonwoven Filter Media (Second Edition), Butterworth-Heinemann, Oxford, UK, pp. 451–519.

[8] Liu, H., Cao, C., Huang, J., Chen, Z., Chen, G. and Lai, Y. (2020). Progress on particulate matter filtration technology: basic concepts, advanced materials, and performances, *Nanoscale,* 12(2), pp. 437–453.

[9] Zhang, S., Rind, N.A., Tang, N. H., Liu, Yin, X., Yu, J. and Ding, B. (2019). Chapter 12 - Electrospun Nanofibers for Air Filtration, in: B. Ding, X. Wang, J. Yu (Eds.), Electrospinning: Nanofabrication and Applications, William Andrew Publishing, Norwich, NY., pp. 365–389.

[10] Guo, Z. and Yang, F. (2017). Surfaces and Interfaces of Biomimetic Superhydrophobic Materials, Wiley, 2017.

[11] Wang, P., Chang, J. and Zhang, L. (2019). Artificially Intelligent Nanomaterials for Environmental Engineering, Wiley, Hoboken, US.

[12] Wei, J. and Li, Y. (2016). Airborne spread of infectious agents in the indoor environment, *American Journal of Infection Control 44 (9, Supplement),* pp. S102–S108.

[13] Centers for Disease Control and Prevention (2020). NIOSH-Approved Particulate Filtering Facepiece Respirators, 2020. https://www.cdc.gov/niosh/npptl/topics/respirators/disp_part/default.html (accessed 20 Jan. 2020).

[14] Liu, D.H.F. and Liptak, B.G. (2018). Air Pollution, CRC Press, Boca Raton, US.

[15] Chuah, C.Y., Goh, K., Yang, Y., Gong, H., Li, W., Karahan, H.E., Guiver, M.D., Wang, R. and Bae T.-H. (2018). Harnessing filler materials for enhancing biogas separation membranes, *Chemical Reviews,* 118(18), pp. 8655–8769.

[16] Ismail, A., Goh, P., Sanip, S. and Aziz, M. (2009). Transport and separation properties of carbon nanotube-mixed matrix membrane, *Separation and Purification Technology,* 70(1), pp. 12–26.

[17] Xu, J., Wu, H., Wang, Z., Qiao, Z., Zhao, S., Wang, J. (2018). Recent advances on the membrane processes for CO_2 separation, *Chinese Journal of Chemical Engineering,* 26(11), pp. 2280–2291.

[18] Jawad, Z.A., (2019). Membrane Technology for CO_2 Sequestration, CRC Press, 2019.

[19] Verstraete, S., Hermia, J. and Vigneron, S. (1994). VOC Separation on Membranes: A Review, in: Vigneron, S., Hermia, J., Chaouki, J. (Eds.), Studies in Environmental Science, Elsevier, Amsterdam, The Netherlands, pp. 359–373.

[20] Petrusová, Z., Machanová, K., Stanovský, P. and Izák, P. (2019). Separation of organic compounds from gaseous mixtures by vapor permeation, *Separation and Purification Technology,* 217, pp. 95–107.

[21] Yeo, Z.Y., Chew, T.L., Zhu, P.W., Mohamed, A.R. and Chai, S.-P. (2012). Conventional processes and membrane technology for carbon dioxide removal from natural gas: A review, *Journal of Natural Gas Chemistry,* 21(3), pp. 282–298.

[22] Yu, Q.L., Hendrix, Y., Lorencik, S. and Brouwers, H.J.H. (2018). Field study of NOx degradation by a mineral-based air purifying paint, *Building and Environment,* 142, pp. 70–82.

SECTION II

Case Studies from Developing World

The Challenges and Intervention of Air Pollution to Health Issue in Cambodia

Dina Heng
Ly Vireak Clinic and Pharmacy MEDICA, Phnom Penh
E-mail: madinaheng@gmail.com

ABSTRACT: Air Pollution is an alarming issue in the globe. World Health Organization (WHO) has shown its concern regarding ill effects of air pollution on public health. Cambodia is not alone, one of the developing countries, which is putting more effort to catch up with the world economic development. An increase in number of factories, rapid infrastructure development and construction activities, and hectic transportation, increase the air pollution. As a result, it is inevitable that adverse health outcomes will affect Cambodian people who are exposed to the increasing high level of polluted air. As evidence WHO, report shows that many diseases including respiratory diseases, heart diseases, fatigue, headaches and anxiety are linked to increased air pollution levels. To support this evidence, according to the Centers for Disease Control and Prevention (CDC), lower respiratory infection is the second top two causing of death in Cambodia; even though the real cause of this disease has not been clearly highlighted. Cambodia has a population of 16, 598,466, with per capita income of US$ 3,760. Life expectancy at birth is Female 71 and Male 67 years, and infant mortality rate is 24/1,000 live births. Many of the top 10 causes of mortality in Cambodia, such as lower respiratory tract infections, Chronic Obstructive Pulmonary Disease (COPD), ischemic heart and neonatal disorders, may be attributed to air pollution. Our community is in the outskirts of Phnom Penh city surrounded by many factories, residential construction sites, new infrastructure building and city rubbish collection site. According to the topic of this program will be enlightened me because many local residents come to pharmacy first before seeing doctors. Therefore, my knowledge sharing with the pharmacist in my pharmacy who is able to provide opportunity to identify and give some advice to patients with relevant health conditions caused by the current air pollution issue. A number of patients who visited my health clinic, have been diagnosed with respiratory infection disease; however, those number has not been used as the case study in order to inform the community about health prevention. All the experience and innovation techniques gained from the workshop will be used as the model of a research study. A case study in my clinic will be planned to be conducted by adopting the possible existing models from the workshop in order to sum up as the optimal findings. Furthermore, *the findings will be shared with the community, schools, garment workers and especially the rubbish collection workers by doing the presentation in one of the public areas. It is worthwhile to contribute the knowledge, finding and suggested prevention methods to all the community stakeholders, as well as to relevant ministries such as the Ministry of Environment, the Ministry of Industry and Handicraft and the Ministry of Health.*

Keywords: Air Pollution, Ambient Air, Health Effect, Intervention.

INTRODUCTION

Air pollution has become an alarming issue in the globe mainly in low-and middle-income countries (World Health Organization, 2016). The ambient air pollution has been attributed to the Global Burden of Disease (GBD) (Brauer *et al.,* 2016). This ambient air pollution alone could cause about 3 million deaths per annum world-wide.

Cambodia is not alone, one of the developing countries, which is putting more effort to catch up with the world economic development. An increase in number of factories, rapid infrastructure development and construction activities, and hectic transportation, increase the air pollution in particular ambient air pollution. As a result, it is evitable that adverse health outcomes will affect Cambodian people who are exposed to the increasing high level of polluted air, particularly in urban areas and in the long-term exposure. As evidence WHO reports shows that many diseases including respiratory diseases, heart diseases, fatigue, headaches and anxiety are linked to the increased air pollution levels.

This paper aims to provide the overview of Cambodian context regarding the linkage of air pollution to public health, the challenges and effects of ambient air pollution, the existing and proposed possible intervention, which help contribute in the public health prevention.

THE CHALLENGES OF AMBIENT AIR POLLUTION

The driving force of air pollution is attributed to urban population explosion and the lack of public transportation in Phnom Penh (Furuuchi *et al.*, 2006). The influx of people from rural areas to Phnom Penh drives more traffic to the city. However, due to the limited public transportation, people have no other option but to drive vehicles by themselves. The dependence on private transport resulted in a double increase in automobiles from 400,000 (1992) to 800,000 (2002) in the city (Ministry of Environment & UNEP, 2009). Therefore, most main roads in the city are highly congested especially during the rush hours including morning, mid-day and late afternoon. Due to the heavy traffic, it is common to take one to two hours just to pass by main intersections or get around markets. When the traffic is heavy, the Particulate Matter (PM) is more released into the air than other time of the day (Zhang & Batterman, 2013). As a result, people who commute daily to work or are regularly outdoors during this time, are more exposed to the Particulate Matter (PM).

THE EFFECTS OF AMBIENT AIR POLLUTION

The long-term exposure to PM can cause acute lung cancer particularly among the elderly (Samet, 2013). According to a non-profit NGO US-based Health Effects Institute funded by the US government and global automotive industry, PM 2.5 pollution in Cambodia is responsible for 72.6 deaths per 100,000 people, based on 2016 data. Globally, the average is 62.5 deaths per 100,000 people (World Health Organization, 2016). The outskirts of Phnom Penh city surrounded by many factories, residential construction sites, new infrastructure building and city rubbish collection site. It is inescapable that people expose to those areas will affect their health through the polluted air. However, there is a lack of study on PM effects on health conditions among Cambodian population especially the rubbish collection workers. Therefore, it is necessary to study the ambient concentration of PM (PM2.5 and PM10) in order to conduct a method to find the disease agent (PM) exists in the transit environment and its health complications. Furthermore, possible solutions will be discussed in the end.

PM is released in the atmosphere when the engine is operated. PM are the particles in the ambient that can cause respiratory diseases, especially lung cancer (World Health Organization, 2016). PM is a complex mixture of organic or inorganic in the forms of solid or liquid particles, which floats in the air and contributes to air pollution (World Health Organization, 2016). These chemical substances can react to each other in the air even before inhalation (Samet, 2013). Even if the amount is small, PM (PM2.5 and PM10) are likely to affect health conditions (World Health Organization, 2006, 2016). Therefore, WHO developed a guideline on PM and stressed the importance of monitoring the ambient concentration of PMs.

When people are commuting in the heavy traffic, commuters directly inhale the contaminated air which contains PM in the form of PM2.5 and PM10 through mouth or nose into the lung (Turpin, 2013). The concentration of PM can be accumulated differently depending on the duration of exposure, exposed period and frequency of their breathing. In general, the majority of people use motorbike more than vehicles without any mask protection. When the traffic is normal, people can travel between 15 to 30 minutes depending on their destinations. However, if they get caught in a traffic jam, people will spend around one to even two hours on the road. Therefore, it is safe to assume that accumulation of PM in the lung can be seen among adults rather than children.

PM contains a mixture of chemical substances that can be carcinogenic to the human (Samet, 2013). Through the inhalation, PM is deeply stored in lung, which can attack lung tissues. Besides, the amount of PM can be accumulated through daily exposure. Therefore, the contaminated lung can develop into a cancer in the long run especially when during later stage of life (World Health Organization, 2016). WHO estimated Cambodian people who die from lung cancer is 62 per 100,000 capita and live with disability is 2369 years of life lost per 100,000 capita (World Health Organization, 2016).

EXISTING INTERVENTION

The ambient air pollution has become a substantial concern in urban areas, particularly in many parts of Phnom Penh, that need to intervene, to prevent acute health risk in the future. However, there is lack of documentation through proper study that could inform the policy makers, health care professionals and community people to understand risk and possible prevention as well as protection among adults and children that mainly commute and are exposed to the contaminated air in the ambient environment.

From the possible and accessible sources, it is highlighted that there is a lack of intervention to tackle ambient air pollution at all levels: upstream, midstream, and downstream. Up to date, particular toxic substance as such PM that highly concentrate in the environment has been studied and founded its harmful effect to people even in the small amount (Bernosky & Vermetter, 2012). In addition, there are information that people could access to, is through some mass media in which it not consistent and is unclear. That information could be found through: website of the Ministry of Environment; Phnom Penh Post; Facebook to name a few. The messages from those social media plus websites were to alert lay people about the condition of toxic air that people could expose to in some areas in Phnom Penh. Connecting to the alert, the primary intervention was to inform people to wear protective mask.

PROPOSED INTERVENTION

Due to limitation of intervention, the government of Cambodia, especially all the relevant ministries: Ministry of Environment, Ministry of Health, as well as city governors, should pay more attention to air pollution in urban areas to better intervene before those challenges will become worse. In this context, three levels of intervention should be applied: primary, secondary and tertiary prevention. The purpose of primary prevention is to avoid exposure to pollutants. This primary prevention should focus on raising awareness among people in the community, informing people to protect themselves from exposing to toxic air. Secondary prevention aims to minimize risk such as reducing the cause of pollutants and so on. The tertiary prevention anticipates solving problems by encouraging people for early diagnose and treatment through providing free health-up at public facilities, as well as subsiding relevant cancer treatment. All of these intervention patterns will be more effective than applying only a particular method of intervention unless it needs involvement from all sectors including public sector, private sector, as well as community itself.

CONCLUSION

Even though, many research studies stated the link between air pollution and ill effects. Cambodia has a limitation of research output. This paper is a preliminary point to open up more studies in this issue of the connection between air pollution and public health in Cambodia context; especially for my private clinic sector. From the forthcoming small-scale study in my clinic, it is informative to educate and raise awareness to the community regarding this global issue.

REFERENCES

[1] Bernosky, J.D. and Vermetter, S.J. (2012). Temporal characteristic of airborne particulate matter in Phnom Penh, Cambodia. *Enviroment Asia,* 5(2), 11–15.

[2] Brauer, M., Freedman, G., Frostad, J., Van Donkelaar, A., Martin, R.V., Dentener, F. and Cohen, A. (2016). Ambient Air Pollution Exposure Estimate of the Global Burden of Disease 2013. *Environ Sci. Technol,* 50(1), 79–88. Doi: 10.1021/acs.est.5b03709.

[3] Furuuchi, M., Murase, T., Yamashita, M., Oyagi, H., Sakai, K.I., Tsukawaki, S. and Hata, M. (2006). Temperature Distribution and Air Pollution in Phnom Penh, Cambodia—Influence of Land Use and the Mekong and Tonle Sap Rivers. *Aerosol and Air Quality Research,* 6(2), 15.

[4] Ministry of Environment and UNEP (2009). Cambodia Environment Outlook (pp. 106). Cambodia: Ministry of Environment.

[5] Samet, J.M. (2013). Introduction. In S. Kurt, A. Cohen and J. Samet (Eds.), Air Pollution and Cancer IARC SCIENTIFIC PUBLICATION (pp. 7–12). Switzerland: World Health Organization. Retrieved from https://www.iarc.fr/en/publications/books/sp161/AirPollutionandCancer161.pdf (International Agency for Research on Cancer).

[6] Turpin, B. (2013). Characterizing Exposures to Atmospheric Carcinogens. In S. Kurt, A. Cohen and J. Samet (Eds), Air Pollution and Cancer IARC Scientific Publication (pp. 29–36). Switzerland: World Health Organization. Retrieved from https://www.iarc.fr/en/publications/books/sp161/AirPollutionandCancer161.pdf (International Agency for Research on Cancer).

[7] WHO Air quality guidelines for particulate matter, ozone, nitrogen dioxide and sulfur dioxides WHO/SDE/PHE/OEH/06.02, 2006.

[8] World Health Organization (2016). Ambient air pollution: A global assessment of exposure and burden of disease Switzerland: World Health Organization.

[9] Zhang, K. and Batterman, S. (2013). Air pollution and health risks due to vehicle traffic. *Sci. Total Environ,* 450–451, 307–316. Doi: 10.1016/J.scitotenv.2013.01.074.

Carbon Footprint: Sector-Wise Emission Analysis of Urban Youth

Pooja Gokhale Sinha
Department of Botany, Sri Venkateswara College, University of Delhi–1100 21
E-mail: pgokhale@svc.ac.in

ABSTRACT: Global climate change due to increased atmospheric concentration of Green House Gases (GHG), particularly species of carbon is a serious environmental challenge. Carbon dioxide (CO_2) is the most potent GHG accounting for up to 77% of total GHG emissions. Successful mitigation of climate change needs a drastic and sustained reduction in global atmospheric CO_2 concentration. Carbon footprint of a person, product or process gives an estimate of CO_2 emitted by it. Calculation of individual carbon footprint is the first step towards combating climate change as it leads to identify major contributing practices for which corresponding mitigation strategies can be designed. Sector and stream-wise carbon footprint of 721 undergraduate students of Delhi of age group 18 to 21 was calculated based on a survey-based questionnaire. Average carbon footprint was calculated as 5.13 tons/annum and was homogenous across streams. Transportation sector was the highest contributor of CO_2 followed by domestic electricity consumption. Statistical analysis indicated that level of awareness regarding the Bureau of energy efficiency (BEE) ratings and the habit of switching off appliances was highly significant amongst the respondents which was not the case in the practice of recycling waste. For any successful mitigation measure, participation of citizens particularly the youth is very important as they are the primary stake holders as well as torch bearers for the future. The study is essentially youth centric, and gives a glimpse into their emission patterns, environmental awareness and sensitivity.

Keywords: Anthropogenic Emissions, Carbon Footprint, Climate Change, Green House Gases (GHG), Sector-wise Analysis, Transportation.

INTRODUCTION

Ruthless anthropogenic activities coupled with increasing population pressure have led to an unprecedented increase in global atmospheric concentration of Green House Gases (GHG). Rapid urbanization and alteration in land use patterns have resulted into climacteric rise in the [GHG] leading to increased global atmospheric temperature or Global Warming. According to the fifth assessment report of IPCC, each of the last three decades have been successively warmer than any preceding decade since 1850, and the period from 1983 to 2012 has been warmest 30-year period in the last 1400 years in the northern hemisphere (IPCC, 2014). Factors that cause or contribute to climate change are known as 'climate forcings' that can be either natural or anthropogenic. Natural factors include volcanic eruptions, alteration in sun's intensity, and very slow changes in the oceanic circulation or land surface. Anthropogenic activities are fossil fuel combustion, industrial activities, emissions from agricultural systems and waste decomposition (Shukla *et al.*, 2003). In an analysis of carbon emission from ten global cities, Kennedy *et al.*, 2010, concluded that power generation, urban design and waste processing are the major determinants of GHG emissions. Studies have been conducted to analyse the factors that govern the input-output of carbon associated with different industrial and domestic processes (Sovacool and Brown, 2010). In India, state-wise carbon emission analysis indicated that electricity generation and transportation are the major sources of CO_2 emission (Ramachandra and Shwetmala, 2012).

Globally, there have been several efforts to combat climate change and a lot of impetus is being given towards quantification as well as reduction in the GHG. Emission reduction is the most pertinent mechanism and has been time and again accepted and reinforced by the international community. In order to bring about emission reduction, the first step is to know how much a country emits and accordingly efforts should be made to create inventories of major GHG, particularly species of Carbon (Druckman and Jackson, 2009). These inventories should be created in such a way that it penetrates deep into the grassroots of the society and is largely participative in nature so that they are accurate and well represented. Calculation of individual carbon footprint is a very accurate method as it involves people and relates their daily practices directly to CO_2 emission.

The term carbon footprint originates from the concept of Ecological footprint, and has been defined and interpreted in different ways depending upon the aim and relevance of the study (Wackernagel and

Rees, 1996; Kleiner, 2007; Abbot, 2008; Andrew, 2008). According to Hammond (2007) it is actually a 'carbon weight' released in grams or tones per person or activity. The definitions of carbon footprint range from being environment centric (ETAP, 2007) to industry centric (Patel, 2006). In a recent study Ramachandra *et al.* (2015), have defined it as the total amount of GHGs impacting the environment, produced both directly and indirectly due to various human activities, expressed in equivalent tons of CO_2. Irrespective of the definition, the topic of grave interest among scientists, policy makers, industrialists and activists. Carbon footprint analysis helps us to determine emissions not only quantitatively but also highlights the sector-wise contribution indicating the impact of human activities. There has been trade-linked country-wise analysis, intra country sector-wise analysis and state-wise analysis of carbon emissions (Hertwich and Peters, 2009; Druckman and Jackson, 2009; Ramachandra *et al.*, 2015). For successful designing and implementation of any mitigation strategy, it is imperative that individual carbon footprint is calculated at local levels, so that any policy is in line with the local developmental needs as well as emission patterns. However, to the best of our knowledge there has been no study to calculate individual carbon footprint of urban youth in India. The study was conducted with the following objectives:

- To calculate individual carbon footprint of undergraduate students, and to observe and analyse stream-wise variation.
- To assess awareness of students towards greenhouse gas emissions and mitigation measures.
- To analyse sector-wise variation in individual carbon footprint.

MATERIALS AND METHODS

An exhaustive questionnaire was designed to calculate individual carbon footprint and assess awareness levels of undergraduate students of the age group 18 to 21 years. It was designed based upon the Low Carbon Lifestyle Tool Kit issued by the Centre for Environment Education, New Delhi (CEE, 2010). The questionnaire consisted of 21 questions, of which six were nominal or awareness based, and 15 were quantitative questions which indicated emission of CO_2 in tonnes per annum associated with a process. Questions covered all major sectors such as transportation, domestic electricity consumption, source of fuel used in cooking, source of water supply and recycling of waste and books etc. The study was conducted over a span of three years from April, 2014 to March, 2017.

The questionnaire was filled up by respondents by participation observation method (Personal Interview) and on an average 20 minutes were given to each participant. Responses were collected and segregated according to streams (Biological Sciences, Physical/Chemical Sciences, Humanities, Languages and Commerce) and type of question (nominal and quantitative questions). In case of quantitative questions, values of CO_2 in tonnes/annum were calculated according to their respective conversion factors (CEE, 2010).

The data obtained was statistical analysed by using software SPSS and descriptive statistics were obtained for various streams as well as sectors. To assess the awareness of students towards greenhouse gas emissions and mitigation measures, known parametric test chi square was used. Stream-wise and sector-wise variation in carbon footprint was analysed using one-way ANOVA. Two-way ANOVA was applied to determine variation in carbon footprint across streams and sector and their interaction.

RESULTS

Stream-wise Analysis

721 students from five different streams—Biological Sciences, Physical/ Chemical Sciences, Humanities, Languages and Commerce participated in the study conducted over a span of three years. Maximum participation of 392 was recorded from Biological Sciences that included students of B.Sc. (Hons.) Botany, Zoology and Biological Sciences and B.Sc. (Prog.) Life Sciences. The number of respondents from Physical and Chemical Sciences, Humanities, Languages and Commerce were 178, 94, 36 and 21 respectively. Average CO_2 emission of all students was calculated as 5.12 ton/annum. Highest carbon emission of 5.37 ton/annum was observed in the students of languages and lowest emission of 4.98 ton/annum was observed in students of Physical and Chemical Sciences. The average emission of Biological Science students was 5.13 ton/annum (Figure 1).

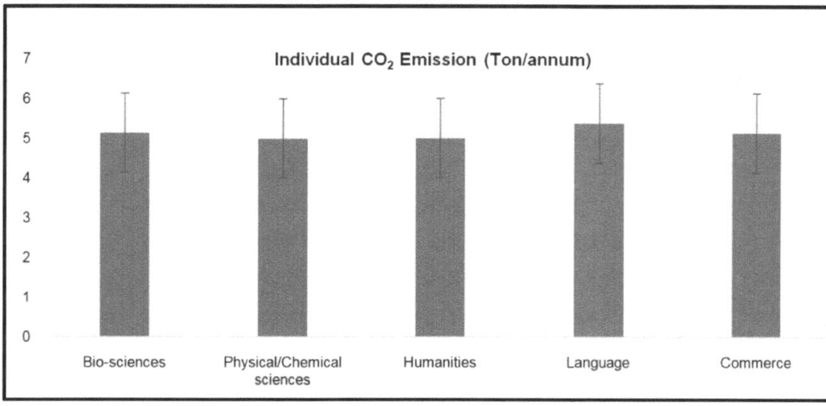

Figure 1: Graph Showing Stream-wise Variation in Carbon Footprint of Respondents

Result of ANOVA to check whether carbon emission shows stream-wise variation. The p-value was observed to be 0.358 which is greater than 0.05 (Table 1). Hence, it was accepted and it is concluded that the difference between carbon emissions across students of different streams is not significant i.e. the carbon emission across different stream is homogenous.

Table 1: Table Showing Stream-wise Variation in Carbon Footprint by One-way ANOVA

Source of Variation	*Sum of Squares*	*Df*	*Mean Square*	*F*	*Sig.*
Between Group	7285472.395	5	1457094.479	1.102	.358
Within Groups	945280942.877	715	1322071.249		
Total	952566415.272	720			

Students' Awareness Towards GHG Emissions and Mitigation Measures

Chi-square analysis of individual awareness towards the Bureau of Energy Efficiency (BEE) star ratings gave a significant value of 0.015 indicating a high degree of awareness that was not stream dependent (Table 2). Testing an individual's habit of switching off appliances, significant value as 0.000 was obtained indicating that there is a significant difference in individual's habit of switching off appliances across different streams. Thereby meaning that among various streams, students' orientation towards switching off appliances varied. Value of 0.102 obtained for the habit of recycling kitchen waste indicates that the practice of recycling waste is not prevalent among students and is independent of the stream of study. Likewise, habit of recycling text books and the practice of switching off ignition at red lights gave significance values of 0.574 and 0.174, respectively. Hence, it may be concluded that practice of recycling old text books and switching off ignition at red lights are independent of the stream of study and are not practised by youth.

Table 2: Chi-square Analysis of Respondents' Awareness Towards Energy Conservation Measures

Parameters	*Awareness Regarding BEE*	*Practice of Switching Off Appliances*	*Practice of Recycling Kitchen Waste*	*Practice of Recycling Old Text Books*	*Practice of Switching Off Ignition at Red Lights*
Significant Values	0.015	0.000	0.102	0.574	0.174

Stream-wise CO_2 emission of respondents by two-way ANOVA gave a value of 0.635 indicating no significant difference in the carbon emissions across different streams (Table 3). However, value as 0.000 was obtained in correlating sector-wise carbon emission of students, indicating that emissions from different sectors are significantly different. For testing interaction effect of streams and sectors, significant value of 0.723 was obtained. Therefore, it can be concluded that the carbon footprint value in the present study is independent of the combination of stream and sector.

Clearly transportation sector is responsible for maximum carbon footprint generation for students of all streams of study (Figure 2). While, waste generation produces minimum carbon footprint.

Table 3: Table Showing Results of Two-way ANOVA

Source	*Df*	*Mean Square*	*F*	*Sig.*
Stream	4	151069.788	.639	.635
Sector	3	8777418.732	37.115	.000
Stream * Sector	12	172654.096	.730	.723

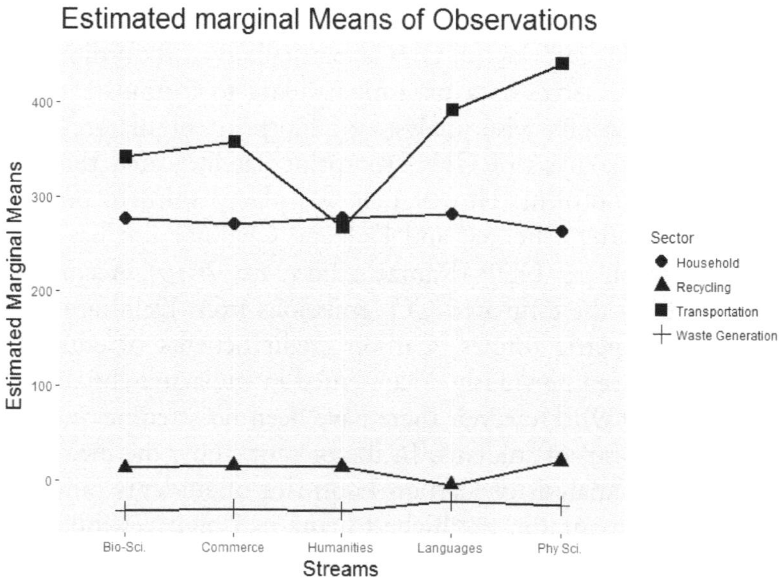

Figure 2: Graph Representing Sector as Well as Stream-wise Variation in Carbon Footprint

DISCUSSION

721 undergraduate students of five streams participated in the survey. Maximum number - 392 of student respondents were recorded in Biological Sciences followed by 178 in Physical Sciences, 94 in Humanities, 36 in languages and 21 in commerce. Average individual carbon footprint of all respondents was calculated as 5.12 ton CO_2/ annum. Measurement of individual carbon footprint is very important as it gives an indication of carbon intensity of a society of a country. In the last few decades there have been several studies reporting country and sector-wise carbon emission. However, at citizen level carbon emission is all the more important as it directly gives an estimation of the relative inputs from different sectors such as transport, lifestyle, household and practice of recycling. The study is important as in future National level mitigation strategies can be devised as there is a fairly good idea on the sectors to be focussed upon.

Stream-wise Analysis of Individual Carbon Footprint

Individual carbon emission of students was 5.12 tonnes of CO_2/ annum. The values showed variation according to the subjects that the students studied. Maximum carbon emission was observed in students of Languages. Inventories of GHGs particularly carbon-dioxide are important also it serves as a preliminary data to formulate policies. There have been city-wise analyses of footprint of major cities of India. Intercity analysis of GHG footprint has indicated that Delhi has maximum footprint. However, it was placed third in per capita CO_2 emission, after Chennai and Kolkata. Chennai also has highest CO_2 eq. emission per GDP (Ramachadra *et al.,* 2015). According to Dhamija (2010) the estimated CO_2 emissions from Delhi are around 15.42 million metric tonnes. Carbon credit benefits of educational institutes have been previously documented in India and abroad (Kafle and Mathur, 2009). However, there have been no attempts indicating individual footprint of students. In the present study, the mean values for stream-wise analysis of Carbon Footprint of students range from 4.98 to 5.37 ton/annum, the highest being in Languages followed by Bio-sciences, Commerce, Humanities and Physical/Chemical Sciences. Thus, it is concluded that different streams had no significant contribution in determining the carbon footprint of the students, thus it independent of the stream of study.

The study analysed students' awareness towards different mechanisms of energy conservation such as Bureau of Energy Efficiency Star rating of appliances (BEE), switching off appliances at plug points when not in use, recycling books and reducing waste generation. Interestingly, significantly higher level of awareness was observed in students' knowledge of BEE star rating of appliances. Also, most of them practised switching off ignition of vehicles at traffic signals. This is indeed a positive indication reflecting upon sensible environmental behaviour. However, the same awareness was not observed in waste segregation and recycling, as well as recycling of used books. Mitigation of global issues like climate change need creation of society that is environment friendly. The first and most important step towards this global goal is to create awareness and sensitize people, particularly the youth about various issues and make them understand the effect of their actions on environment. The observation highlights the need to manage climate change with concerted efforts from not only scientists but also social scientists, as the issue is now more of a social behaviour. The transition to a very low carbon economy needs elementary changes in technology, regulatory frameworks, infrastructure, business practices, consumption patterns and lifestyles. There is a need for a target to be set which aids the local and national governments to frame climate change policies and regulations

Sector-wise emission analysis revealed that among the various sectors analysed, maximum carbon emissions was attributed to transportation sector followed by household appliances, indicating that indirectly fossil fuel combustion as well as dependence on conventional fuels is a significant contributor to carbon footprint. The IPCC in its report has already cautioned that if no appropriate mitigation measures are practise the global carbon emissions due to transportation sector could reach up to 12 Gt CO_2 eq by 2050 (IPCC, 2014). Not only appropriate measures need to be devised, they have to followed and practised in a sustained manner. Long-term and short-term strategies should be formulated keeping in mind local needs and limitations. On a global scale the per capita carbon emission of Indians is much lower than that of westerners. However, there is a steady increase in people using personal transport that is adding up to their carbon footprint (Leather *et al.*, 2011). According to estimates, CO_2 emission for Delhi, for the year 2007–08 was around 7.66 MMT using the top-down approach, whereas using the bottom-up approach it was estimated to be 8.17 MMT (Dhamija, 2012). India has the second largest populations of the world and its gross domestic product has almost doubled between 1995 and 2005 (IMF, 2008). In this very duration, there was a surge in the sale of motorised two-as well as four wheelers (SIAM, 2008). This strong motorization has led to increasing concerns about local and long-range air pollution and its impacts on climate change. Indeed, already by the year 2000, India was among the ten countries with highest exhaust pollutants from the road transportation sector and road fuel consumption approximately doubled every ten years since 1980 (Baidya and Borken, 2009). The transportation sector is the most active sector in terms of contribution to increased carbon emission amongst youth. With urbanisation, comes the ascend of new wealthy and middle classes, spurring demand for goods and services. Suddenly, corporate India urgently needed better transportation and infrastructure, from freight facilities to public transportation for commuting employees, and the newly prosperous wanted their own two wheelers and vehicles. Concomitantly the Indian automotive industry took off to meet the demands of both businesses and individuals. Today, India is among the fastest growing automotive markets in the world. In India, Delhi tops the list of being the city with highest number of car ownerships which according to some estimates have already crossed the 1 crore mark. People often drive on more unnecessary journeys be it to the supermarket or the shops. If few walk or cycle or use public transport or even use CNG as a fuel, it will reduce carbon footprint. Despite the countless benefits of public transport, the modern generation has a higher crook towards private vehicles in order to prevent time. The study highlights that the young age group is aware about switching off ignition at red lights and practise it, however we suggest a switch to public transport such as DTC buses would further reduce their emissions and thus the carbon footprint significantly.

The next sector which draws our immediate attention is the Power Consumption sector. The Government of India set up Bureau of Energy Efficiency (BEE) on 1st March 2002 under the provisions of the Energy Conservation Act, 2001 (bee.gov.in, 2002). The mission of the Bureau of Energy Efficiency is to assist in developing policies and strategies with a thrust on self-regulation and market principles, within the overall framework of the Energy Conservation Act, 2001 with the primary objective of reducing energy intensity of the Indian economy. India is the world's third largest producer and fourth largest consumer of electricity. The main contribution to the growth in the total electricity generation is through power generation by coal fired plants and non-conventional renewal energy sources (RES) whereas the contribution from natural gas, oil and hydro plants has decreased in last five years (2012–2017). According to Pahauri and Spreng (2002), there have been significant rise in the energy consumption of Indian household in the last few decades. They have attributed this rise to increase in per capita expenditure, rise in population and higher energy intensity of sectors such as food and agriculture. Similar observations have been reported by Chunekar *et al.* (2016), according to whom India has seen a phenomenal rise in residential electricity consumption (REC), which has increased by around 50 times since 1971. The per-capita consumption of the consumers in Delhi is more than 1561 units per annum as against the national average of 1010 units (FY 2014–15). Delhi, being an urban place with high load density, has seen the electricity consumption increasing from 20040.22 MUs in 2003–04 to 30759.888 MUs in 2016–17. The increase in percentage terms is 53.49% for Delhi for the past fifteen years (Rao *et al.*, 2009). Power sector has become a serious challenge for the

Government as it is the primary driver for all round development of the territory. The study highlights that the students are aware about BEE and practise switching off electrical appliances. Electrical appliances like air conditioners and others have a significant contribution in increasing the carbon footprint. Customers often focus on the cost of owning the appliance and tend to overlook the cost of using the appliance. The Government of India (Bureau of Energy Efficiency, Ministry of Power) has introduced the Standards and Labelling Program in May 2006 to provide the users with information related to energy efficiency of electrical appliances. Under this program the manufacturers are required to place a label showing how much electricity the appliance will consumer under certain conditions. The program is currently running for refrigerators, air conditioners, televisions, geysers, tube lights and fans among the household appliances. From this study we suggest that the masses have to be more sensitized about the BEE rating and its importance in reducing the carbon footprint.

Waste is an imperative contributor to carbon emissions. Minimizing waste can lead to saving of unnecessary big emissions and lower land fill requirements, resulting in reduction of pollution in air and land. Waste not only discharges CO_2 and methane into the atmosphere, but also acts as an air, groundwater and soil pollutant. In order to reduce the waste generated the three R's—Reduce Reuse and Recycle were introduced in order to create a sustainable life whereby less amount of waste enters into landfills thereby reducing the carbon footprint. The present study reveals that the practice of recycling waste is independent of the stream of study. People however need to be more aware about segregation of waste at source and practise recycling of all kinds of waste.

CONCLUSION

There is a need to curtail our CO_2 emissions and the first step towards it is to know how much we emit. The present study calculates individual carbon footprint based on their lifestyle, transportation and waste recycling behaviour. In order to reduce the carbon footprint simple steps such as walking and use of public transport over private vehicles, turning off lights when not in use, replacing incandescent light bulbs with compact fluorescent or LEDs, buying used products and reselling or recycling items can be adopted by people.

ACKNOWLEDGEMENT

The author acknowledges the STAR STATUS College Scheme by the Department of Biotechnology (DBT), Ministry of Science and Technology, Government of India, under which the present study was conducted.

REFERENCES

[1] Abbott, J. (2008). Report-What is carbon footprint. 19 February 2008. The Edinburg Centre for Carbon Management, 2008.

[2] Andrew, J.E. (2008). What is carbon footprint? An overview of definitions and methodologies. Vegetable industry carbon footprint study—discussion papers and workshop (26 September). Sydney: Horticulture Australia Limited.

[3] Baidya, S. and Borken, K.J. (2009). Atmospheric emissions from road transportation in India. Energy Policy, 37:3912–22.

[4] Bee (2002). https://beeindia.gov.in/content/about-bee

[5] CEE (2010). http://www.ceeindia.org/cee/pdf_files/Toolkit_English_final.pdf

[6] Chunekar, A., Varshney, S. and Dixit, S. (2016). Residential Electricity Consumption in India: What do we know? Prayas Energy Group, Pune, India.

[7] Dhamija (2010). Executive summary of inventorization of greenhouse gases—sources and sinks in Delhi. Climate Change Agenda for Delhi 2009–12, Government of Delhi; 2010.

[8] Druckman, A. and Jackson, T. (2009). The carbon footprint of UK households 1990–2004: A socio-economically disaggregated, quasi-multi-regional input–output model. Ecological Economics, 68 (7), 2066–2077.

[9] ETAP (2007). The Carbon Trust Helps UK Businesses Reduce their Environmental Impact, Press Release, http://ec.europa.eu/environment/etap/pdfs/jan07_ca rbon_trust_initiative.pdf

[10] Hammond, G. (2007). Time to give due weight to the 'carbon footprint' issue. Nature, 445(7125), 256–256.

[11] Hertwich, E. and Peters, G. (2009). Carbon Footprint of Nations: A Global, Trade-Linked Analysis.

[12] IMF (International Monetary Fund), (2008). World Economic Outlook Database, April 2008 Edition. Available at: /http://www.imf.org/external/pubs/ft/weo/ 2008/01/weodata/index.aspxS,

[13] IPCC (2014). Sims R., R. Schaeffer, F. Creutzig, X. Cruz-Núñez, M. D'Agosto, D. Dimitriu, M.J. Figueroa Meza, L. Fulton, S. Kobayashi, O. Lah, A. McKinnon, P. Newman, M. Ouyang, J.J. Schauer, D. Sperling and G. Tiwari, (2014): Transport. In: Climate Change 2014: Mitigation of

Climate Change. Contribution of Working Group III to the Fifth Assessment Report of the Intergovernmental Panel on Climate Change [Edenhofer, O., R. Pichs-Madruga, Y. Sokona, E. Farahani, S. Kadner, K. Seyboth, A. Adler, I. Baum, S. Brunner, P. Eickemeier, B. Kriemann, J. Savolainen, S. Schlömer, C. von Stechow, T. Zwickel and J.C. Minx (eds.)]. Cambridge University Press, Cambridge, United Kingdom and New York, NY, USA.

[14] Kafle and Mathur, J. (2009). Energy and Buildings, Vol. 41, (1), 133–137.

[15] Kennedy, C. *et al.* (2010). Methodology for inventorying greenhouse gas emissions from global cities. *Energy Policy*, 38, 4828–37.

[16] Kleiner, K. (2007). The corporate race to cut carbon. Nature, 3, 40–43.

[17] Narasimha Rao *et al.* (2009). An overview of Indian Energy trends: low carbon growth and development challenges. *India Prayas Energy Group.*

[18] Pachauri, S. and Spreng, D. (2002). Direct and indirect energy requirements of households in India. *Energy Policy*, 30(G), 511–23.

[19] Patel, J. (2006). "Green sky thinking". Environment Business (122), 32.

[20] Ramachandra, T.V. and Shwetmala (2012). Decentralised Carbon Footprint Analysis for Opting Climate Change Mitigation Strategies in India. *Renewable and Sustainable Energy Reviews*, 16, 5820–5833.

[21] Ramachandra, T.V., Sreejith, K. and Aithal, Bharath H. (2015). GHG footprint of major cities in India, *Renewable and Sustainable Energy Reviews* 44 2015 473–495, http://dx.doi.org/10.1016/j.rser.2014.12.036

[22] Shukla, P.R., Sharma, S.K., Garg, A. *et al.* (2003). Climate change vulnerability assessment and adaptation. In Climate change and India: vulnerability assessment and adaptation. Universities Press Pvt. Ltd., India.

[23] SIAM (2008). Automobile domestic sales trends. Available at: /http://www.siamindia.com/scripts/domestic-sales-trend.aspxS, accessed 12.05.2008.

[24] Sovacool, B.K. and Brown, A. (2010). Twelve metropolitan carbon footprints: A preliminary comparative global assessment. *Energy Policy*, 38(9), 4856–69.

[25] Wackernagel, M. and Rees, W.E. (1996). Our ecological footprint: reducing human impact on the Earth. BC. Canada: New Society Publishers Gabriola Island, Canada.

ABBREVIATIONS

ANOVA	–	Analysis of Variance
BEE	–	Bureau of Energy Efficiency
GHG	–	Green House Gases
IPCC	–	Intergovernmental Panel on Climate Change

Environmental Impact of Coal-Fired Power Plant on Local Air Quality: Case Study of Southern Java, Indonesia

Widyarani[1*], Dindin Makhmuddin[2], Andri Fachrur Rozie[3], Diana Rahayuning Wulan[1] and Sudaryati Cahyaningsih[1]

[1]Research Unit for Clean Technology, Indonesian Institute of Sciences (LIPI), Komplek LIPI Bandung, Jalan Sangkuriang, Gedung 50, Bandung 40135, Indonesia

[2]Pusat Unggulan Pengembangan Iptek Terapan, Padjadjaran University

[3]Research Centre for Informatics, Indonesian Institute of Sciences (LIPI)

*E-mail: widyarani@lipi.go.id

ABSTRACT: Air quality was monitored on the area surrounding a coal-fired power plant (CPP) in Palabuhanratu Bay, in the southern part of Java Island, Indonesia. The measurement was performed at three settlement areas to the north, east, and west of the CPP, as well as the CPP area itself. The monitored parameters were total suspended particulate (TSP), PM_{10}, NO_2, SO_2, and CO. The measurement was performed quarterly since the beginning of power plant operation in 2014 until 2019. The concentration of all pollutants at the CPP area was higher than the other locations and showing an increasing trend. For settlement areas around the CPP, a weak increasing trend at the beginning until the 33rd month of CPP operation was observed for PM_{10}, SO_2, and NO_2. Data from local health facilities shows a higher admitted respiratory health problem compared with the admission before CPP operation. Despite the weak correlation, attention should be given on both increases in air pollution parameters and admission of respiratory problem.

Keywords: Coal-Fired Power Plant, Southern Java, Air Quality, Air Pollution, Monitoring, Particulate Matter, Respiratory Problem.

INTRODUCTION

Electricity generation is a source of atmospheric environmental pollution. For instance, in 2014, electricity generation contributed to 1% of PM_{10} emission, 12% of NOx emission, and 68% of SO_2 emission in the USA (USEPA, 2018). In 2010, more than 60% of world electricity was generated from fossil fuel while the share of renewable sources was only 20%. In 2020, it is projected that electricity from renewable sources will account for 30%, but fossil fuel will still dominate at around 55% (EIA, 2019).

Coal combustion in coal-fired power plants can emit components such as particulate matters, sulphur oxides (SOx) and nitrogen oxides (NOx), polyaromatic hydrocarbon, and heavy metals, which affect air quality and are linked to several health problems. Particulate matters (PM_{10} and $PM_{2.5}$) are linked to respiratory problems, lung cancer, heart attack, and congestive heart failure. NOx is linked to asthma and chronic pulmonary disease. NOx, SOx, and CO are linked to stroke (Prehoda and Pearce, 2017; Shah *et al.*, 2015).

In Indonesia, from the 65 Gigawatts of installed power plant capacity in 2018, only 15% came from renewable sources. On the other hand, coal contributed to 45% of the primary energy source for electricity generation. Despite the plan to increase renewable energy sources contribution to 25% in 2025, most electricity in Indonesia is still generated from fossil fuel (Ministry of Energy and Mineral Resources, 2019).

Since 2014, a coal-fired power plant (CPP) has been operated on Palabuhanratu Bay, in the southern coast of Java Island, Indonesia. The CPP comprises of three generating units, each of 300 Megawatts capacity. The coal used in the CPP has a heating rate of 2064 kcal/kWh and is consumed at 13500–14400 metric tonnes per day.

The objective of this paper was to evaluate the air quality in the settlement area around Palabuhanratu CPP and how it affects the health condition. Measurement of air quality was performed quarterly in three settlement areas around the CPP, as well as the CPP area itself. Admission to the clinics at the community health centre was used as an indicator of health condition.

MATERIALS AND METHODS

Air Quality Sampling and Analysis

The study took place at the area surrounding Palabuhanratu Coal-fired Power Plant (106°32' E, 07°01' S), which is located on Sukabumi Regency, Indonesia. Sampling points were settlement areas at the north, east, and west of the CPP, as well as the CPP area itself (Figure 1).

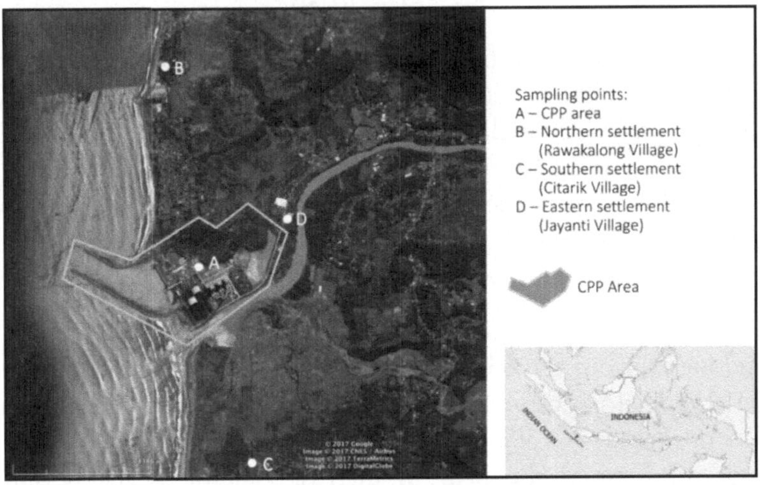

Figure 1: Coal Power Plant Location and Air Quality Sampling Points (Base map was retrieved from Google Earth)

Table 1: Parameter and Measurement Methods[a]

Parameter	*Sampling Time*	*Sampling Equipment*	*Analytical Method*
Total Suspended Particulate (TSP)	24 h	Hi-volume sampler	Gravimetric
PM_{10}	24 h	Hi-volume sampler	Gravimetric
Sulphur dioxide (SO_2)	24 h	Impinger gas sampler	Pararosaniline
Nitrogen dioxide (NO_2)	24 h	Impinger gas sampler	Saltzman
Carbon monoxide	24 h	NDIR analyser	NDIR

[a] Sampling and analytical methods were in accordance with Indonesian Air Quality Standard and Indonesian National Standard.

Samples were taken once every quarter of a year, starting from the start of CPP operation in March 2014 (designated Month-1) through

November 2019 (Month-69). Measured parameters with the applied sampling and analytical methods are presented in Table 1.

Health Condition Data

The incidence of respiratory and other diseases was approached by admission to the clinics at the community health centre ('Puskesmas'). Data was obtained from Puskesmas Palabuhanratu for the period of March 2014 through June 2019 and from Puskesmas Citarik for the period of March 2014–April 2015 and August 2017–November 2018. Puskesmas Palabuhanratu was located at the north of the CPP and served five villages in Palabuhanratu Sub-district namely Palabuhanratu, Citepus, Cibodas, Buniwangi, and Cimanggu. Puskesmas Citarik was located at the north of the CPP and served five villages in Palabuhanratu Sub-district namely Citarik, Cikadu, Tonjong, Pasirsuren, and Jayanti.

Calculation and Data Analysis

The total population of Palabuhanratu Sub-district had grown from 104,231 in 2013 to 112,809 in 2017 (BPS Kabupaten Sukabumi, 2014, 2018). Population data for years 2014, 2015, and 2016 were estimated using linear regression. Admission was calculated as:

$$\text{Admission per 1000 population} = \frac{\text{Total admission for a particular disease}}{\text{Population in the service are}} \times 1000$$

The change of values over time was plotted with linear regression to test the correlation. ANOVA and Kruskal-Wallis tests were used to compare central values. The Kruskal-Wallis test was used when the assumption of normality was not fulfilled. The p-value less than 0.05 was regarded as significant. The data was processed using Microsoft Excel 2016 and Jamovi v.1.1.9.0 (The jamovi project, 2019).

RESULTS AND DISCUSSION

Figure 2 shows the concentrations of the total suspended particulate (TSP). At the start of the operation, TSP concentrations in all locations were 23–29 $\mu g/m^3$. Figure 2a shows TSP concentration at the CPP area increased by 1.4 $\mu g/m^3$ per month ($R^2 = 0.94$). The increase of

TSP concentration is also shown at the northern (Figure 2b) and southern (Figure 2c) settlements, however, the correlation coefficient is too low to make a firm conclusion. Comparison of central values does not show any significant differences between locations.

Figure 3 shows the concentration of particulate matter sized 10 μm or less (PM_{10}). At the start of the operation, PM_{10} concentrations in all locations were 11–14 μg/m^3. Figure 3a shows PM_{10} concentration at the CPP area increased by 0.3 μg/m^3 per month (R^2 = 0.64). Figures 3b-d do not show any clear trend of changes of PM_{10} concentration. Comparison of central values shows there are significant differences between locations, except between northern and eastern settlements. PM_{10} concentration at the southern settlement was lower than the other locations.

Figures 4, 5 and 6 show the concentrations of SO_2, NO_2, and CO, respectively. Similar to PM_{10}, the concentration of these gases at the CPP area shows an increasing trend (R^2 = 0.81–0.85). On the other hand, there is no clear trend of changes of these gases at the other locations. The exception is CO concentration at the southern settlement, which increased by 5.8 μg/m^3 per month (R^2 = 0.74). Comparison of central values shows there are significant differences between locations, except between northern and eastern settlements. Concentrations of SO_2, NO_2, and CO at the southern settlement were lower than the other locations.

Dispersion of air pollutants from a point source such as CPP stack is influenced by stack height and weather condition, as well as the chemical reactions that may occur with other particles or gases and coagulation, breaking and sedimentation of particles in the atmosphere (Cooper and Alley, 1994). From Figures 2–6, it can be deduced that concentrations of pollutant at the CPP were higher than the other locations and showing an increasing trend. The settlement areas around the CPP do not show a clear increasing trend. However, for PM_{10}, SO_2, and NO_2, there seemed to be a weak increasing trend at the beginning of CPP operation that peaked at Month-33.

Figure 7a shows the admission of respiratory problems in Puskesmas Palabuhanratu from March 2014 through June 2019. Respiratory problems include common cold, asthma, influenza, tuberculosis, upper respiratory tract infection, and lower respiratory tract infection.

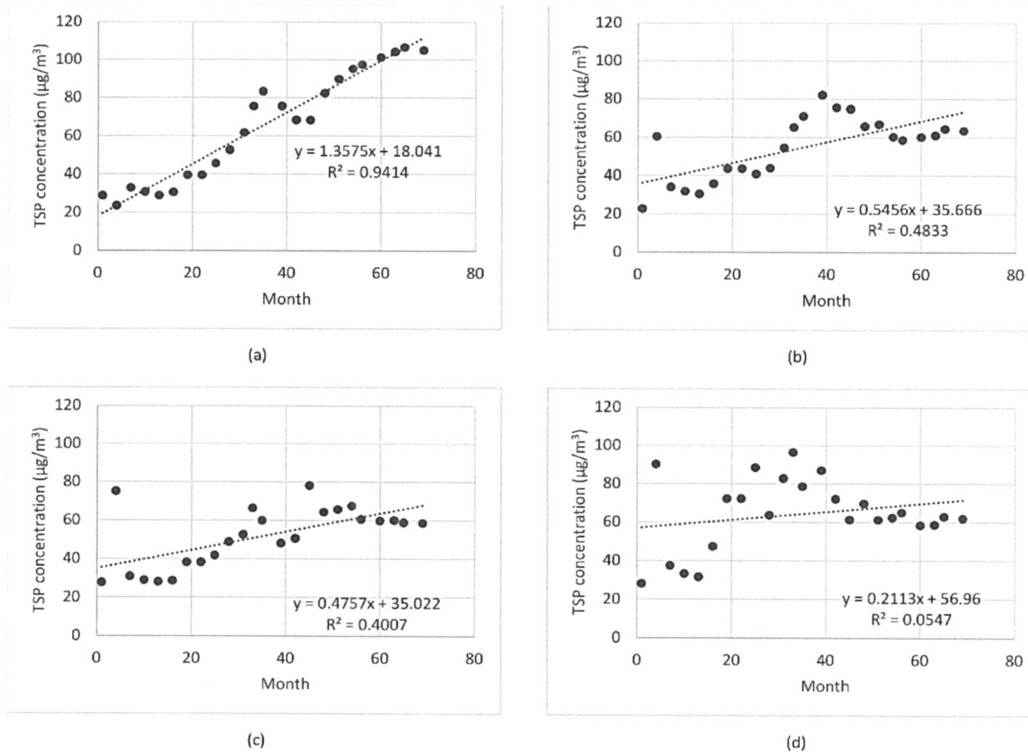

Figure 2: TSP Concentration in the CPP Area (a), Northern (b), Southern (c), and Eastern (d) Settlements. (Month Counted from the Start of the CPP Operation)

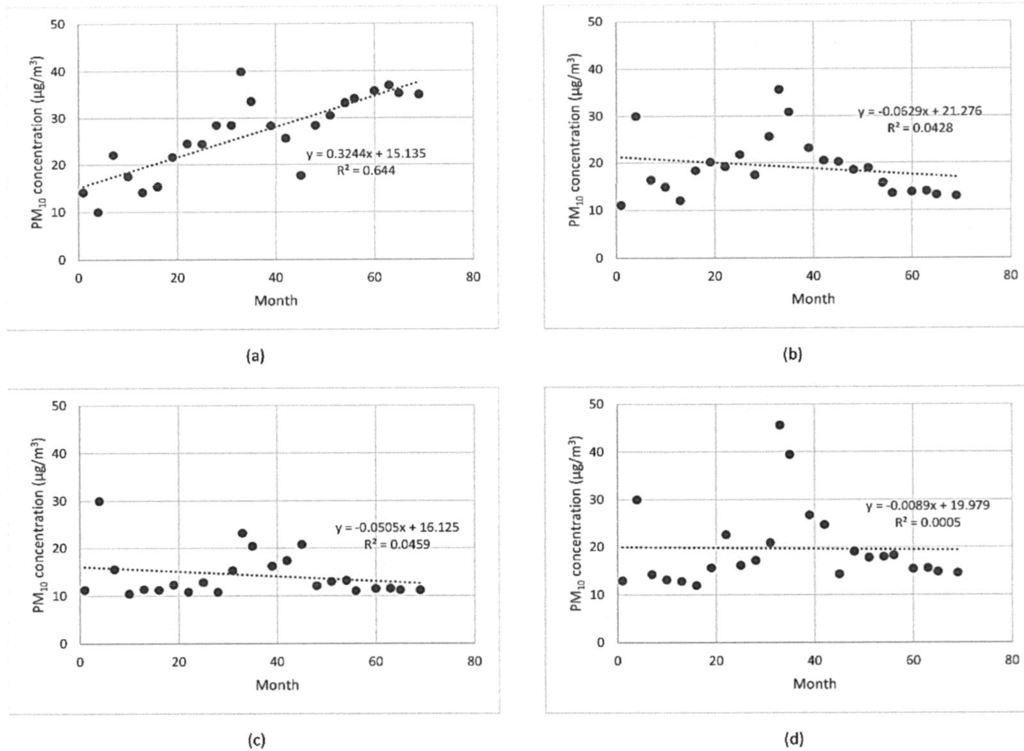

Figure 3: PM_{10} Concentration in the CPP Area (a), Northern (b), Southern (c), and Eastern (d) Settlements. (Month Counted from the Start of the CPP Operation)

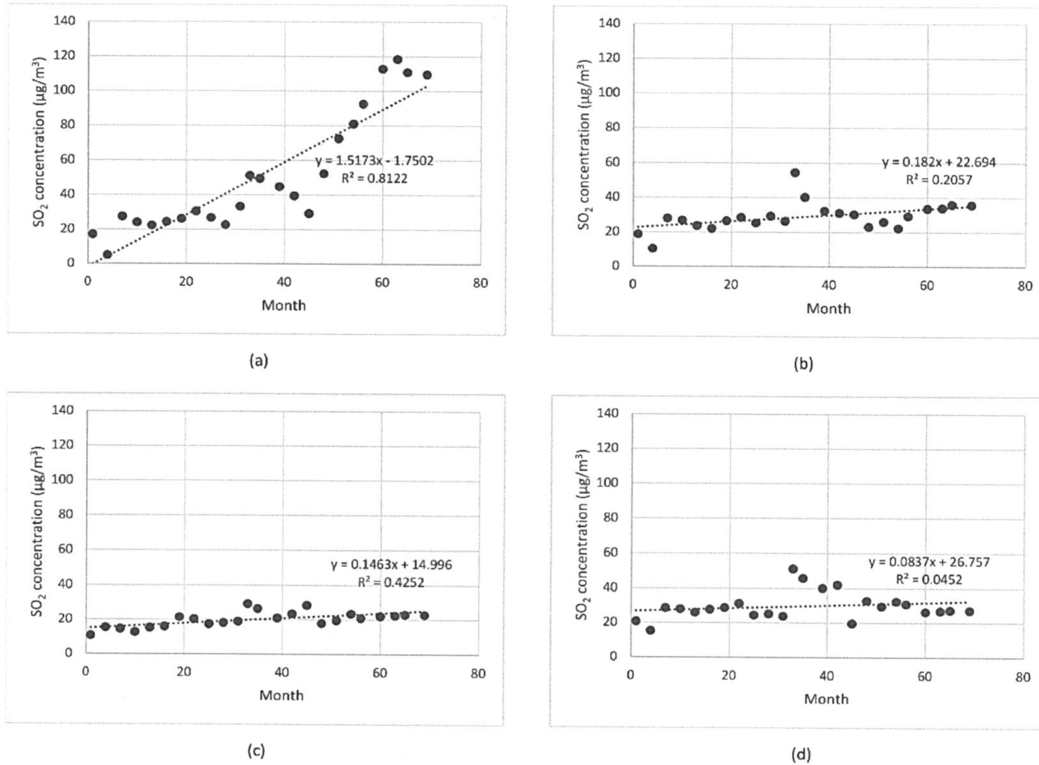

Figure 4: SO_2 Concentration in the CPP Area (a), Northern (b), Southern (c), and Eastern (d) Settlements. (Month Counted from the Start of the CPP Operation)

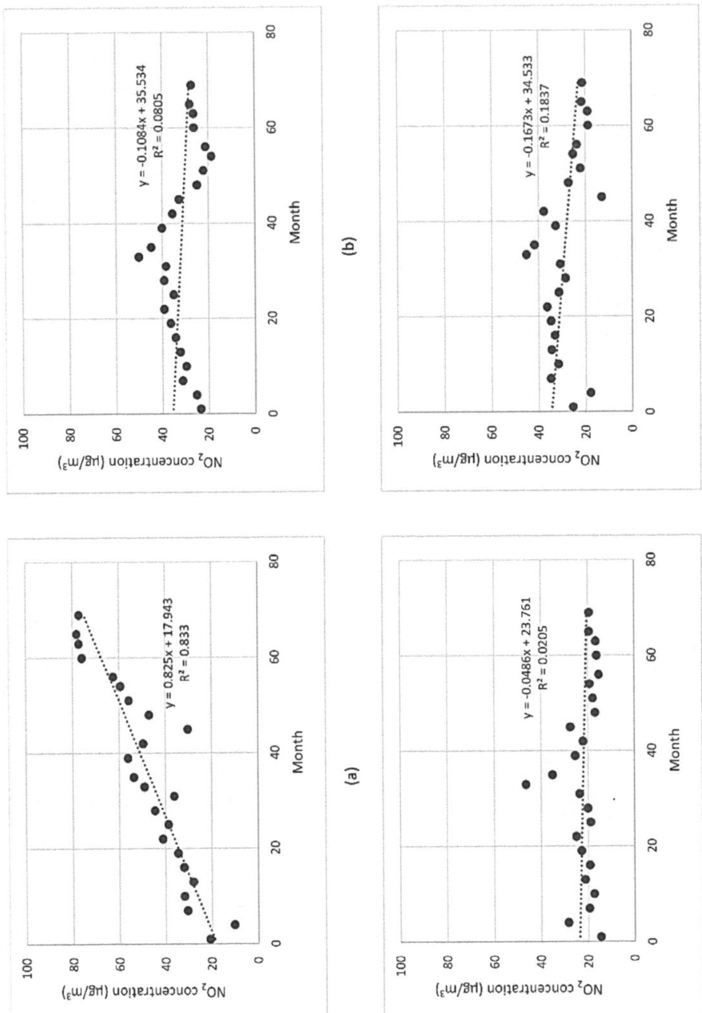

Figure 5: NO_2 Concentration in the CPP Area (a), Northern (b), Southern (c), and Eastern (d) Settlements. (Month Counted from the Start of the CPP Operation)

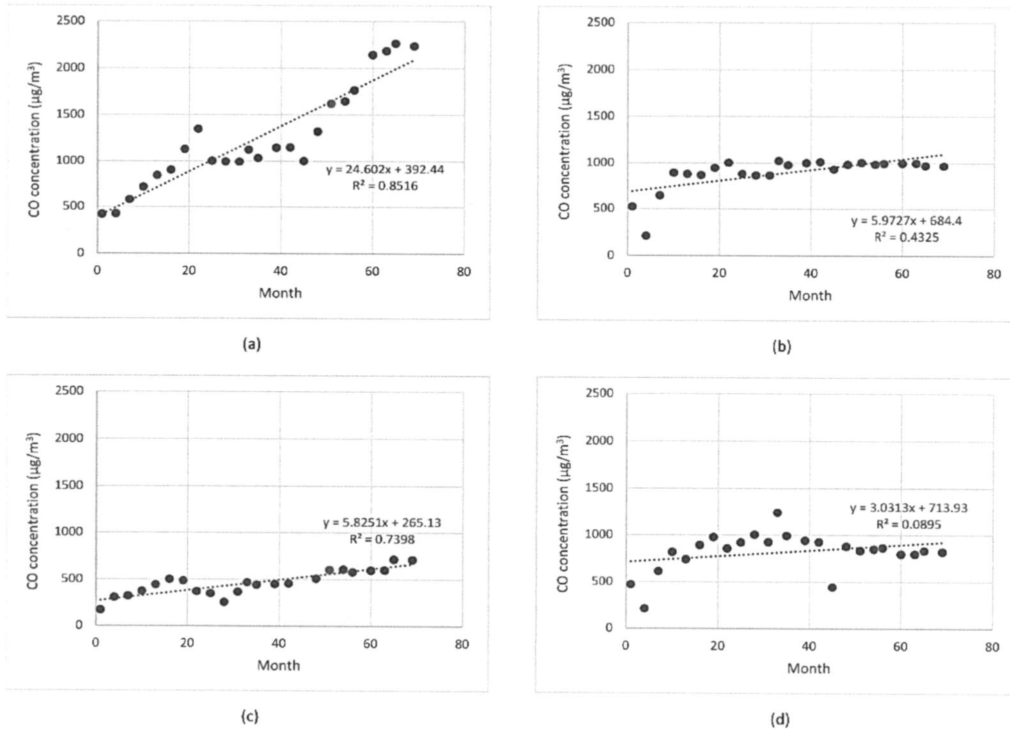

Figure 6: CO Concentration in the CPP Area (a), Northern (b), Southern (c), and Eastern (d) Settlements. (Month Counted from the Start of the CPP Operation)

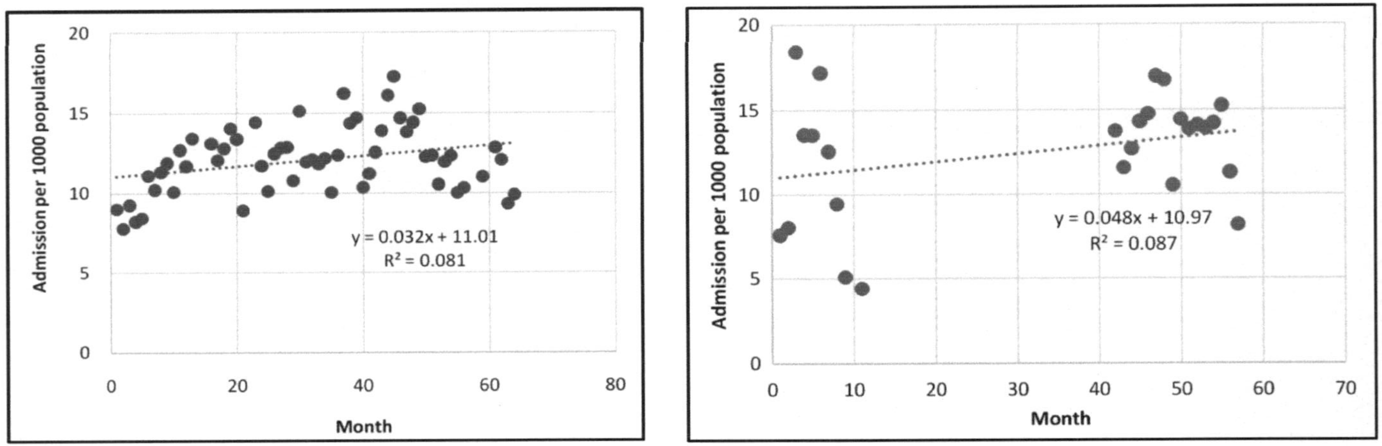

(a) (b)

Figure 7: Admission of Respiratory Problem per 1000 Population at Puskesmas Palabuhanratu (a) and Citarik (b). (Month Counted from the Start of the CPP Operation)

Figure 7a does not show any clear change of admission rate during the study period. A closer examination of admission rate during the dry season (June-August), which is the peak of respiratory problem incidents, is shown in Table 2. Table 2 shows that the average admission of respiratory problem per 1000 population was increased from 9.2 in dry season 2014 to 12.9 in dry season 2016, then decreased again to 11.6 in dry season 2018. The admission rate in 2015 and 2016 was twice the rate in 2013, before the CPP operation.

Table 2: Average Admission per 1000 Population at Puskesmas Palabuhanratu during the Dry Season (June–August)

Admission per 1000 Population	*Year*						
	2013	*2014 (Month 4–6)*	*2015 (Month 16–18)*	*2016 (Month 28–30)*	*2017 (Month 40–42)*	*2018 (Month 52–54)*	*2019 (Month-64)*
Respiratory problem	6.3	9.2	12.6	12.9	11.4	11.6	9.9
Myalgia	2.4	2.8	3.4	2.6	4.1	3.1	2.4

Starting from 2014, Indonesia has implemented a universal healthcare scheme. This scheme enables healthcare access for part of the population (mostly low income, informal sector) who previously did not have any form of health insurance. This might contribute to the increase of admission since some people who previously would not go to doctors or clinic due to financial reason might eventually go after the scheme was implemented. However, compared to other illness such as myalgia (Table 2), admission of respiratory problems shows a higher increase during the same period.

Figure 7(b) shows the admission of respiratory problems in Puskesmas Citarik from March 2014 through November 2018. Unfortunately, due to missing data between May 2015 and July 2017, no conclusion can be drawn.

CONCLUSIONS

Air pollutant concentrations at the CPP were higher than the surrounding settlement area. On the other hand, concentrations at the southern settlement were lower than the other locations. Despite the weak correlation, the increase of both air pollution parameters and admission of the respiratory problem should be considered as a mandatory concern. Thus, a periodic measurement should be performed to monitor the status of air pollution and community health in the area.

ACKNOWLEDGEMENTS

The authors are grateful to PT PLN (Persero), Java-Bali Generation Unit for their support that enabled this study.

REFERENCES

[1] BPS Kabupaten Sukabumi (2014). Kecamatan Palabuhanratu Dalam Angka 2014. BPS Kabupaten Sukabumi, Sukabumi, Indonesia.

[2] BPS Kabupaten Sukabumi (2018). Kecamatan Palabuhanratu Dalam Angka 2018. BPS Kabupaten Sukabumi, Sukabumi, Indonesia.

[3] Cooper, C.D. and Alley, F.C. (1994). Air Pollution Control: A Design Approach, 2nd ed. Waveland Press, Prospect Heights, IL, USA.

[4] EIA (2019). International Energy Outlook 2019 [WWW Document]. URL https://www.eia.gov/outlooks/ieo/pdf/ieo2019.pdf (accessed 1.23.20).

[5] Ministry of Energy and Mineral Resources (2019). Handbook of Energy & Economic Statistics of Indonesia [WWW Document]. URL https://www.esdm.go.id/assets/media/content/content-handbook-of-energy-and-economic-statistics-of-indonesia-2018-final-edition.pdf (accessed 1.23.20).

[6] Prehoda, E.W. and Pearce, J.M. (2017). Potential lives saved by replacing coal with solar photovoltaic electricity production in the U.S. Renew. Sustain. *Energy Rev.* 80, 710–715. https://doi.org/10.1016/j.rser.2017.05.119

[7] Shah, A.S.V., Lee, K.K., McAllister, D.A., Hunter, A., Nair, H., Whiteley, W., Langrish, J.P., Newby, D.E. and Mills, N.L. (2015). Short term exposure to air pollution and stroke: systematic review and meta-analysis. *BMJ 350*. https://doi.org/10.1136/bmj.h1295

[8] The Jamovi Project (2019). jamovi. (Version 1.1) [Computer Software]. URL https://www.jamovi.org.

[9] USEPA (2018). 2014 National Emissions Inventory, version 2 Technical Support Document [WWW Document]. URL https://www.epa.gov/sites/production/files/2018-07/documents/nei2014v2_tsd_05jul2018.pdf (accessed 1.23.20).

Monitoring and Analysis of CO_2 and NO_2 Levels: A Case Study of the Capital City of Mauritius

K. Boodhoo[1] and M.R. Lollchund[2]

[1]Department of Chemistry, University of Mauritius

[2]Department of Physics, University of Mauritius

[1]E-mail: kishore.boodhoo@uom.ac.mu

ABSTRACT: Nowadays, with the advent of motor vehicles and industries which release gases into the atmosphere, the quality of air in some regions of the world has deteriorated to an extent that in some cities like Mexico City the air has become unbreathable. To prevent such unwanted air quality problem, it is important to monitor and analyse the levels of carbon dioxide (CO_2) and nitrogen dioxide (NO_2) in different cities of the world. This paper focuses on Port-Louis, the capital city of Mauritius. The period of monitoring started from February 2015 for CO_2 while from August 2015 for NO_2. The in-situ measurements lasted for at least six months at four chosen sites (State Bank of Mauritius (SBM) Building, Air Mauritius Building, Mauritius Telecom Tower and Sun Trust Building). Analysis of the collected data revealed that the highest CO_2 concentration was 950 ppm while for NO_2 it was 110 ppb. Both highest records were during the morning traffic peak hours. Overall, CO_2 concentrations varied between 400 and 950 ppm whereas those for NO_2 varied between 40 and 110 ppb. A spectral analysis using the Empirical Ensemble Mode Decomposition (EEMD) method was also carried out to understand the periodicity in the data as well as the overall trend in the levels of the gases. This analysis confirmed the correlation of gas concentrations with traffic flow density and revealed a linear decrease in the gas's concentration in air. On the whole, the air in the capital of Port-Louis was noted to be of good quality.

Keywords: Air Quality Monitoring, Modelling, Carbon Dioxide, Nitrogen Dioxide, Spectral Analysis.

INTRODUCTION

Over the last three decades, greenhouse gases emissions have increased by an average of 1.6% per year with carbon dioxide (CO_2) emissions from the use of fossil fuels which is growing at a rate of 1.9% per year, provided no actions are taken (Pachauri and Meyer, 2014). Globally, there has been a net increase of 100 ppm in the carbon dioxide levels in 2007 compared to preindustrial times, giving rise to the phenomenon of enhanced greenhouse effect, and leading to global warming. The greenhouse gases mainly include water vapour, carbon dioxide, methane and nitrous oxide (Solomon, 2007). Global warming has significantly influenced the environment and economic prospects of a nation. According to UN's Intergovernmental Panel on Climate Change (IPCC), a reduction of 4 Wm^{-2} of outgoing infra radiation is expected if the concentration of CO_2 is doubled and other parameters are kept constant.

Nowadays, there is a high demand for energy because of rapid population and economic growth, especially in emerging market economies. Consequently, there is even more consumption of fossil fuels leading to higher greenhouse gas emissions. This change in climatic conditions will become drastic in the middle of the century. The IPCC (Pachauri and Meyer, 2014) has predicted that there could be a rise in the average global air temperature up to 5.8°C by 2100, if there is no control on the emissions of greenhouse gases. The effects will be disastrous where the ice caps will be melting at a faster rate and the global rainfall will dramatically increase leading to massive floods in some parts of the world. Although there will be higher humidity in the atmosphere, some areas will become even drier. For example, with the rise in sea level, there is intrusion of saltwater within the freshwater aquifers, leading to a definite loss of these resources. In addition, with the rise in temperature there is coral bleaching and with this phenomenon the natural habitats of the aquatic species are destroyed and hence the population of fish is drastically reduced. Other devastating effects include more intense cyclones and spread of diseases over longer span of time due to mosquitoes. Keeping the air quality acceptable has become an important task for decision makers as well as for non-governmental organizations (Patania *et al.*, 2009). Hence it is imperative to monitor the levels of carbon dioxide and nitrogen oxides in highly populated and highly industrialized areas in various parts of the world, including Mauritius.

An increase in the mean atmospheric temperature will provide more energy to the air and water, leading to extreme weather conditions where more intense rain showers and cyclones will be recorded. In addition, with the increase in air temperature, deeper layers of the ocean would become warmer over centuries giving way to sea level rise. It is predicted that there will be a half a meter rise in sea level by 2100. Low-lying areas in countries such as Bangladesh and Small Island developing states (SIDS) will get flooded. Already it is observed that there has been a 10–25 cm rise over the last 100 years. This is due to the melting of glaciers, but mostly from thermal expansion of seawater.

PREDICTED EFFECTS ON HUMAN HEALTH

Extreme heat waves will be felt in summers but fewer prolonged cold periods in winters. With the increase in the number of very hot days in temperate zones, very young and very old people as well as those suffering from respiratory diseases, heart disease and high blood pressure will get affected. With global warming, mosquitoes causing malaria, dengue and chikungunya will manifest longer during the year.

SOURCES OF CARBON DIOXIDE

Use of Coal

Coal is widely available as an energy reserve and is cheap to produce. Coal reserves are predicted to last for the next 200 years. Small proportion of ancient plant matter, that was covered over by water and could not be converted to carbon dioxide during that period, is converted to coal instead. When burnt, not only carbon dioxide and water are emitted, but also sulphur dioxide, fluorides, uranium and other radioactive metals, and heavy metals. Despite this disadvantage, coal is still used to produce electricity.

Petroleum and Natural Gas

They are primarily hydrocarbons, produced by cracking process of crude oil. One typical example is gasoline which is convenient, safe to handle and cheap to produce. In ocean sediments and permafrost, there is a significant amount of natural gas held in methane hydrates (chlathrates). More oil is found in tar sands and oil shales, but it is expensive to exploit these sources of natural gas.

On the other hand, high levels of nitrogen oxides contribute to photochemical smog, a situation where the air composition includes ozone, a carcinogenic agent. More precisely, nitrogen dioxide (NO_2) is responsible for the generation of smog, the most common form of air pollution which appears as a yellowish-brownish-gray haze in many big cities throughout the world. Smog is a term that signifies smoke and fog.

Ozone is produced due to photochemical reactions involving nitrogen oxides (NO_x gases) and unburnt hydrocarbons that are released from internal combustion engines. Another source of hydrocarbons is due to the evaporation of solvents, liquid fuels and other organic compounds. Collectively the latter are known as volatile organic compounds. The overall reaction is as follows:

$$\text{VOCs} + NO + O_2 \longrightarrow \text{Mixture of } O_3 + HNO_3 + \text{Organics}$$

The mechanism for the production of ground-level ozone is produced from NO_2 as follows:

$$NO_2 \xrightarrow{UV} NO + O$$

$$O + O_2 \longrightarrow O_3$$

NO_x GASES

They comprise of two gases, namely nitric oxide, NO, and nitrogen dioxide, NO_2. In clean air, the levels of NO_x are low and are due to reaction between nitrogen and oxygen during lightning flashes. They become a pollutant when the same reaction occurs at a high temperature in a combustion engine and are released to a much higher concentration:

$$N_2 + O_2 \longrightarrow 2NO$$

The higher the temperature the more NO is produced since reaction is endothermic. NO produced in such a way is known as thermal NO. The mechanism for its formation consists of two steps:

$$O_2 \longrightarrow 2O$$

$$N_2 + O \longrightarrow NO + N$$

The rate of determination of reaction is the second step.

NO is also produced to a lower extent from the oxidation of nitrogen atom present in the fuel itself; hence it is known as fuel NO. Nitric oxide is also emitted from electric power plants. The nitric oxide in air gets oxidized to nitrogen dioxide, NO_2 and the process may take minutes to hours. The yellow colour of the smog is due to the presence of NO_2. The latter absorbs visible light near 400 nm, removing the purple component while allowing the yellow light to be transmitted.

For the generation of smog, there are several conditions that must be met:

(a) High traffic flow to emit high levels of NO and volatile organic compounds.
(b) Abundant sunlight to allow photochemical reactions to proceed.
(c) Little movement of the air mass to prevent dilution of reactants.

International agreements have been signed with the government of Mauritius to curb the emissions of carbon dioxide. The latter is more and more involved into renewable sources of energy such as wind and sun.

For a preliminary study, the city of Port-Louis is selected since it is being everyday highly subjected to exhausts from motor vehicles nearly round the clock. Recent figures show that an annual average daily traffic (AADT) of 90,000 enters the capital (Dorsami and Puchooa, 2014).

AIM AND OBJECTIVES

The aim of this research work was to study the levels of carbon dioxide and nitrogen dioxide at four sites in the capital city of Mauritius, Port-Louis. The objectives were

1. To monitor the levels of these gases at the four sites and study their variations over specific periods of time (peak hours, night, week-ends, etc.)
2. To model their trends with time.

MATERIALS AND METHODS

Sites Description and Local Emission Sources

The city of Port-Louis (Figure 1) is the country's economic, cultured, political centre and most popular city with a population of around

148,000 (Statistics Mauritius, 2011). Moreover, Port-Louis has the main office and commercial land. Almost all government ministries and departments are in the capital. The hottest month is January where the average temperature is around 33° C, while the coldest month is July, with an average temperature of around 20° C. The wettest month is in March with an average rainfall of 221 mm of rain.

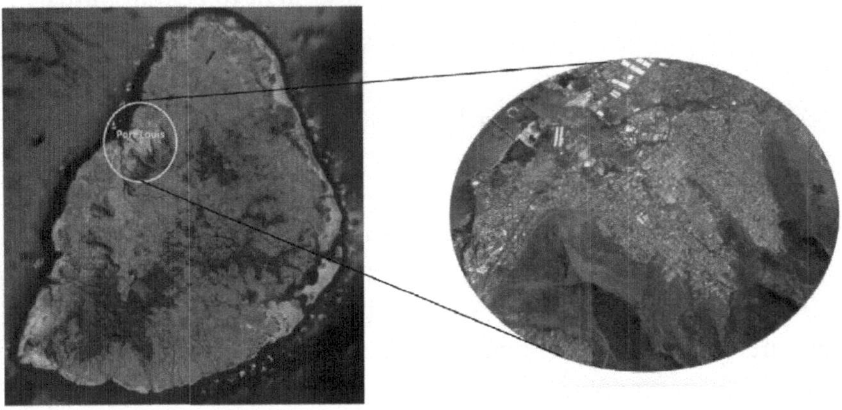

Figure 1: Location of the City of Port-Louis

There is severe traffic congestion at the entrances to Port-Louis in the morning and at the exits in the evening, caused mainly by work trips. Within Port-Louis, there is congestion in the morning and evening. Table 1 below shows the projected traffic flow:

Table 1: Projected Traffic Flow in Port-Louis from 2015 to 2020 (Dorsami and Puchooa, 2014)

Year	*Traffic Growth Rate*	*Motorcycles*	*Cars*	*Class B Heavy Vehicles*	*Class C Heavy Vehicles*	*Buses*	*Total Vehicles*
2015	6%	363	3628	195	9	270	4465
2016	6%	385	3846	207	9	286	4733
2017	6%	408	4077	219	10	303	5017
2018	6%	432	4322	232	10	321	5317
2019	6%	458	4581	246	11	340	5636
2020	6%	485	4856	261	11	360	5963

For a small island of Mauritius and above all, for a small city like Port-Louis, with such a high number of motor vehicles visiting the latter, there is bound to have very high levels of CO_2 and NO_2 in the region. More frightening still, is the projected number of vehicles visiting the capital in the few years to come.

In order to monitor the levels of CO_2 and NO_2 four sites were chosen (Figure 2) as follows:

1. State Bank of Mauritius (SBM) Building
2. Air Mauritius Building
3. Mauritius Telecom Tower
4. Sun Trust Building.

Figure 2: Locations where the Detectors were Placed (1. SBM, 2. Air Mauritius, 3. Mauritius Telecom Tower, 4. Sun Trust building)

The reason for choosing these sites was that they are highly frequented areas of the capital. It should be mentioned that simultaneous continuous measurements of carbon dioxide have been made since February 2015 at the four sites. Similarly, nitrogen dioxide measurements have been made since August 2015. Mobile monitoring system for measuring the concentration of the gases, which provide quality data continuously 24 hrs. daily during the measurement campaign, was installed on the ground floor of the buildings.

RESULTS AND DISCUSSIONS

CO_2 Levels

As mentioned earlier, carbon dioxide is mainly produced by the combustion of fossil fuels. In clean air, the concentration of the latter is about 0.04%. Despite this very small concentration, it is a highly potent greenhouse gas owing to its anti-symmetric and bending vibrations absorbing and emitting in the thermal IR window.

Carbon dioxide levels were measured using CO_2 data-logger model CO_2 from Extech instruments and distributed by MicroDAQ.com Ltd, USA. This apparatus monitors for CO_2 concentrations, measures temperature as well as relative humidity (RH). The meter updates readings every 30 seconds and records readings for long time environment monitoring. The memory capacity is 15999 points (5333 RH, temperature and CO_2 records).

Time Series

Figures 3, 4 and 5 below are examples of time series measurements of carbon dioxide on a particular day.

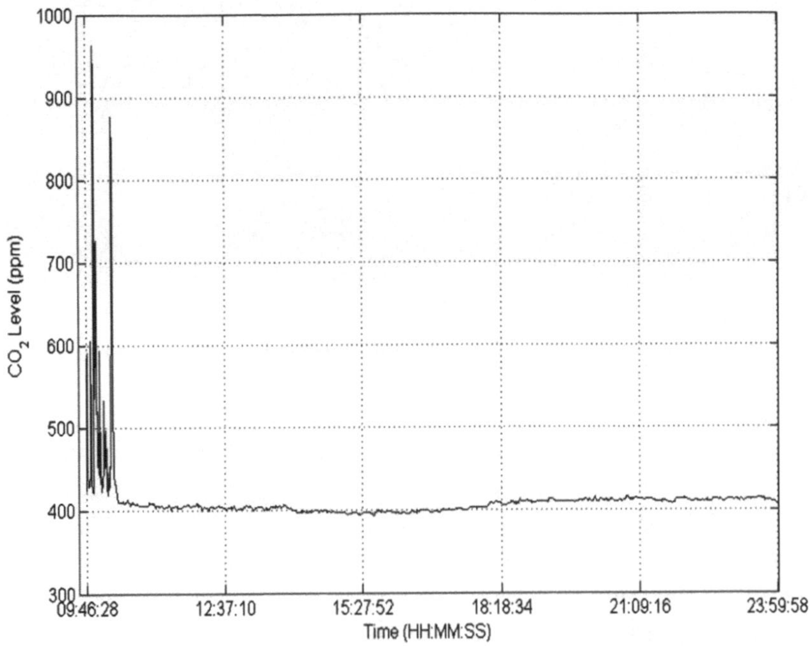

Figure 3: CO_2 Levels at SBM Building on 30 March 2015

It can be observed at SBM Building during the rush hours in the morning, before 10 am, the levels of carbon dioxide are very high, reaching nearly 950 ppm, compared to those in the afternoon. This high level of CO_2 can be attributed to the high inflow of traffic to the capital. After 10 am the levels of CO_2 have gone down to about 400 ppm throughout the rest of the day and up to midnight. This clearly demonstrates that there has been quick diffusion of the gas owing to the fact that Port-Louis is along the coastline.

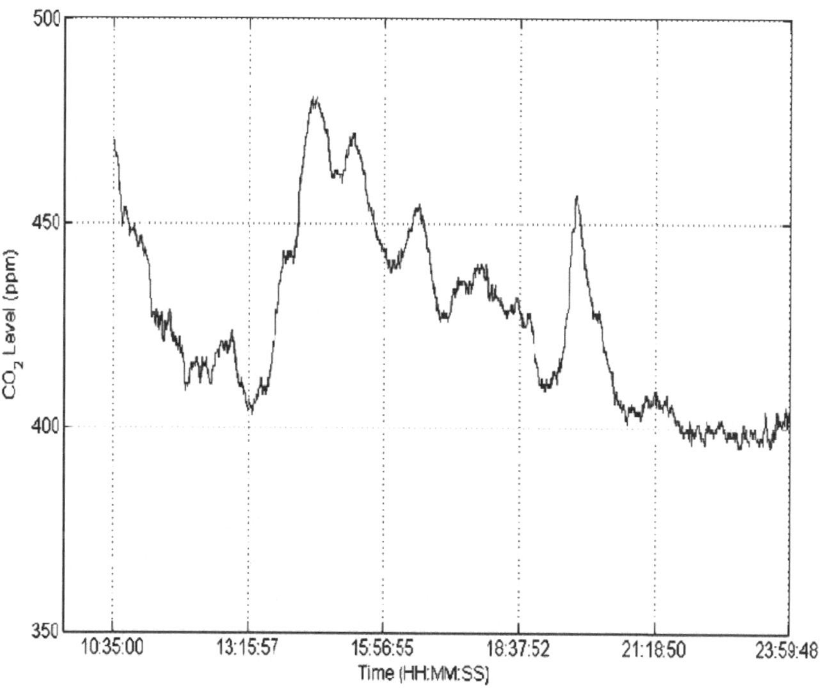

Figure 4: CO_2 Levels at Sun Trust Building on 01 April 2015

At Sun Trust building, in the early hours of the day, the levels of CO_2 were quite high, about approximately 475 ppm. However, in the afternoon, after about 3 pm, the levels can be observed to start increasing and reached the peak levels at about approximately 475 ppm at nearly 4 pm. Again, this can be explained by the fact that at this time of the day there is high flow of traffic since the offices get closed. Thereafter, the levels of carbon dioxide start to decrease. Interestingly, the levels of that particular greenhouse gas started to increase again at 8 pm and came back down to 400 ppm within an hour.

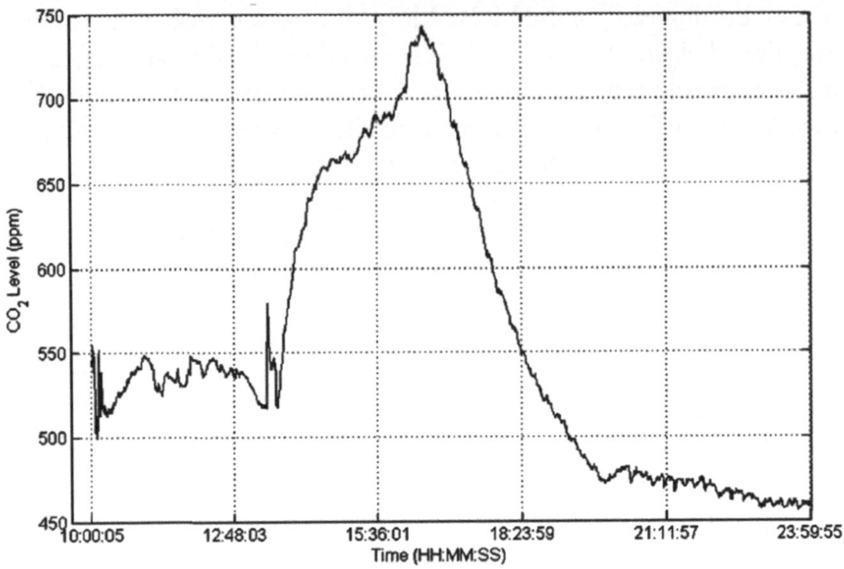

Figure 5: CO_2 Levels at Mauritius Telecom Building on 01 April 2015

At Telecom tower, strangely, the levels observed for carbon dioxide were opposite to what we have seen for the other two sites, in other words, in the morning the levels were maintained under 550 ppm while in the afternoon, they showed a maximum just under 750 ppm. After 6 pm, the values started to decrease and reached 450 ppm at midnight. Again, this scenario can be explained by high vehicular traffic in the afternoon when people are heading home. A plausible reason for the long time period for carbon dioxide to reach 450 ppm can be attributed to the fact that there is a slow diffusion of the air around that building since it is a highly congested area and air circulation is not favoured.

Spectral Analysis Using the EEMD Method

EMD basically detaches non-linear oscillatory patterns of higher frequencies from those of lower frequencies in a set of data. The method is straightforward. For a given time series signal EMD represents the signal in a set of basis functions (simpler signals), which are termed as intrinsic mode functions (IMF), using local temporal and structural characteristics of the data. An IMF should satisfy the following properties: (1) zero mean; and (2) the number of local extrema equals the number of zero crossings ± 1 (Huang *et al.*, 1998; Coughlin & Tung, 2005; Molla *et al.*, 2006).

The IMFs are obtained by the following procedure:

- Connect the maxima in using smooth lines and denote this curve as upper envelope, $E_u(t)$. Similarly, connect the maxima in $x(t)$ using smooth lines and denote this curve as lower envelope, $E_l(t)$.
- Compute the mean of these envelopes:

$$m(t) = 1/2(E_u(t) + E_l(t)) \quad \ldots (1)$$

It should be noted that $m(t)$, at any temporal location, is nearly equal to the lowest frequency component in the data.

- Subtract the mean of the envelopes from $x(t)$ and store the result in $h_1(t)$.

  That is

$$h_1(t) = x(t) - m(t) \quad \ldots (2)$$

  This results into isolating the higher frequency components from the lower frequency components in the original signal. Ideally $h_1(t)$ is the first IMF.

- If $h_1(t)$ does not satisfy the properties of an IMF, then it is treated as a "new" signal and is re-entered in the process of computing the IMFs. This iteration, known as the sifting process, is repeated until the IMF is obtained.
- Repeat the process until all IMFs, $h_k(t)$, are obtained. At this final stage (say after n iterations), a constant or monotonic function $r_n(t)$ remains which is termed as the residual. It represents any trend within the original series.

It is intuitive to observe that the original signal can be reconstructed by summing all IMFs and the residual using,

$$\sum_{k=1}^{n} h_k(t) + r_n(t) \quad \ldots (3)$$

The IMF components and the residual have physical significance.

The EEMD algorithm for the time series signal $x(t)$ can be described as follows (Wu & Huang, 2009).

- Generate $x^i(t) = x(t) + w^i(t)$ where $w^i(t)$ (i = 1,..., N) are different realizations of white Gaussian noise.
- Each $x^i(t)$ (i =1,..., N) is fully decomposed using the EMD algorithm described above to obtain their IMFs ($h_k^i(t)$), where k = 1,..., n.
- Assign $\overline{h_k(t)}$ as the k-th mode of $x(t)$, which is obtained as the average of the corresponding IMFs using,

$$\overline{h_k(t)} = 1/N \sum_{i=1}^{N} h_k^i(t) \quad \ldots (4)$$

It should also be mentioned that the white noise introduced in EEMD helps in the separation of different timescales in noisy data, but has no involvement in the final IMFs (Wu *et al.*, 2014). The added noise cancels out in the ensemble average. The EEMD method is very effective in extracting signals contained in the data, and represents a major enhancement of the EMD method (Wu & Huang, 2009; Sharma & Kaur, 2014; Wu *et al.*, 2014).

The EEMD method was applied to long term data measured at SBM. Figure 6 displays the resulting time-series of the data and its complete decomposition by EEMD in terms of IMFs are shown in Figure 7 and a residual in figure 8. The minimum value of this time-series is 560 ppm, maximum is 701 ppm, and mean is 595 ppm with a standard deviation of 14.1 ppm. It can also be observed that the EEMD decomposition separates the time series into six IMFs and a trend.

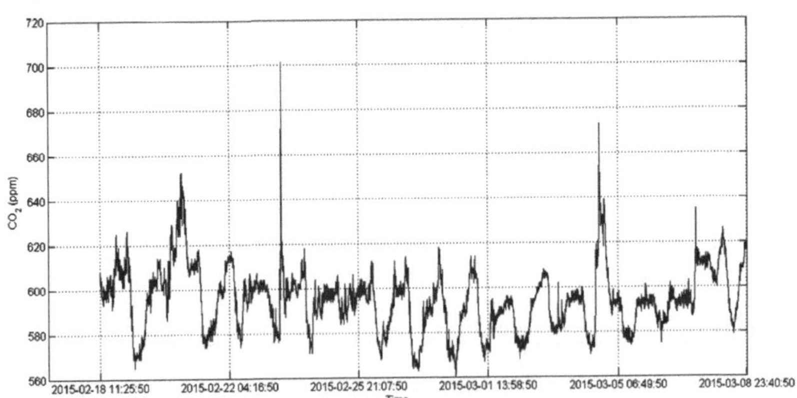

Figure 6: Long Term CO_2 Data Measured at SBM Building

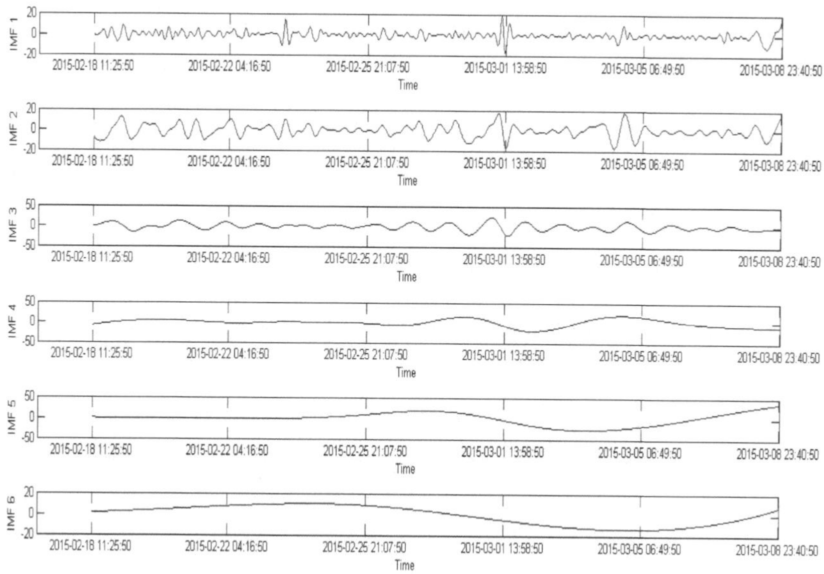

Figure 7: IMF for CO_2 Data Measured at SBM Building

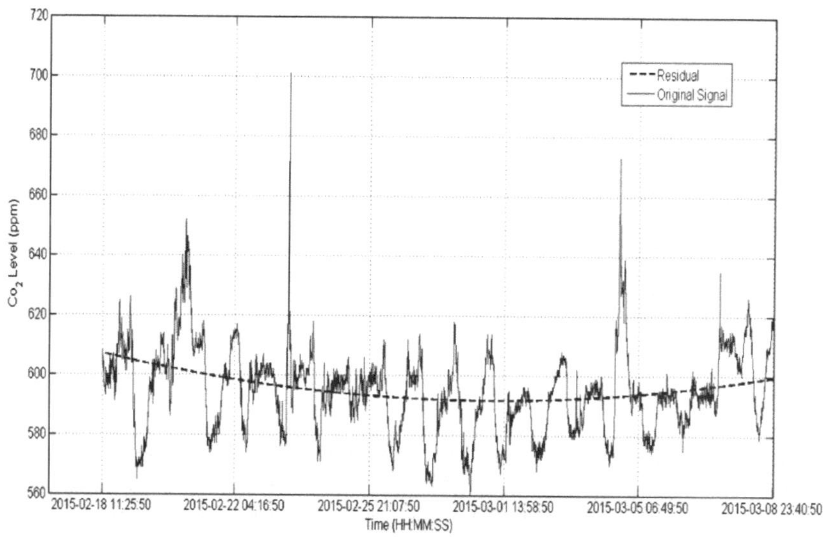

Figure 8: The CO_2 Data and its Residual Obtained by EEMD

NO_2 Levels

The detector used was from Chi'an Tech Instruments Co. Ltd. It is portable and includes data logging function. Unfortunately, only 2 detectors were obtained and they were placed only on SBM building.

Figure 9 below shows the time series of the levels of the nitrogen dioxides over about three months of sampling.

Time Series

It can be observed that over the period from August 2015 'to November 2015 there has been a linear decrease in terms of NO_2 concentrations. The maximum concentration noted was about 110 ppb whereas the minimum concentration was about 40 ppb. In this range, it is fair to say that the air is of good quality since the limit allowed for NO_2 in the atmosphere is 100 ppb. This decrease in the NO_2 levels can be attributed to the fact that there is land breeze and sea breeze which work in synergy to reduce the NO_2 concentration in air.

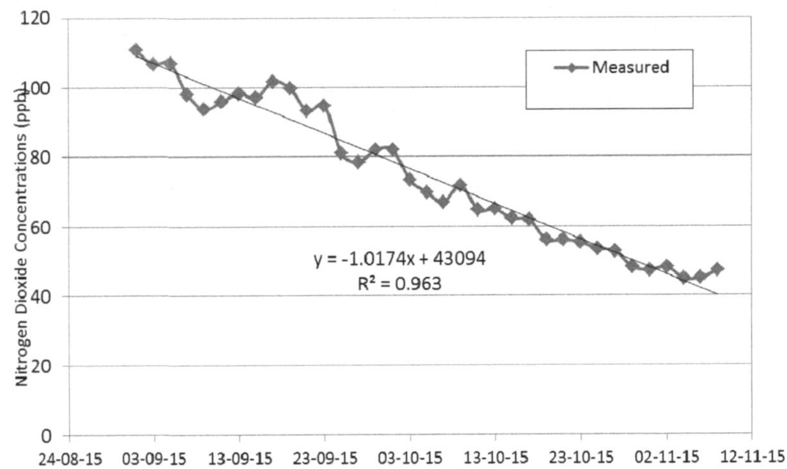

Figure 9: Time Evolution of NO_2 Concentrations in Air on SBM Building

CONCLUSIONS

Measurements and treatment of data related to the concentrations of carbon dioxide and nitrogen dioxide revealed that the air quality in the region of Port-Louis is relatively good compared to polluted cities like Mexico City. The major source of these gases is from the combustion engines in motor vehicles. Despite the huge number of vehicles visiting the capital, the air quality is still within the permissible limits thanks to the action of land breeze and sea breeze. However, we need to be very careful given that the number of motor vehicles keeps on increasing.

REFERENCES

[1] Coughlin K. and Tung K.K., (2005). Empirical mode decomposition of climate variability", WSPC, chapter7. http://depts.washington.edu/amath/old_website/research/articles/Tung/journals/coughlin-tungHHT05.pdf.

[2] Dorsamy, C. and Puchooa, C. (2014). Alleviating Traffic Congestion along the M1 corridor—An economic perspective. In Road Development Authority. Report, http://rda.govmu.org/english//DOCUMENTS/ECONOMIC%20PERSPECTIVE%20ARTICLE.PDF, pp. 1–11.

[3] Huang, N.E., Shen, Z., Long, S.R., Wu, M.C., Shih, H.H., Zheng, Q., Yen, N.C., Tung, C.C. and Liu, H.H., (1998). The empirical mode decomposetion and the Hilbert spectrum for nonlinear and non-stationary time series analysis, *Proc. Roy. Soc.,* A, 454, pp 903–995.

[4] Molla, M.K.I., Rahman, M.S., Sumi, A. and Bamik, P. (2006). Empirical mode decomposition analysis of climate changes with special reference to rainfall data. *Discrete Dynamics in Nature and Society*, Vol. 2006, pp. 1–17.

[5] Pachauri, K.R. and Meyer L. (2014). IPCC climate change 2014 synthesis report, Geneva, Switzerland.

[6] Patania, F., A. Gagliano, A., F. Nocera, F. and Galesi, A. (2009). Air quality in street canyons: a case study. *WIT Transactions on Ecology and the Environment*, Vol. 123, WIT Press, England.

[7] Sharma, B. and Kaur, S. (2014). Distinction between EMD & EEMD Algorithm for pitch detection in speech processing. *International Journal of Engineering Trends and Technology (IJETT),* 7(3), pp 119–125.

[8] Solomon, S. (2007). Climate Change 2007—The Physical Science Basis. "Contribution of Working Group I to the Fourth Assessment Report of the IPCC". Cambridge, UK. https://www.ipcc.ch/report/ar4/wg1

[9] Statistics Mauritius (2011). Census 2011. http://statsmauritius.govmu.org/English/ Census and Surveys/Pages/census/Census-2011.aspx

[10] Wu, Z. and Huang, N.E. (2009). Ensemble empirical mode decomposition: A noise-assisted data analysis method. *Advances in Adaptive Data Analysis,* 1(1), pp. 1–41.

[11] Wu Z., Huang N.E. and Wallace J.M. (2014). Adaptive and local analysis of climate data. *Engineering Sciences*, 12(2), pp. 41–45.

Study of Urban Area Air Quality in Yangon, Myanmar

Moe Myitzu
Department of Research and Innovation, Yangon
E-mail: moemyitzu@gmail.com

ABSTRACT: Particulate Matters sampling for both coarse and fine ones have been carried out during the period from March 2011 to March 2012. Air sample set (course and fine) were collected at the urban area (Food and Environmental Monitoring Division, Department of Atomic Energy), Bahan Township, in Yangon, Myanmar. Air sample set (course and fine) was collected at urban area using GENT PM-10 Stacked Filter Unit with Nuclepore polycarbonate filters, period was 24 hours and frequency was once a week. The ambient mass and element concentration in Air sample were determined. The particulate mass concentration was determined by Micro balance and Elemental analysis of particulate matter (PM) on filter determined using Wavelength Dispersive X-ray Fluorescence Spectrometer (S8 Tiger, BRUKER). The Black Carbon concentration of Air filters were determined by Smoke Stain Reflectometer (M-43 D). PM concentration of air filters recorded the highest concentration in the winter while the lowest concentration was record in the rainy season.

Keywords: Particulate Matter, Coarse, Fine, Air Filter, Elemental Analysis, Black Carbon, XRF Analysis.

INTRODUCTION

Air pollution has become a matter of global concern because of its impact on human health and environmental quality. Air pollution is the biggest environmental rick for public health (Santoso *et al.*, 2011). Air pollution contributes to many health problems such as cardiovascular disease, liver and blood disease, headache, anxiety, eye/nose/throat irritation, asthma, nervous system disorders, lung cancer, problems of the reproductive system and other chronic and long term diseases. The health risks are associated with the total suspended particulate matter (TSPM) (Athanasios *et al.*, 2008). Particulate matter consists of solid particle of organic and inorganic substance suspended in the air (Kate *et al.,* 2015). The components of particulate matter are Sulphur, Nitrogen, Sodium, Chlorine, Black Carbon and some toxic elements. The small particle in the air reach the heart and lungs and can result in heart attack, asthma, decreased lung function, coughing, wheezing, bronchitis and ultimately a reduction in life expectancy.

There are two main types of air pollution—ambient air pollution (outdoor pollution) and household (or indoor) air pollution which refers to pollution generated by household combustion of fuels (caused by burning fuel such as coal, wood or kerosene) using open fires or basic stoves in poorly ventilated spaces. Both indoor and outdoor air pollution can contribute to each other, as air moves from inside buildings to the outside, and vice versa. A major component in Air pollution is particulate matter (PM) which can be course and fine. Coarse particles can be regarded as those with a diameter ranging from 2.5 to 10 µm ($PM_{2.5\text{-}10}$); and fine particles are those less than 2.5 µm ($PM_{2.5}$) (Graizia *et al.,* 2001).

X-ray fluorescence (XRF) spectrometry is an elemental analysis technique with broad application in science and industry. XRF is based on the principle that individual atoms, when excited by an external energy source, emit X-ray photons of a characteristic energy or wavelength. By counting the number of photons of each energy emitted from a sample, the elements present may be identified and quantitated (James *et al.*, 2012). The identification of the elements by X-ray methods is possible due to the characteristic radiation emitted from the inner electronic shells of the atoms under certain conditions. The emitted quanta of radiation are X-ray photons whose specific energies permit the identification of their source atoms (James *et al.*, 2012). The method is fast, accurate and non-destructive, and usually requires only a minimum of sample preparation. Applications are very broad and include the metal, cement, oil, polymer, plastic and food industries, along with mining, mineralogy and geology, and environmental analysis. XRF is also a very useful analysis technique for research and pharmacy. Spectrometer systems can be divided into two main groups: energy dispersive systems (EDXRF) and wavelength dispersive systems (WDXRF). The elements that can be analyzed and their detection levels mainly depend on the spectrometer system used. The elemental range for EDXRF goes from sodium to uranium (Na to U). For WDXRF it is even wider, from beryllium to uranium (Be to U). The concentration range goes from (sub) ppm levels to 100%. The elements with high atomic numbers have better detection limits than the lighter elements. The precision and reproducibility of XRF analysis is very high. Very accurate results are possible when good standard specimens are available, but also in applications where no specific standards can be found. The measurement time depends on the number of elements to be determined and the required accuracy, and varies between seconds and 30 minutes. The analysis time after the measurement is only a few seconds. X-Ray Fluorescence (XRF) is widely used to the filter samples of airborne particulate matters (Watson *et al.*, 1999).

Yangon served as the capital of Myanmar until 2006. Yangon is Myanmar's most popular city and its most important commercial center. Many construction sites, industries including tobacco, Garment factories are located in Yangon. Thus, suspended particulate matter is a main pollution and it may become a problem for the public health.

The main purpose of this study was to known sources of particulates and reduction the air pollution of urban area.

MATERIALS AND METHODS

Samples' Collection

Particulate Matters sampling for both coarse and fine have been carried out during the period from March 2011 to March 2012. Air sample set (course and fine) were collected at the urban area (Food and Environmental Monitoring Division, Department of Atomic Energy), Bahan Township (16.8006°N, 96.1655°E), in Yangon, Myanmar. Figure 1 is the sampling site.

In this research, both fine filters (for fine particles smaller than 2.5 µm ($PM_{2.5}$) and coarse filters (for coarse particles between 2.5–10 µm ($PM_{2.5-10}$), 47 mm diameter Nuclepore Polycarbonate filters are used. "GENT" Stacked Filter Unit (SFU) is used for sample collection. The sampler operated at a flow rate of 17 liters per min, period was 24 hours and the frequency was once a week. The air is drawn through the sampler by means of a diaphragm vacuum pump, which is enclosed in a special housing together with a needle valve, vacuum gauge, flow meter, volume meter, time switch (for interrupted sampling) and hour meter. It consists a double (staked) filter cassette (which should be loaded with two different Nuclepore filters) (Maenhaut *et al.,* 1994).

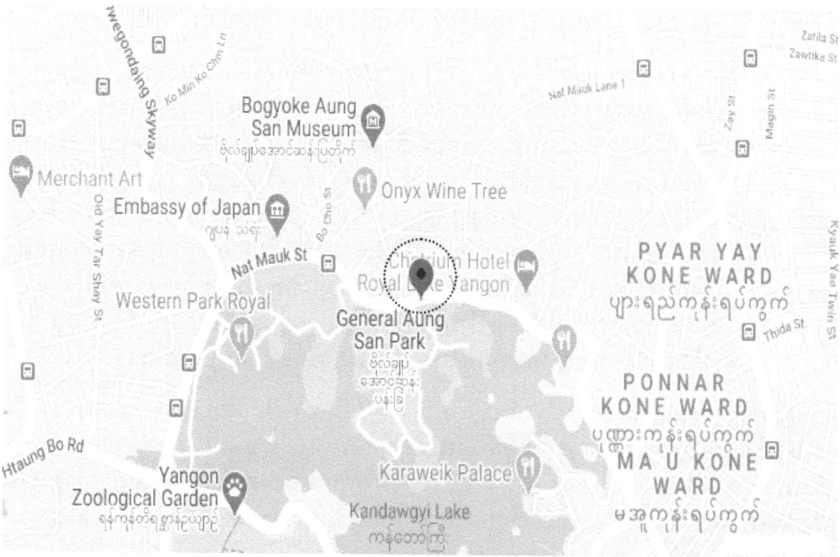

Figure 1: Sampling Site (red/circle), Bahan Township, Yangon

Analytical Technique

The mass of particulate matters on both coarse and fine filter were determined by weighing the filters before and after exposure. Then they were divided by the volume of air passing through the filters to obtain the concentration of $PM_{2.5}$ and $PM_{10-2.5}$ ($\mu g/m^3$) respectively.

The Black Carbon concentration of Air filters were determined by Smoke Stain Reflectometer (M-43 D). The digital model consists of an LCD display meter unit and a measuring head, which is connected to the unit. The unit also carries the main switch (at the back of the meter), coarse and fine sensitivity controls and the zero control. The Reflectometer must be used on a flat, level surface with its LCD digital display facing forwards.

WDXRF analytical method can be applied to determine the trace elements on air filters. The WDXRF method is characterized by excellent spectral resolution, which minimize peak overlaps (Watson *et al.*, 1999). In this research work, S8 Tiger WDXRF (4K) in National Analytical Laboratory was used to determine the concentration of toxic elements which can cause both short- and long-term medical problems. Standard Reference Material (SRM 2783, Air Particulate on Filter Media) were used for the calibration of common and toxic elements contained in various fraction of airborne particulate matter. A unit of SRM (2783) included two loaded filters and two blank filters.

WD-XRF measurements were performed with a wavelength-dispersive S8 Tiger x-ray spectrometer (Bruker, Germany) provided with Spectra Plus software. The system is equipped with (i) a rhodium x-ray tube, a 4 kW generator, an eight-position crystal changer, a 56-position sample chamber, (ii) a scintillation counter (detector) and (iii) a flow-proportional counter (detector). The spectrometer operates under vacuum conditions and a 34 mm channel mask. The samples were placed in a sampler holder (Bruker) with a diameter of 34 mm and covered with an aluminum non-backscattering cup (Bruker).

In X-ray tubes, a high voltage of 10–60 kV is applied between the cathode and the anode to produce X-ray. It is used to accelerate the electrons. The electrons are emitted from a heated wolfram filament, the cathode. They are accelerated by the high voltage and collide with the anode. The anode consists of a metal (Rh). When colliding with the target, the accelerated electrons interact with the electrons in the metal, releasing part of their energy as photons. These electromagnetic waves, called radiation, with a wavelength of 0.1 to 2 A° are emitted, besides continuous radiation also X-rays with a characteristic wavelength. The X-rays are generated in high vacuum. The X-rays leave the beryllium window at the front of the tube.

A scintillation counter measures ionizing radiation. The scintillator consists of a transparent crystal that fluoresces when struck by ionizing radiation. A sensitive photomultiplier tube measured the light from the crystal. When radiation strikes the scintillator a flash of light is produced. The photomultiplier amplifies this signal and measures the intensity. The intensity of the light flash depends on the energy of the charged particles. Cesium Iodine (CsI) in crystalline form is used as the scintillator for the detection of photons and alpha particles; sodium iodide (NaI) containing a small amount of thallium is used as a scintillator for the detection of X-rays.

A proportional counter is a type of gaseous ionization detector. An inert gas is used to fill the tube, with a quench gas added as a stabilizer. A common proportional gas mixture is 90% argon, 10% methane known as P-10 gas. An incoming ionizing particle, if it has sufficient energy, liberates electrons from the atomic orbitals of the gas atom, leaving an electron and positively charged atom (ion pair). The electron drift under the influence of the electric field of the proportional counter toward the electrode (anode). At the same time, the positive ions move toward the cathode. On their way to the cathode, the drifting electrons gain enough energy to create further ion pairs. If the operating voltage is chosen carefully, an avalanche process occurs that exponentially increase the total number of electrons in the detector. This amplified signal is read out; it is proportional to the intensity of X-rays.

RESULT AND DISCUSSION

Average concentration of PM mass for 12 months (April, 2011) to (March, 2012) and average values of Black Carbon concentration are shown in Table 1. Elemental Concentration of $PM_{2.5}$ is summarized in Table 2.

Table 1: Average Concentration of PM Mass and Black Carbon Concentration (μg/m^3)

No.	*Time*	*Particulate Matter (μg/m^3)*		*Black Carbon (μg/m^3)*	
		Coarse	*Fine*	*Coarse*	*Fine*
1.	April, 2011	28.25	22.50	2.55	6.29
2.	May, 2011	26.50	21.50	2.00	7.20
3.	June, 2011	23.50	22.12	3.05	5.23
4.	July, 2011	18.25	16.13	3.05	4.46
5.	August, 2011	12.87	22.77	4.52	3.42
6.	September, 2011	11.25	19.56	7.01	4.81
7.	October, 2011	15.09	17.43	5.02	6.22
8.	November, 2011	23.62	21.65	4.40	5.66
9.	December, 2011	30.31	15.24	4.42	8.35
10.	January, 2012	32.23	15.85	3.24	9.00
11.	February, 2012	37.64	23.28	4.48	7.29
12.	March, 2012	33.31	20.02	3.52	6.55

Table 2: Elemental Concentration of $PM_{2.5}$ (ng/m^3)

No.	*Elements*	*Concentration (ng/m^3)*		
		April, 2011	*September, 2011*	*February, 2012*
1.	Al	51.7	34.5	60.9
2.	As	12	–	15
3.	Ca	2360	1920	5900
4.	Cr	–	–	5.2
5.	Cu	4.8	–	–
6.	Fe	68	120	202
7.	K	145	230	270
8.	Mg	35	65	47
9.	Mn	7	–	–
10.	Na	65	42	152
11.	Ni	7	–	–
12.	Pb	24	–	35
13.	S	215	521	123
14.	Si	546	421	1600
15.	Ti	–	–	10.93
16.	Zn	96	–	134

According to the values in Table 1, the mass concentration on fine filter of February (winter) is highest concentration (23.38 $\mu g/m^3$) and October (rainy) is the lowest concentration (17.43 $\mu g/m^3$) and the mass concentration on coarse filter of February (winter) is also highest concentration (37.64 $\mu g/m^3$) and September(rainy) is the lowest concentration (11.25 $\mu g/m^3$). The average value of concentration of mass in Air show seasonal significant variation.

The elemental concentration of Ca (Calcium) and Si (Silicon) are high value in three seasons, 2360 (ng/m^3), 1920 (ng/m^3) and 5900 (ng/m^3) and 546 (ng/m^3), 421 (ng/m^3) and 1600 (ng/m^3) respectively. They could be from construction site. Al, Ca, Fe and Si were could be from the contaminating road dust.

CONCLUSION

The result of particulate mass concentration of air filters was within the World Health Organization (WHO) recommended standards.

These results showed that, air pollution recorded the highest concentration in the winter season (November to February) and the lowest concentration was recorded during the Rainy season (June to October). The concentrations of Black Carbon on filters were 3–9 µg/m^3 and 2–7 µg/m for fine and coarse ones, respectively. The results showed that, the main sources of air pollution in Myanmar were inefficient modes of construction, vehicles and industrial zone. Since X-ray fluorescence analysis can analyze a sample non-destructively and quickly, it can be applied to a wide range of uses such as manufacturing and quality control. Recently, the quantitative analysis of trace elements became possible because of high-sensitivity technologies such as filtering, which eliminates background interference, and thin film methods. In the future, X-ray fluorescence analysis will become more widespread particularly in measuring hazardous metals in materials and soils.

ACKNOWLEDGEMENTS

The author would like to acknowledge the Food and Environmental division, Atomic Energy Department for the use of GENT air sampler for sampling, micro balance for weighing and Smoke Stain Reflectometer for black carbon measurement. The author also thanks to National Analytical Laboratory for the use of WDXRF spectrometer for elemental analysis.

REFERENCES

[1] Adams, Kate (2015). Particulate matter Components, Sources and Health: Systematic Approaches to Testing Effects, *Journal of the Air & Waste Management Association*, pp. 544–558.

[2] BRUKER AXS GmbH (2011). S8 Tiger X-ray Spectrometer Operation Manual, Karisruhe, Germany, pp. 2.1–2.8.

[3] Gutheri, James M. and Ferguson, Jeffrey R. (2012). Overview of X-ray Fluorescence, University of Missouri Research Reactor, Columbia.

[4] He, Kebin (2001). The Characteristic of $PM_{2.5}$ in Beijing, China, *Atmospheric Environment*, 35, pp. 4959–4970.

[5] https://archaeometry.missouri.edu/xrf_overview.html

[6] Maenhaut, W. (1994). The "GENT" Stacked Filter Unit (SFU) Sampler for the Collection of Atmospheric Aerosols in Two Size Fractions: Description and Instructions for Installation and Use, pp. 249–263, Gent, Belgium.

[7] Marcazzan, Grazia M.; Vaccaro, Stefano; Valli, Gianluigi and Vecchi, Roberta (2001). Characterisation of PM_{10} and $PM_{2.5}$ particulate matter in the ambient air of Milan (Italy), *Atmospheric Environment* 35 (2001), pp. 4639–4650.

[8] Santoso, M. (2011). Preliminary study of the Ambient Air Pollution in Serpong, Indonesia, Atmospheric Pollution Research 2, pp. 190–196.

[9] Valavanidis, Athanasios (2008). Airborne Particulate Matter and Human Health: Toxicological Assessment and Importance of Size and Composition of Particles for Oxidative Damage and Carcinogenic Mechanisms, *Journal of Environmental Science and Health*, pp. 339–362.

[10] Watson, J.G. (1999). Elemental Analysis of Airborne Particle, *Amsterdam, The Netherlands,* pp. 67–96.

Case Study of Quantifying the Carbon Emission of the Crematoria to Identify Carbon Sinkable Tea Vegetation Requirement to Counterpoise the Carbon Foot Print of Crematoria

Thammita A.S. Anuruddha[1], E.A.N.K. Edirisinghe[2], T.L. Wijerathne[3] and N.S.K. Samarasinghe[4]

[1,2,4]National Engineering Research and Development Center, Srilanka

[3]Srilanka Tea Research Institute

E-mail: [1]anuruddhalk@gmail.com; [2]nandana@nerdc.lk; [3]thushari.wijeratne@gmail.com; [4]nalaka1020@gmail.com

ABSTRACT: Anthropogenic emissions which harm terrestrial ecosystems are one of the major concerns to up-keep the environmental sustainability. This paper discusses a study of the volumetric carbon emission of the crematoria and attempting to identify the replantation of commercial crop (in this case, TEA, Camellia sinensis) requirement to make the "environmental effect" neutral caused by its carbon sequestration. In the modern world, the usage of crematoria is rapidly increased due to the limited availability of burial grounds. Making crematoria comply with emission standards, is not much concerned particularly in third world countries. National Engineering Research and Development Center (NRDC) is the pioneer developer of several versions of crematorium solutions. It transfers technology to interested commercial fabricators of crematoria. This study is based on emission data of NERDC cremator unit developments. However, this estimation of Carbon Dioxide emissions was done based on the latest version which complies with the EURO standards. The possibility of counterpoising these emissions using tea plantations in different regions of Sri Lanka were explored based on different levels of compliance to the recommendations of the Tea Research Institute (TRI) considering the reported values of their carbon sequestration potentials.

Keywords: Crematoria Emission, Carbon sink of TEA, Camellia sinensis, TRI, NERDC, CO_2 Emission Counterpoise.

INTRODUCTION

In global climate index 2019 published by German watch, Sri Lanka has been consecutively placed twice in the second position which implies that the Sri Lanka is the second most affected country in terms of natural disasters that occur as a consequence of climatic-changes. Being an island country in the Indian Ocean, located close to the south-east of India, controlling of carbon emission is very important to preserve the both countries' local climatic stability.

Nowadays, the land shortage is a significant problem for highly populated countries. Monaco, Japan, Sri Lanka, Bangladesh are some of the highly populated countries, which suffer from such limited land availability difficulties.[3] Less availability of burial grounds forces the increased use of crematoriums for the funerals of their loved ones. Cremation of corpses becomes the most viable solution all-around in highly populated countries. However, due to religious and cultural grounds, some communities do not prefer to cremate corpses.

In functionality-wise, crematoria are not far different from incinerators. Although crematoria are also combustors, in legal/regulatory point of view, these facilities are not considered as incinerators.[6] In Sri Lanka, a constituted law is not still available for crematoria. The National Engineering Research and Development Centre (NERDC) transfers the latest low emission technology to licensed companies and other entities such as NGOs and those who are interested in fabricating crematoria and incinerators. There are few suppliers who are importing crematoria from other countries, also available. Local governments are purchasing and installing crematoria according to their regional demand. An island-wide crematoria survey was done by NERDC in 2018. There are 265 crematoria in operation around the island including few imported units from foreign countries. At least one corpse is cremated daily in each facility makes 265 corpses per day and 68,900 per year which makes heavy emission of carbon variations to the atmosphere. In this study, CO_2 emission was calculated in the cremation process and the required volume of possible commercial crops was also calculated where the cultivation of crops in new areas can counterpoise the environmental effect by absorbing the amount of

CO_2 of crematoria all over the island. It is suggested that the local governments support and encourage the commercial growers to replant such amount of vegetation in the island.

Carbon sequestration of crop plants i.e. the long-term storage of atmospheric CO_2 in plant biomass is a hot topic in the current era where it is considered as a method of mitigating the impacts of climate change. Therefore, the carbon sequestration potentials of different plant species have been determined in the world at different levels.

Tea (Camellia sinensis) is one of the major commercial crops cultivated in a vast extent of land in Sri Lanka. Its carbon sequestration potential has also been quantified in different elevations.[8,9] As the tea industry contributes to 0.7% of the gross domestic product of Sri Lanka (Central Bank, 2018). Therefore, promoting tea plantations would be beneficial both in terms of the economy as well as the environment. It thereby becomes a suitable candidate for counterpoising the environmental effects for the emissions of crematoria in Sri Lanka. The estimated carbon sequestration potentials of 30 years old tea plants in different elevations are given in Table 2.

Table 1: Estimated Carbon Sequestration Potential of Different Plant Species

Plant/Land Use Type/Ecosystem	*Carbon Sequestration Potential (MT of C $ha^{-1}yr^{-1}$)*	*Reference*
Teak	1.10–466.40	Abayasiri and Ranasinghe, 2000
Albizia	5.88	Kaye *et al.*, 2000
Mesic savannas	2.80	Williams *et al.*, 2004
Smallholder agro-forestry	1.50–3.50	Montagnini and Nair, 2004
Silver oak	2.09	Niranjana and Viswanath, 2005
Rubber	7.69	Tillekeratne, 2007
Shaded coffee	5.30	Anonymous, 2008
Coconut	8.60–26.90	Ranasinghe and Thimothias, 2012
Mahogany	15.66	Perera and Ranasinghe, 2013
Shaded tea	2.3–6.7	Wijeratne *et al.*, 2014

Table 2: Carbon Sequestration Potential of Thirty Years Old Tea Plants in Different Elevations in Sri Lanka[8]

Region	*Type of tea*	*Bush Densities (# of bushes ha^{-1})*	*C Sequestration (MT C $ha^{-1} yr^{-1}$)*
LC	SD	8000	1.71
LC	VP	12500	1.99
MC	SD	8000	0.78
MC	VP	12500	1.02
UC	SD	8000	0.26
UC	VP	12500	0.27

Legends: (LC = low country, MC = mid country, UC = up country, VP = vegetatively propagated plants, SD = seedling plants)

Accordingly, the tea plantations in low country (LC) region shows the highest carbon sequestration potential, followed by mid-country and it was lowest in the up-country region.

The objective of this case study is to quantify the carbon emission of crematoria which are installed in Sri Lanka. Then it proposes a reasonable way to neutralize carbon emission in a profitable manner by replanting a commercial crop in abandoned or bare lands in the country. In this case that will be abandoned tea estates.

MATERIALS AND METHODS

According to the survey done by NERDC in 2018, 265 cremator units are installed and in operation around the island. It is considered that at least one corpse would be cremated per day for five days a week. According to a recent survey, almost 50% of annual deaths are cremated. Tuesdays and Fridays are not used to cremating of corpses since long ago due to local customs and religious beliefs that have been long standing in the society. Liquid Petroleum gas is being used to operate crematorium and all calculations were based on that assumption. Less than ten units of old-styled gasifier crematoria are still available which uses biomass and those have been omitted in this study. Operation of a crematorium generates a substantial amount of greenhouse gases. Burning of cadavers, casket and the LP gas contribute to the total CO_2 production. Other than the direct CO_2 emissions, Electricity is also used during operation of Crematoria as the system consists of electrical Flue gas blowers and Burner fans. These electrical Machines contribute to increasing carbon footprint by consuming electricity from the min grid. Electricity is a direct result of the burning of fossil fuel.

RESULTS AN DISCUSSIONS

Analysis of CO_2 emission in the course of the operation of a crematorium is given below.

Estimation of CO_2 Emission during the Combustion of Corpse and the Casket

For calculations an empirical formula for the human body was generated based on the following chemical compositions (% by weight) of the human body.[10]

Table 3: Element Percentage of Corpse by Weight

Chemical Element	*Percentage of Corpse by Weight*
Oxygen	65%
Carbon	18%
Hydrogen	10%
Nitrogen	3%
Cadmium	1.5%
Phosphorous	1.2%
Potassium	0.2%
Sulfur	0.2%
Chlorine	0.2%
Sodium	0.1%
Magnesium	0.05%

3.2 Combustion of a Corpse

Transition of main elements are presented by the following equation after the combustion of human body.

$$C\,H_{6.33}O_{2.66} + 1.25(O_2 + 3.76\,N_2) \rightarrow CO_2 + 3.16\,H_2O + 4.7N_2 \quad \ldots (1)$$

Where, average weight of a corpse is around 80 Kg,

CO_2 Emission from Corpse

$$= \frac{\text{Molecular Weight of } CO_2 \text{ Emitted}}{\left(\begin{array}{c}\text{Molercular Weight of Empirical}\\ \text{Formula for the Corpse}\end{array}\right)} \times \text{Average Weight of a Corpse} \qquad \dots (2)$$

Where,

$$\frac{44}{60.89} 80 \text{ Kg} = 57.8 \text{ Kg}$$

Assuming chemical composition of biomass for casket, empirical formula $C_{4.2}H_{5.8}O_{2.6}N_{tr}$ (moisture free basis)[15] was assumed. (tr = trace amount)

Combustion of a Casket

Combustion of casket is modelled by the following equation,

$$C_{4.2}H_{5.8}O_{2.6} + 4.35(O_2 + 3.76N_2) \rightarrow 4.35CO_2 + 2.6H_2O + 16.36N_2 \qquad \dots (3)$$

Assuming, weight of the casket	= 34 Kg+-
Moisture content	= 30% Wet basis
Ash generation	= 1% by weight
Dry matter from the casket	= 23.8 Kg

$$CO_2 \text{ Emission from Casket} = \frac{\text{Molercular Weight of } CO_2 \text{ Emitted}}{\begin{array}{c}\text{Molecular Weight of Casket Material}\\ \text{Casket Dry Matter Weight}\end{array}} \qquad \dots (4)$$

Where,

$$\frac{191.4}{97.8} 23.8 = 46.57 \text{ Kg}$$

CO_2 Emission through LP Gas Combustion

Assuming burning of Butane

$$C_4H_{10} + 6.5\ (O_2 + 3.76N_2) \rightarrow 4CO_2 + 5H_2O + 24.4N_2 \qquad \dots (5)$$

Average time for the burning process is 1.5 h, during which 32 Kg of LPG is consumed.

$$\begin{pmatrix}\text{Total } CO_2 \text{ Weight Generated}\\ \text{during the Process}\end{pmatrix} = \frac{\text{Molecular Weight of } CO_2 \text{ Emitted}}{\text{Moercular Weight of Butane Content}}\ \text{LPG Weight Used for Burning} \quad \ldots (6)$$

Where,

$$\frac{176}{58 \times 32} = 97.1 \text{ Kg}$$

Total Direct CO_2 emission during the corpse burning process is 201.5 Kg.

Average Power requirement for the crematorium is 14 KW. Average time taken for the burning process is 1.5 hours.

Thermal power plants are the primary power generation source in Sri Lanka. It is estimated to be 67%[11] of the total generation capacity.

Average amount of CO_2 emissions per KW/h is 0.71 Kg.[12]

Therefore,

$$\begin{pmatrix}\text{Number of Units Required}\\ \text{for the Burning Process}\end{pmatrix} = \text{Power Required for Burning Process} \times \text{Duration of Burning Process} \quad \ldots (7)$$

Where,

$$14 \times 1.5 = 21 \text{ KW/h}$$

$$\begin{pmatrix}\text{Indirect } CO_2 \text{ Emission through}\\ \text{Electricity Consumption}\end{pmatrix} = \begin{pmatrix}\text{Power Consumption during}\\ \text{the Burning Process}\end{pmatrix} \times \begin{pmatrix}\text{Average } CO_2 \text{ Emitted}\\ \text{per KW/h during Generation}\end{pmatrix} \quad \ldots (8)$$

$$21 \times 0.71 = 111.825 \text{ Kg}$$

Therefore,

$$\begin{pmatrix}\text{Total } CO_2 \text{ Emission during}\\ \text{the Burning Process of a Corpse}\end{pmatrix} = \begin{pmatrix}CO_2 \text{ Emitted during Burning}\\ \text{of Corpse and Casket}\end{pmatrix} \times (CO_2 \text{ Emitted During Power Generation}) \quad \ldots (9)$$

Where,

$$201.5 + 14.91 = 216.41 \text{ Kg } CO_2\text{/Corpse}$$

Table 4: Breakdown of CO_2 Emission during the Burning of a Corpse

Contributor	*CO_2 Emission/Kg*
Corpse	57.8
Casket	46.57
Gas volume	97.1
Electricity usage	14.91
Total emission	216.41

Number of crematoria in Sri Lanka is approximately 265 and an average of 1 corpse is burnt per day in each unit.

Cremation takes place only during 5 days of a given week (customarily Funeral rituals are not carried out on Tuesdays and Fridays in Sri Lanka). It is considered that there are 52 weeks per year.

$$\begin{pmatrix}\text{Number of corpses burnt during the period}\\ \text{of a year based on the above assumptions}\end{pmatrix} = \begin{pmatrix}\text{Number of}\\ \text{Corpses}\end{pmatrix} \times \begin{pmatrix}\text{Number of}\\ \text{Crematoria}\end{pmatrix} \times \begin{pmatrix}\text{Number of Operating}\\ \text{Days Per Week}\end{pmatrix} \times \begin{pmatrix}\text{Number}\\ \text{of Weeks}\end{pmatrix} \quad \ldots (10)$$

Where,

$$1 \times 265 \times 5 \times 52 = 68900 \text{ MT yr}^{-1}$$

Therefore,

$$\text{Total amount of } CO^2 \text{ emission per year} = CO^2 \text{ emission per corpse} \times \text{Calculated average deaths per year} \quad \ldots (11)$$

Where,

$$216.41 \times 68900 = 14{,}911 \text{ MT yr}^{-1}$$

It clearly shows that when, in compliance with the Tea Research Institute of Srilanka (TRI) recommendations, the CO_2 absorption potentials of the tea plantations increase tremendously and the tea plantations become more and more environmentally friendly. Therefore, the local governments should take initiatives to provide support for the commercial tea growers considering their compliance to the TRI recommendations.

Table 5: CO_2 Emissions Due to Cremations from 2015 to 2018[13]

Year	*No. of Deaths*	*CO_2 Emission/ Metric Tons*
2015	65,807	14,242
2016	65,382	14,149
2017	69,911	15,129
2018	69,749	15,094

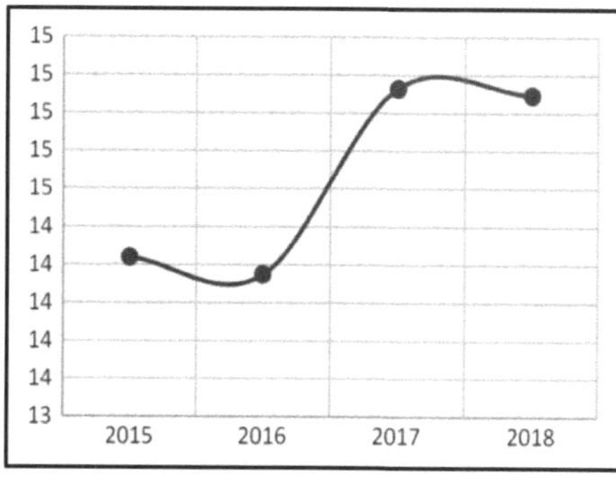

Figure 1: CO_2 Emission Due to Cremation

Legend: CO_2 emissions per year = (No. of Deaths) × (216.41 CO_2 emission per corpse)

Table 6: Estimation of CO_2 Absorption by Tea Plantations in Different Elevations

Scenario	*Region*	*CO_2 Absorption ($MT\ ha^{-1} yr^{-1}$)*	*Required Land Extent (ha)*
100% tea + 100% HS + 100% MS	LC	24.42	610.70
	MC	12.82	1,162.78
	UC	8.60	1,734.74
100% tea + 50% HS + 50% MS	LC	15.60	955.84
	MC	8.06	1,850.60
	UC	4.78	3,117.59
100% tea + 0% HS + 0% MS	LC	6.78	2,198.15
	MC	3.29	4,530.57
	UC	0.97	15,369.00

Legend: (1 MT of C = 3.67 MT of CO_2 equivalents)

(LC = low country, MC = mid country, UC = up country, HS = high shade, MS = medium shade)

CONCLUSION

The total emissions of CO_2 due to the burning of corpses in Sri Lanka, 14,911 MT y^{r-1} could be counterpoised by commercial tea plantations in different elevations without much effort. The required land extent of tea plantations to counterpoise these emissions is lowest in the low-country, followed by mid-country and it is highest in the up-country tea growing elevations. Further, the required land extent to counter-poise the emissions reduce tremendously with the increase in compliance with the TRI recommendations especially in terms of in-corporations and management of both high and medium shade trees.

Although tea cannot be cultivated in all regions in Sri Lanka and all the crematoria are owned and controlled by local governmental bodies, they can come into agreements to exchange funds among them and transfer funds to the respective local governments where ever tea can be cultivated and thereby replantation can be encouraged using such funds. This will save the environment while benefiting the commercial growers as well whilst ensuring environmental protection resulting in sustainability.

REFERENCES

[1] Green, Julia K.; Seneviratne, Sonia I.; Berg, Alexis M.; Findell, Kirsten L.; Hagemann, Stefan; Lawrence, David M. and Gentine, Pierre, Large influence of soil moisture on long-term terrestrial carbon uptake, *NATURE*, Vol. 565, 24 January 2019, pp. 476–479.

[2] Suwanmontri, Chanon; Kositanont, Charnwit and Panich, Noppaporn, Carbon Dioxide Absorption of Common Trees in Chulalongkorn University, *Modern Applied Science;* Vol. 7, No. 3; 2013, pp. 1–4.

[3] UN world population prospects 2019 https://population.un.org/wpp/Download/Standard/Population (accessed 20/01/2020).

[4] Keenan, Trevor F.; Prentice, I. Colin; Canadell, Josep G.; Williams, Christopher A.; Wang, Han; Raupach, Michael and Collatz, G. James (2016). Recent pause in the growth rate of atmospheric CO_2 due to enhanced terrestrial carbon uptake, https://www.nature.com/articles/ncomms16137 (accessed 20/01/2020).

[5] Effects of Thinning on CO_2 Absorption Capacity of Rhizophora apiculata Plantations in Can Gio Mangroves, Mangrove ecosystems technical reports, Vol. 6, pp. 47–49.

[6] Mari, Monste *et al.* (2010), Toxic emissions from crematories: A review, *Environmental International,* 36 (2010), pp. 131–137.

[7] Edirisinghe, E.A.N.K.; Warahena, A.; Pieris, W.K.R. and Salgadu, A.R.C., Energy Efficient Cremator Solution in Conformity with Environmental Regulations—an Investigation with a Case Study, *NERDC Research Symposium* (2017).

[8] Wijeratne, T.L.; Costa, De; W.A.J.M. and Wijeratne, M.A. (2014). Carbon Sequestration Potential of Tea Plantations in Sri Lanka as an Option for Mitigating Climate Change; A Step towards a Greener Economy, pp. 205–212. In A. P. Keerthipala (ed.), *Proceedings of the 5th Symposium on Plantation Crop Research*—"Towards a green Plantation Economy". Sugarcane Research Institute, Uda Walawe, 70190, Sri Lanka.

[9] Wijeratne, T.L., Sugathapala, A.W.L.C., Pathirana, V.P.R.P.V., Jayawardhene, A.D.I.U., Sivapriyadharshani, A. and Sidhakaran, V., Review on Climate change research and development of carbon budget for tea plantations, Tea Research Institute of Sri Lanka, 238th meeting of the experiments and extension forum, July 2019.

[10] Anne Marie Helmenstine, Chemical Composition of the Human Body https://www.thoughtco.com/chemical-composition-of-the-human-body-603995 (accessed 20/01/2020).

[11] Abayasiri, S. and Ranasinghe, D.M.S.H.K. The contribution of forest plantations (Teak) in Sri Lanka in acting as a Carbon sink to reduce greenhouse gases in the atmosphere. Abstract presented in: *Proceedings of the Sixth Annual Forestry and Environment Symposium 2000.* Department of Forestry and Environmental Science, University of Sri Jayawardenepura, Sri Lanka: 29 (2000).

[12] Kaye, J.P., Resh, S.C., Kaye, M.W. and Chimner, R.A. Nutrient and carbon dynamics in a replacement series of Eucalyptus and Albizia trees. Ecology, 81(12): 3267–3273 (2000).

[13] Williams, R.J., Hutley, L.B., Cook, G.D., Russell-Smith, J., Edwards, A. and Chen, X. Assessing the carbon sequestration potential of mesic savannas in the Northern Territory, Australia: Approaches, uncertainties and potential impacts of fire. View point. *Functional Plant Biology,* 31: 415–422 (2004).

[14] Scurlock, J.M.O. and Hall, D.O., The Global Carbon Sink: a grassland perspective, *Global Change Biology,* 229–233 (1998), 4.

[15] Tillman, David A., etc., Wood combustion, principles, processes and economics, Academic Press, USA.

[16] Anonymous (2008). Available online http://www.coffeehabitat.com/2008/03/coffee-and-carb.html. Last cited 27.07.2009.

[17] Ranasinghe, C.S. and Thimothias, K.S.H. (2012). Estimation of carbon sequestration potential in coconut plantations under different agro-eco-

logical regions and land suitability classes. *Journal of National Science Foundation Sri Lanka*, 40(1): 77–93.

[18] Perera, L.S.A. and Ranasinghe, D.M.S.H.K. (2013). Carbon Sequestration Potential of Privately Managed Swietenia macrophylla Plantations in the Wet and Intermediate Zones of Sri Lanka. *Proceedings of International Forestry and Environment Symposium 2013.* Department of Forestry and Environmental Science, University of Sri Jayewardenepura, Sri Lanka.

[19] Central Bank (2018). *The Annual Report for the year 2018.* Central Bank of Sri Lanka, Colombo, Sri Lanka.

The Impacts of Indoor Air Pollution on Human Health and the Challenges to Control it in Peri Urban Areas

(Case Study of Mazyopa Compound-Lusaka District)

Clavel Mulenga

Public Health Inspector, Lusaka City Council, Lusaka

E-mail: clavelmulenga2@gmail.com

ABSTRACT: According to WHO, Poor housing triggers several negatives impact on environmental health, human health and general performance of national development. Housing in peri urban areas of the world cities example in Lusaka, is a serious problem for the government and Local Authority, but primarily for the individual family. Mazyopa Settlement has a total population of about 3,880: the number of households is estimated at 766 and the average household size is 5 (LCC, 2006). Indoor Air Pollution occurs when certain air pollutants from particulate, microbes and gases contaminate the air of the indoor areas. These air pollutants can cause respiratory diseases, cancer, eyes and Skin irritation etc. Causes of Indoor Air Pollution in informal housing are: Toxic products e.g. asbestos, lead etc., fuels for cooking, inadequate ventilation, high temperature and humidity are few of the primary causes of indoor air pollution in homes. It is clear from the evidence gathered in my research on: "Impacts of poor housing conditions and their effects on human health" that the predominant health problems is upper respiratory infection in these informal settlements which is directly linked to Indoor Air Pollution as a result of poorly ventilated houses, built without adequate window area and are overcrowded with more than the required number of occupants. Further these, houses are so congested owning to the fact that they are closely built to each other living no room for air circulation leading to none air movement creating polluted air for the residents of these communities. The above-mentioned pollutants are also present by natural introduction or by human activities. Most microbes find their way in these houses due to sick members of the family and continue to be present due to poor ventilation of these house. While as chemical pollutants are as a result of man's activities for example the use of charcoal for cooking in a home or the painting of the interior walls with paint containing Lead. These too once introduced in the indoor air it pollutes the air quality and persist because the buildings are poorly constructed and ventilated. Challenges still remain as efforts to assess health risks associated with indoor air pollution are limited by insufficient information about the number of people exposed, the pattern and severity of exposures, and the health consequences of exposures. An overall strategy should be developed to investigate indoor exposures, health effects, control options, and public policy alternatives in these informal settlements.

Keywords: Indoor Air Pollution, Health, Air Pollutants, Upper Respiratory Infection.

INTRODUCTION

One of the basic human needs is shelter and civilization has led to a significant fraction of the world's population migrating to urban cities. This population migration has subsequently put the pressure on the housing deficit faced by urban cities of the most if not all developed and developing countries, intern it has resulted into rapid growth of squatter camps or informal settlements in the urban cities. The type of dwellings built in these settlements are rated Poor housing and conceived to retain air pollutants indoors for longer period, exposing the occupants. Poor housing triggers several negatives impact on environmental health, human health and general performance of national development (WHO Report, 2003).

Housing in peri urban areas of the world cities example in Lusaka, is a serious problem for the government and Local Authority, but primarily for the individual family. These houses are built without approved building plans from the Local Authority. They do not conform to the standard design and construction provided in the Public Health Act CAP 295 (building regulations) of the Laws of Zambia and other related laws. Mazyopa Settlement has a total population of about 3,880: the number of households is estimated at 766 and the average household size is 5 (LCC, 2006). The housing structures are informal, without approved layout plans and authorization from the local authority. The houses are built very close to each other. A few houses are made of concrete blocks with iron/asbestos roofing sheets. It's the congestion, overcrowding, dampness and poor ventilation of households that are primary cause of Indoor Air Pollutions which amount to a greater percentage of the factor causing acute respiratory infections i.e. Upper and Lower Acute Respiratory Infection in these informal settlements. People living in these shelters use synthetic chemicals e.g. paint and have relied on crude fuels for their cooking and heating needs. The combustion of these crude biomass fuels such as crop residues, animal dung and wood—generates smoke that adversely affects the health of occupants.

This paper will not discuss the health effects or impacts of the many different pollutants which can be found in indoor air singularly, rather will group broad categories of adverse health effects and describe the relevant indoor exposures which may give rise to these health effects with bias to respiratory effects seen in Mazyopa Compound.

Indoor air pollution occurs when certain air pollutants from particles and gases contaminate the air of the indoor areas. These air pollutants can cause respiratory diseases or even cancer. Removing these pollutants can improve the quality of your indoor air. **IAP** is ubiquitous, and takes many forms, ranging from smoke emitted from solid fuel combustion, especially in households in developing countries, to complex mixtures of volatile and semi-volatile organic compounds present in buildings (Junfeng (Jim) Zhang* and Kirk R. Smith, 2016).

Everyday exposure to multiple chemicals, most of which are present indoors, has contributed to increasing prevalence of Upper Respiratory Infection such as flu, whooping cough, asthma, autism, childhood cancer, medically unexplained symptoms, and perhaps other illnesses as evidenced in many informal settlements with poor housing structures. Carbon monoxide (CO) poisoning cases are reported, mainly as a consequence of inadequate ventilation of appliances. Concerns have also been reported about exposure to nitrogen dioxide (NO_2) emitted from gas stoves (Zhang J, Smith KR, Ma Y *et al.*, 2000). Compared to combustion of solid fuels, however, gaseous fuels in simple devices emit substantially smaller amounts of pollution, including particulate matter (PM), CO, eye irritating volatile organic compounds (*e.g.* aldehydes), and carcinogenic compounds such as benzene and 1, 3-butidiene and polycyclic aromatic hydrocarbons

(Zhang J and Smith KR, 1999). Daily exposure to products of incomplete combustion poses both acute and chronic health risks. A recent (Zhang J and Smith KR, 1996). World Health Organization (WHO) report estimated that indoor smoke from solid fuels ranked as one of the top ten risk factors for the global burden of disease, accounting for an estimated 1.6 million premature deaths each year (WHO. World Health Report, 2002). Among all environmental risk factors, it ranked second only to poor water/sanitation/hygiene (WHO. World Health Report, 2002). The existing literature provides strong evidence that smoke from solid fuels is a risk factor for acute respiratory infections (ARI), chronic obstructive pulmonary disease (COPD) and lung cancer (from coal smoke) (Smith KR, 2002). Evidence from 13 studies in developing countries indicates that young children living in solid-fuel using households have two to three times more risk of serious ARI than unexposed children after adjustment for potential confounders including socio-economic status (Smith KR *et al.*, 2000). An evaluation of eight studies in developing countries indicates that women cooking over biomass fires for many years have two to four times more risk of COPD than those unexposed after adjustment for potential confounding factors.

CAUSES OF INDOOR AIR POLLUTION IN INFORMAL HOUSING

Before you comprehend the effects of indoor air pollutions you must first understand the causes as well as what we can do improve our quality of air both indoor and outdoor.

These causes can be categorised in three:

1. *Biological Causes*: these are sources that encourage biological contaminates such as moulds, mites, and other microbes in the air may cause diseases.
2. *Chemical Causes*: these are sources of contaminates the give raise to chemical pollutants in the air such Toxic products include asbestos, lead, formaldehyde, tobacco smoke and Fuels for cooking, *Millions of people around the world prepare meals using traditional methods (i.e. wood charcoal, coal, dung, crude Fuel) on open fires.* Smoking is one obvious source of indoor air pollution, and can give rise to emissions of fine particulates, polyaromatic hydrocarbons (PAHs) and dioxins. Furniture made from

fibreboard has been associated with exposure to formaldehyde, although steps have been taken by the manufacturers to minimize this problem.

3. *Physical Causes:* Although not a direct source of any pollutant, inadequate ventilation, overcrowding, dampness, Congestion, High temperatures and humidity are all primary causes of Indoor Air Pollution. Further, dust particulate matter is that air pollutant. Natural sources, such as dust mites, can also cause health effects. The problem has been exacerbated by modern luxuries such as carpets and central heating, and reduced ventilation brought about by the desire to insulate properties.

EFFECTS OF INDOOR AIR POLLUTIONS ON HUMAN HEALTH

The effects of IAP on Human Health have been groups as:

1. **Effects on the Respiratory System**: Several effects on the respiratory system have been associated with exposure to IAP. These include acute and chronic changes in pulmonary function, increased incidence and prevalence of respiratory symptoms, acute in pre-existing respiratory symptoms, and sensitization of the airways to allergens present in the indoor environment. Combustion products, ETS (ETS is a complex mixture of pollutants whose source is primarily cigarette Smoking), respirable suspended particles, CO, nicotine, nitrogen oxides, formaldehydes and biological contaminants are the main agents associated with respiratory health effects indoors (Berglund B *et al.*, 1984).
2. **Allergy and Other Effects on the Immune System:** Allergic asthma and extrinsic allergic alveolitis (hypersensitivity pneumonitis) are the two most serious allergic diseases caused by allergens in indoor air. House dust mites, pets, insects, and moulds in the indoor environment are important causes of allergic asthma and rhino conjunctivitis. Outdoor allergens such as pollens and moulds may penetrate into the indoor environment through open windows, doors, or ventilation systems. The airborne allergens vary with seasons, weather conditions, geographical location, and the local indoor environment (European Concerted Action, 1988)
3. **Cancer and Effects on Reproduction:** An increased risk of developing lung cancer has been linked to exposure to

environmental tobacco smoke (ETS) and to radon decay products. A few indoor air pollutants, notably asbestos, radon and environmental tobacco smoke (ETS) have been associated with cancer (European Concerted Action, 1989).

4. **Effects on the Skin and Mucous Membranes in the Eyes, Nose and Throat:** ETS, Respirable suspended particles, GO, nicotine, nitrogen oxides. The major sites of irritative changes caused by ETS are the eyes and nasopharynx. Eye and conjunctiva irritation, nasal discomfort, sore throat, sneezing, and cough are frequently reported symptoms (European Concerted Action, 1989).
5. **Sensory Effects and other Effects on the Nervous System:** Complex mixtures of organic chemicals e.g. in indoor air also have the potential to invoke subtle effects on the central and peripheral nervous system, leading to changes in behaviour and performance. There are still major difficulties in predicting sensory effects and effects on the nervous system of chemicals at commonly encountered indoor concentrations, especially the early, subtle and reversible effects (European Concerted Action, 1990).
6. **Effects on the Cardiovascular System:** Cardiovascular effects have only infrequently been described as being associated with exposure to IAP. Only exposures to ETS and carbon monoxide (CO) have been implicated in cardiovascular symptoms, and in changes in Cardio Vascular Disease (CVD) morbidity and mortality.

For some groups of effects, clear relationships with exposure to IAP have been reported in the world literature. Among these are respiratory disease (particularly among children), allergy (particularly to house dust mites) and mucous membrane irritation (particularly due to formaldehyde). Large numbers of people have been, and are still being affected (B. Berglund, 1991).

PUBLIC HEALTH RELEVANCE OF IAP

The public health relevance of the effects of IAP varies, not only from substance to substance, but also from country to country, depending on the presence of specific local sources and climatic influences. According to recent findings, over 2 million death occur every single day due to indoor air pollution. Young children, infants and the foetus are known to be more susceptible to adverse effects of lead at a given blood lead level than adults. Smoking and asbestos exposure have been shown to act synergistically in the causation of lung cancer; smokers suffer from COPD more than nonsmokers, and to the extent that COPD patients are more susceptible to pollutants than others, smoking causes some people to be more susceptible. It is not clear to what extent smokers who do not have COPD are more susceptible to effects of other pollutants than non-smokers.

RELATIONSHIP OF HOUSING CONDITION TO INDOOR AIR POLLUTION

Analysis Diagram

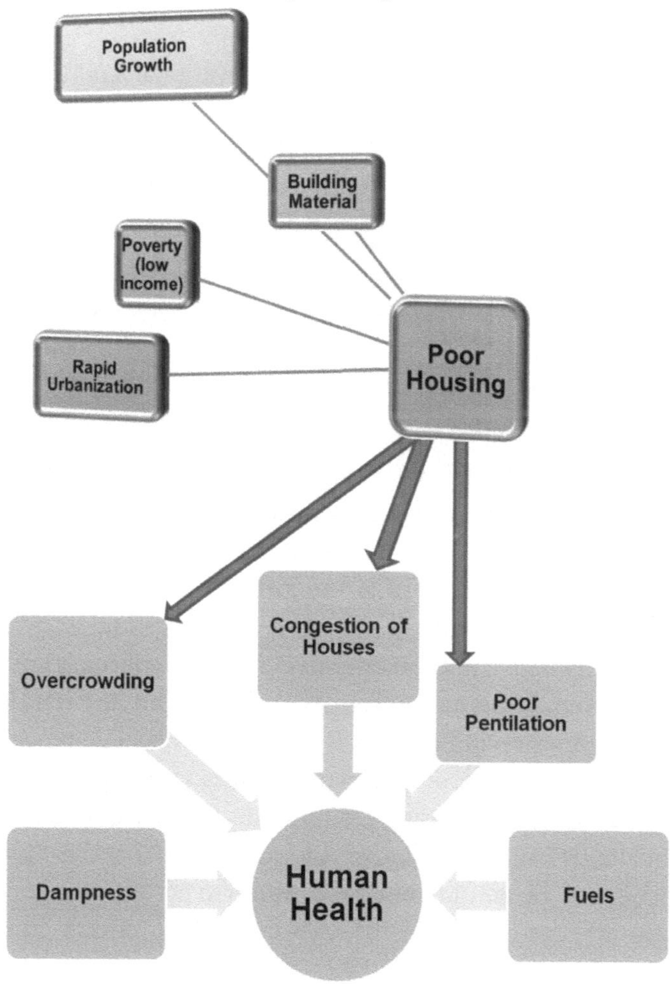

The "Sick Building Syndrome"

The "Sick Building Syndrome" is a term used to describe the reduced comfort and health status of occupants in a particular building or part of it where the occupants complain about indoor air quality and manifest symptoms which they assign to that reduced quality, Symptoms reported in SBS have typically included mucous membrane and eye irritation, cough, chest tightness, fatigue, headache and malaise (NATOICCMS Pilot Study, 1991).

MATERIALS AND METHODS

Methods of studying health Impacts of indoor pollutants can be grouped into three broad categories:

1. Human studies, subdivided into observational and experimental studies.
2. *Animal studies:* which can be subdivided into a number of categories depending on their length (acute, sub chronic, and chronic) or end-point (morbidity, mortality, carcinogenicity, irritation, etc.), and
3. Vitro studies; in which effects of pollutants on cell or organ cultures are studied. These studies have the advantage that they are less costly than animal studies, and that results can generally be obtained in a shorter period of time.

However, this study is an observational study focussing on epidemiology of indoor air pollutants are mostly. The study sort to collect electronic data, data from interviews and other literature and observe the epidemiological tendencies of Indoor Air Pollutants, their resultant health impacts on humans. Through a thorough and systematic investigation, qualitative and quantitative data was collected and used to paint a clear public health concern with regard the impacts of Indoor Air Pollution on human health in Peri Urban areas. The process of risk characterization for indoor pollutants occurs through several phases: hazard identification, exposure assessment, dose-effect evaluation, and finally qualitative and quantitative risk assessment. The final product of this process may be an individual risk estimate per exposure unit or the evaluation of the incidence of the concerned effects in a given population (B. Berglund *et al.*, 1991).

RESULTS

It is clear from the Table 1 below that most of the respondents fell in the age group of 20 to 24 years. Very few were aged between 40 to 44 years.

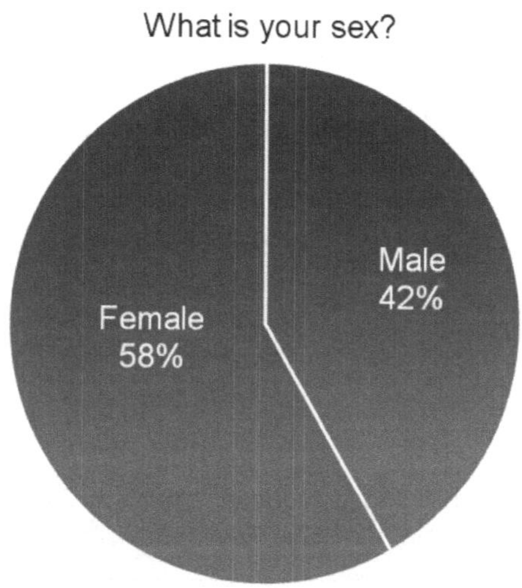

Figure 1: Percentage Distribution of Respondents by their Sex

Source: *Field Data*: As observed from Figure above, most (58 percent) of the respondents that participated in the study were females and remaining 42% were males.

Table 1: Number and Percent Distribution by Age

Age-Group	*Number*	*Percent*
20–24	38	40
40–44	24	25
50–54	34	35
Total	**96**	**100**

Source: Field Data.

It is clear from Table 2 that most (42 percent) of the respondents indicated that they lived on plots with between 3–4 houses on the plot or in the yard and 35 percent of the respondent indicated they had between 5–6 houses on the plot and 23 percent of the respondent had 1–2 housing structures on the plot. This information clearly shows serious housing congestion which prevents free air flow required in any settlement. Posing serious health risks on residents.

Table 2: Number and Percent Distribution of Respondents on Number of Houses per Plot

Number of House per Plot	*Number*	*Percent %*
1–2	22	23
3–4	40	42
5–6	34	35
Total	**96**	**100**

Source: Field Data.

As can be observed from Table 3 below, 90 percent of the respondents indicated that more than 5 households occupy a room of a housing units. Very few (6 percent) respondents indicated that less than 5 household occupy a room. This Denoted serious overcrowding in these houses.

Table 3: Number and Percent of Household per Dwelling

Households Per Room	*Number*	*Percent*
1–5	6	6.25
5–10	90	93.75
Total	**96**	**100**

Source: Field Data.

It is clear that in terms of the relation of window sizes to the floor area, most (73 percent) of the housing units had window sizes that were abnormal in relation to the floor area.

Table 4: Relation of Window Size to Floor Area

Windows Size to Floor	*Number*	*Percent %*
Below 1/10 (not normal)	70	73
1/10 (normal)	22	23
Above 1/10 (normal)	4	4
Total	**96**	**100**

Source: Field Data.

It was revived that the highest 77% of the respondents lived in houses that have not been Damp proofed and 23% lived in damp proofed homes. Suffice to not that Mazyopa is in a near a stream so the more of it is water logged.

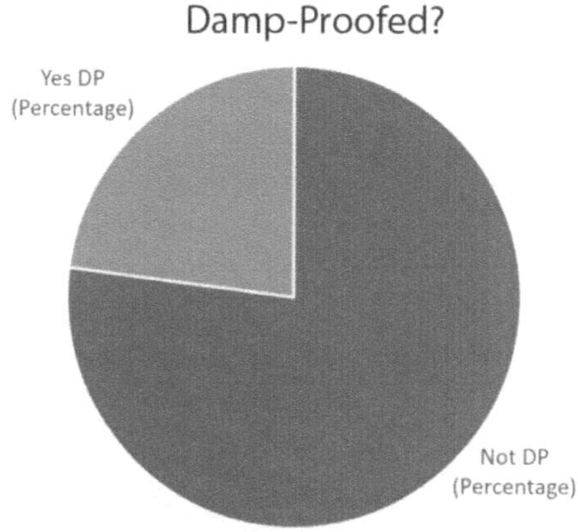

Figure 2: Damp-Proofed Floors
Source: Field Data.

It is clear from Table 5 that most of the respondents use Charcoal for cooking 40 percent, followed by other means 35 percent and then stove as the least 25 percent.

Table 5: Number and Percent Distribution by Fuels Used in Cooking

Fuels Used in Cooking	*Number*	*Percent %*
Charcoal	38	40
Stove	24	25
Other means (kerosene, biogas)	34	35
Total	**96**	**100**

Source: Field Data.

It is evident from the Table 6 that the most (42 percent) prevalent disease in Mazyopa Compound was Acute Respiratory, followed by malaria (33 percent).

Table 6: Information on Most Prevalent Diseases in Mazyopa Compound

Type of Disease	*Number*	*Percent %*
Diarrhoea	24	25
Malaria	32	33
Acute respiratory infection	40	42
Total	**96**	**100**

Source: Field Data.

Table 7: Frequency of the ARI Diseases MAZYOPE

How often do you suffer from these diseases?	*Number*	*Percent*
Weekly	39	41
Monthly	40	42
Yearly	17	18
Total	**96**	**100**

Source: Field Data.

Table 8: Showing Top Six Diseases in Mazyopa in 2019

Disease	*August*	*September*	*October*	*November*	*December*
Acute respiratory infections	2165	840	4099	2173	1936
Pneumonia	5820	174	97	202	200
Whooping cough	961	821	462	105	425
TB	47	36	95	37	98
Malaria	36	32	24	32	36

Source: Disease aggregation forms for Chipata Clinic.

DISCUSSION

The world over tend to focus on air quality in streets and other public spaces. In fact, a significant proportion of most people's exposure to many pollutants occurs in the home and in the workplace. The latter is a health and safety matter and outside the scope of this discussion. However, the generation of pollution in the home is an important issue that needs to be discussed whenever considering the effects of harmful substances in the air on our world population.

It is clear from the evidence gathered in this paper that the Peri-Urban Areas like Mazyopa compound are gripping with perpetual Health problems like Acute Respiratory Infections, both Upper Respiratory Infection i.e. whooping cough, flu, etc. and Lower Acute Respiratory Infection e.g. pneumonia, as a result of Indoor Air Pollution in this area. This is attributable to:

- The poor ventilations in these houses, as can be seen from the Table 4 where 73% of the respondents lived in houses which had window openings below the required area for the rooms with regard the $1/10^{th}$ of the Floor area for the room.

- Table 3 shows that 94% of respondent to the survey lived in over-crowded rooms 5–10 persons per room more than the required number of occupants required in relation to floor area per square metre per head ratio.
- Figure 2, shows clearly that the houses in Mazyopa are not Damp proofed and they are built on a water logged area. Mazyopa compound lays on a stream. Dampness in homes is one area of concern. Reduced Ventilation in homes causes moisture to be trapped inside causing dampness which in turn creates a favorable environment for house dust mites and moulds. It enhances the growth of mites and fungi that produce substances sensitive to susceptible individuals.
- Table 5, clearly proves to us the sources of IAP pollutants such as CO Carbon monoxide, Sulphur oxides SO, and nitrogen (N) which are by products of these Fuels used in cooking and heating of Homes. 40% of the respondent said they used Charcoal as fuel for cooking in their homes and 35% used Kerosene and other fuels for cooking indoors and 25%. The first two groups represent 70% of those exposed to the by-products air pollutants.
- Table 2, the findings showed that the housing units are crusted together affirming that the area is congested. On average a plot of 10 meter squared would house 3 to 4 house structures. These houses are one to two roomed with an average of 4–5 people in a one roomed. Which affect the free flow of air circulation. The houses in Mazyopa compound are so congested owning to the fact that they are closely built to each other living no room for air circulation. Again, that contributes to none air movement retaining polluted air in the settlement and houses for long. Free flow of air helps in the dilutions of Air pollutants indoor but the non-movement due to congestion creates a conducive environment for the high exposure by residents to these air pollutants. Very evident from these findings that houses are overcrowded. See Table 2 Congestion, back to back dwelling and overcrowding in these poorly designed and constructed houses enable the transmission of Acute Respiratory Infections, tuberculosis, Meningitis etc.

The above factors are the primary causes of retention of Indoor Air pollutants naturally introduced or by human activities indoors giving high exposure to occupants. Most microbes find their way in these houses due to sick members of the family and continue to be present due to poor ventilation of these house. While chemical pollutants are as a result of man's activities for example the use of charcoal for cooking in a home or the painting of the interior walls with paint containing Lead, once introduced in the indoor air it pollutes the air and persist because the buildings are poorly constructed and ventilated.

Impacts of IAP on Human Health

In terms of the prevalent diseases, most (42 percent) of the respondents indicated that the most prevalent disease in Mazyopa Compound was Acute Upper Respiratory, these were experienced mostly on a monthly basis. Poor ventilated and light houses increase spread of acute respiratory infection. (Pneumonia, influenza, whooping cough etc.) Lung cancer is a very serious disease with a high fatality rate; however, the number of people affected is much lower than the number of people contracting respiratory disease or allergies, or experiencing irritation effects due to exposure to indoor pollution. Several infectious diseases are known to be transmitted from one person to the next when infectious agents are propelled into the air by coughing, sneezing, singing, talking, etc. The major reason why airborne infections are much more likely to spread indoors than outdoors is, that the dilution in outdoor air is usually so great that chances of inhaling enough infectious droplet nuclei to become infected are negligible. Also, people spend much more time indoors than outdoors, The respiratory system of young children and those exposed to other air pollutants especially the chemical pollutants and Elderly people with impaired pulmonary function and/or weakened defense systems are considered to be more susceptible to environmental insults, as an insult of a given size will affect them more. In addition, patients already suffering from Chronic Obstructive Pulmonary Disease (COPD) are considered to be more susceptible than healthy individuals. The Clear case informal settlements like mazyopa.

Indoor Air Pollutants Prevalent in Mazyopa Compound

Microbiological Agents, which include bacterial, virus, protozoa etc. that may be transmitted via air droplets in the environment where the source constantly continues to discharge and there is no free flow of air.

Carbon monoxide occurs as the result of poorly ventilated and maintained combustion sources (gas boilers, fires etc) but the burden of morbidity and mortality is probably under-estimated in official figures.

Asbestos and Man-Made Mineral Fibres (MMMF)—a common material in older dwellings, but usually causing low level exposure unless disturbed likely to causes asbestosis or cancers.

Lead—mainly in old lead paint and water pipes/solder joints. Lead causes brain and nerve damage, kidney failure.

Formaldehyde—is another leading cause of indoor air pollution, found in paints, sealants and wood preservatives.

Tobacco Smoke—that comes from outdoor and indoor areas can also be an indoor air pollution and causes respiratory irritation, pneumonia. Other indoor air pollutants (e.g. carbon dioxide nitrogen oxides, volatile organic compounds and biomass burning products)

Effects of indoor air pollution can be life threating. Kids and old age people are more prone to the effects of indoor air pollution. These effects are dependent on the pollutant one is exposed to for example.

Interventions to Mitigate the Impacts of IAP on Human in Mazyopa Compound Homes

This disease burden is highly preventable by appropriate building design and building use further the upgrading of the settlement. Properly ventilate these houses in informal settlement of the world cities and the upgrading of this settlements. Use of environmentally friendly chemicals in inside the houses these include cleaning materials, pesticides, paints and other fumigants used to preserves house furniture. The use of clean energy in cooking and for those can't afford to adopt cooking outside when using fuels that produce toxic element that pollution the environment. Finally, the good hygiene practices.

CONCLUSIONS

Housing condition plays a critical role in the spread, control and prevention of many diseases, especially in the transmission of communicable diseases. If in poor conditions or state, residents will be exposed to disease pathogens, pollutants and hazards that poses threats to human health and survival. Therefore, homes can both protect from disease or facilities negative health effects.

It was evident from the findings that being an information settlement, Mazyopa Compound is a good example of housing units that retain

Indoor Air Pollutants exposing the occupants especially the children and women. It is clear that the common diseases experienced are Acute Respiratory Infection which are perpetuate by Indoor Air Pollution as a result of overcrowding in these houses, Poor ventilation of the house, congested housing structures and use of fuels indoors for cooking. Factors that promote the spread of Acute Respiratory Infection both Upper Respiratory Infection and Acute Lower Respiratory Infection. Respiratory infections may spread in indoor environments when specific sources of infectious agents are present, or smaller indoor mixing volumes allow infectious diseases to spread more easily from one person to the next. In Mazyopa most homes use unvented kerosene heaters, which elevated levels of sulphur dioxide (SO,) indoor. As the human respiratory system is the organ directly affected by air pollution, the potential respiratory health effects of indoor and outdoor air pollution have been widely investigated.

Effects of indoor air pollution can be life threating e.g. respiratory irritation, pneumonia, brain and nerve damage, kidney failure and asbestosis or cancer a large segment of the study population which is a true presentation of the population is exposed to ETS, to combustion products from unvented combustion appliances, and to biological contaminants in the mazyopa houses. Effects of exposure to IAP on the respiratory system of Humans are well documented, and sizeable public health benefits can be achieved when houses in mazyopa are well ventilated, the use of fuels for cooking indoors are discouraged and the reduction of factors that encourage the proliferation of infectious agents in buildings or by air irradiation or minimization of crowding or reducing Dampness. Although official efforts to control air pollution have traditionally focused on outdoor air, it is now apparent that elevated contaminant concentrations are common inside private homes.

Concerns about potential public health problems due to indoor air pollution are based on evidence that urban residents typically spend more than 90 percent of their time indoors, concentrations of some contaminants are higher indoors than outdoors, and for some pollutants personal exposures are not characterized adequately by outdoor measurements. Among the more important indoor contaminants associated with health or irritation effects are passive tobacco smoke, radon decay products, carbon monoxide, nitrogen dioxide, formaldehyde, asbestos fibres, microorganisms, and aeroallergens.

Challenges still remain as efforts to assess health risks associated with indoor air pollution are limited by insufficient information about the number of people exposed, the pattern and severity of exposures, and the health consequences of exposures. An overall strategy should be developed to investigate indoor exposures, health effects, control options, and public policy alternatives in these informal settlements.

ACKNOWLEDGEMENTS

I wish to acknowledge the assistance and support rendered to me by Rusangu University-Lusaka Campus Management, Lusaka City Council and the Ministry of Health, Chipata Clinic staff in the gathering of data both literature and field source data to compile this paper.

REFERENCES

[1] WHO (2003). Report.

[2] Lusaka City Council (LCC), (2006). informal settlement report.

[3] Berglund, B., Brunekreef, B., Knoppel, H., Undvau, T., Maroni, M. and Mblhave, L. (1991). Indoor Air Quality and Its Impact on Man, Report # 10 *Effects of Indoor Air Pollution on Human Health.* European Concerted Action, Brussels – Luxembourg.

[4] Berglund, B., Berglund, U., Lindvall, T., Spengler, J. and Sundell, J. Indoor air quality—containing papers from the Third International Conference on Indoor Air Quality and Climate, Stockholm, Sweden, 1984. *Environment International, 12* (1 986) (special issue).

[5] Conserve Energy Future website: https://www.conserve-energy-future.com/causes-and-effects-of-indoor-air-pollution.php

[6] European Concerted Action, "Indoor Air Quality and its Impact on Man": Report No. 1: Radon in indoor air. Commission of the European Communities. Luxembourg, 1988.

[7] European Concerted Action, "Indoor Air Quality and its Impact on Man": Report No. 3: IAP by NO, in European countries. *Commission of the European Communities, Luxembourg,* 1989.

[8] European Concerted Action, "Indoor Air Quality and its impact on Man": Report No. 4: Sick Building Syndrome—a practical guide. Commission of the European Communities, Luxembourg, 1989.

[9] European Concerted Action, "Indoor Air Quality and its Impact on Man": Report No. 7: IAP by formaldehyde in European countries. Commission of the European Communities, Luxembourg, 1990

[10] Jun Feng (Jim) Zhang and Smith, Kirk R. (2016). *Indoor air pollution: a global health concern.* University of Medicine and Dentistry of New.

[11] Lusaka City Council (2006). Informal settlement report.
[12] NATOICCMS Pilot Study on Indoor Air Quality: Epidemiology and medical management of building-related complaints and illnesses. Report on a meeting held in Oslo 19–21 August 1991 (in press).
[13] Smith, K.R. Indoor air pollution in developing countries: recommendations for research. *Indoor Air* 2002; 12: 198–207
[14] Smith, K.R., Samet, J.M., Romieu, I. and Bruce, N. Indoor air pollution in developing countries and acute lower respiratory infections in children. *Thorax* 2000; 55: 518–32.
[15] World Health Organization: Indoor air quality: Radon and formaldehyde. Environmental Health Series No. 13, Copenhagen, 1986.
[16] World Health Organization Indoor air quality: Biological contaminants. *World Health Organization Regional Publications European Series* No. 31, Copenhagen, 1990.
[17] WHO. *World Health Report: Reducing Risks, Promoting Healthy Life.* Geneva: WHO, 2002.
[18] WHO. *World Health Report: Reducing Risks, Promoting Healthy Life.* Geneva: WHO, 2002.
[19] Zhang, J., Smith, K.R., Ma, Y. *et al.,* Greenhouse gases and other pollutants from household stoves in China: A database for emission factors. *Atmos Environ* 2000; 34: 4537–49.
[20] Zhang, J., Smith, K.R. Emissions of carbonyl compounds from various cook stoves in China. *Environ Sci Technol.,* 1999; 33: 2311–20.
[21] Zhang J, Smith KR. Emissions of carbonyl compounds from various cook stoves in China. *Environ Sci Technol* 1999; 33: 2311–20.
[22] Zhang, J., Smith, K.R. Hydrocarbon emissions and health risks from cookstoves in developing countries. *J. Expos Anal Environ Epidemiol,* 1996; 6: 147–61.

ABBREVIATIONS

ARI – Acute Respiratory Infection
COPD – Chronic Obstructive Pulmonary Disease
CO – Carbon Monoxide
CVD – Cardio Vascular Disease
ETS – Environmental Tobacco Smoke
IAP – Indoor Air Pollutions
LCC – Lusaka City Council
LARI – Lower Acute Respiratory infection
PM – Particulate Matter
URI – Upper Respiratory Infection
WHO – World Health Organisation

APPENDIX

General Matrix for the Evaluation of the Impact of Indoor Pollutants on the Community

Health Impact	*Severe Effects*	*Minor Effects*
87% of people affected in Mazyopa	Death Cancer Serious Disease	Mild Disease Discomfort Annoyance Reduced productivity
>10% of the population of Mazyopa compound	Large exposure-Large Health Impact	Large Exposure-Limited Health impacts
< 10% of the population of Mazyopa Compound	Limited exposure-Large health impact	Limited exposure-Limited Health Impacts

Air Pollution Know Your Enemy—A Case Study of Harare, Zimbabwe

Sikayi Gloria Winnie

Health Department, City of Harare,
Rowan Martin Building, Pennefather Avenue,
P.O. Box 596 Harare, Zimbabwe
E-mail: wgsikayi@gmail.com

ABSTRACT: Air pollution is one of the major causes of environmental pollution. Air pollution is a real enemy that threatens human life by causing and aggravating respiratory infections, heart diseases and cancers. It poses several negative effects on the flora and fauna. The major air pollutants monitored in Harare, Zimbabwe were Suspended Particulate Matter (SPM), Sulphur dioxide (SO_2), and Nitrogen oxides. This paper seeks to give an insight on the burden of air pollution in Harare, Zimbabwe and the interventions that can be employed to mitigate air pollution. The monitoring of pollutants was done from eight strategic sites around the city. Sulphur dioxide levels were observed to be generally above the WHO annual mean of 50 $\mu g/m^3$. The levels of SO_2 were highest in the heavy industrial sites. The highest Nitrogen dioxide levels were recorded at a site along the busiest major road, which leads to the Beitbridge border post. Nitrogen dioxide levels in all areas were low (annual means of between 19.72 $\mu g/m^3$ and 32.92 $\mu g/m^{3)}$ and within the WHO annual mean of 40 $\mu g/m^3$. The highest Suspended Particulate Matter was recorded at Beatrice Road Infectious Diseases Hospital sited between Mbare high-density suburb and the heavy industry and the lowest levels of SPM were recorded in the low density areas with no industrial activity. This case study is based on retrospective study of the information collected over five years from 2002 to 2006. The trends show that air pollution was increasing. There was no further information available beyond 2006 as the City of Harare faced enormous financial challenges to maintain and replace the aged air pollution equipment. Thus, as a matter of urgency, air pollution monitoring needs to be revived and should be improved and upgraded from the old methods to the new technological methods to be in lined with the international applied global methods.

Keywords: Air Pollution, Air Pollutants, Sulphur Dioxide, Nitrogen Oxides, Suspended Particulate Matter, Strategic Sites, Monitoring.

INTRODUCTION

Air pollution is indeed a global enemy as it has caused a lot of human suffering. Air pollution is responsible for morbidity and mortality from diseases such as respiratory infections, heart diseases and cancers. In the year 2016, ambient air pollution was responsible for 4.2 million deaths. Worldwide, ambient air pollution is estimated to cause about 16% of the lung cancer deaths, 25% of the chronic obstructive pulmonary disease deaths, about 17% of ischemic heart diseases and stroke, and about 26% of respiratory infection deaths.[5]

Air pollution, originating in cities, impacts human health causing morbidity and mortality associated with lung infections. Other respiratory conditions including, asthma, emphysema and bronchitis are also primary health problems associated with human exposure to pollution. Children and the aged are especially susceptible to air pollution related illnesses.

Air pollution has caused severe deleterious effects globally that include the increase in global warming and climate change. Sulphur dioxide (SO_2) and Nitrogen dioxide (NO_2), do not only cause morbidity in humans but are also the primary causes of acid rain when they react with water in the atmosphere. Acid rain damages buildings, automobile coatings and vegetation. Carbon monoxide (CO_2) prevents blood from carrying oxygen causing mortality in humans. Another pollutant, Suspended Particulate Matter (SPM), has adverse effects on the appearance of buildings and retards vegetation growth by blocking the stomata.

REGIONAL AND GLOBAL TRENDS OF AIR POLLUTION

A major share of the global emissions of air pollutants results directly from urban activities such as energy production, manufacturing and transportation. According to WHO data 2017, 92 nations were evaluated to determine which countries were most polluted. Of these 92 countries, three nations, Pakistan, Qatar and Afghanistan had the highest levels of pollution. According to conserve-energy-future.com, Pakistan had an average concentration of particulate matter, $PM_{2.5}$ of 101 $\mu g/m^3$. This was the highest rate of pollution in the world. Qatar had an average concentration of $PM_{2.5}$ of 92 $\mu g/m^3$ and Afghanistan had an average concentration of $PM_{2.5}$ of 84 $\mu g/m^3$ making it the third most polluted city in the world.[4]

The Status of Air Pollution in Africa

Increased activity in key social and economic sectors is contributing significantly to the air pollution burden in Africa. Unsustainable patterns of consumption and production of energy resources by transport and industry have been the leading sources of outdoor pollution. The Moroccan industry for example burns 1 million tons of fuel each year, generating 2 million tons of CO_2. The extensive use of coal based energy sources explains South Africa's proportionally larger CO_2 emissions. Mining and cement production in Zimbabwe, Morocco, Zambia and South Africa among others are also contributing significantly to the region's air pollution mainly through dust emissions and CO_2 from the combustion of coal.[8]

The transport sector is also being recognized as the highest polluter in key African cities such as Nairobi, Cairo, Johannesburg, Cape Town and Dakar. There has been a doubling effect of the motor vehicle fleets in Zimbabwe and Botswana over the past 10 years.[8] Transport systems are emitting tons of atmospheric pollutants mainly oxides of Nitrogen and Sulphur. The use of imported second hand vehicles and poor road networks and congestions during peak hours has contributed enormously to the air pollution burden.

The household sector has also contributed to indoor air pollution through the use of firewood, charcoal and kerosene. Traditional cooking using fire in Africa is significant enough to cause air pollution. Clearing of land for agriculture through the burning of forests cannot be ignored as it contributes to the greenhouse gas emissions. Dust is a small but an important contributor to atmospheric pollution. Wind-blown dusts from mines, cement manufacture and sand blasting activities cannot be also ignored.

Air Pollution in Zimbabwe

In Zimbabwe like the rest of the world, air pollution still remains one of the pressing environmental problems. The majority of energy production in Zimbabwe is from non-renewable energy. Zimbabwe has four power plants that depend on thermal power for the generation of electricity, Hwange, 920 Megawatts (MW), Harare (125 MW), Bulawayo (120 MW) and Munyati (120 MW). Zimbabwean coal is known to contain 2–3% Sulphur, and burning of fossil fuels leads to the emission of Sulphur dioxide.[7]

People living close to mines and factories suffer dust emissions quietly. In rural areas and parts of urban areas, indoor pollution has affected many people using wood, cow dung, charcoal and kerosene for cooking, heating and lighting. Other household cleaning products such as insect repellants, air fresheners and detergents are all causes of indoor air pollution in cities causing difficulty in breathing.

Another contribution to air pollution in Zimbabwean industries comes from the use of old technology and equipment. Companies cannot replace their old machinery due to the economic meltdown.

Zimbabwe experienced a rapid expansion in the vehicle population from 2009. Since the introduction of the multi-currency regime, commonly referred to as dollarization in 2009, Zimbabwe has imported pre-owned vehicles worth over US$ 5 billion. In 2018 alone, vehicle imports reached nearly 330,000 by November.[6]

Data on air pollution in third world cities is lacking.[6] Most towns and cities in Zimbabwe do not have air pollution monitoring systems and records to show the impact of air pollution on their citizens. This report reviews the status of air pollution in Harare, the capital city of Zimbabwe over a 5-year period from 2002 to 2006. This is the period when Harare last monitored its air pollutants. The data was collected mainly from the City of Harare, Health Department's annual reports. The main sources and constituents of air pollution in Harare, are shown in Table 1.

Table 1: Sources and Main Constituents of Air Pollution in Harare, Zimbabwe

Constituent	*Sources*
SO_2	Burning of fossil fuels in manufacturing industries (boilers), thermal power station, vehicle emissions
NO_2	Burning of fossil fuels in manufacturing industries (boilers), fertilizer manufacturing, thermal power station, vehicle emissions

(Contd...)

(Table 1 contd...)

Constituent	*Sources*
SPM	Dust from cement manufacturing, fertilizer manufacturing, base mineral grinding, sand blasting, tobacco processing, quarrying, concrete premixing, detergent manufacturing, construction, dust roads, vehicle emissions, power plants, yard sweeping
Smoke	Boilers, Burning of tyres in home industries, occasional veld fires, occasional fires from the municipal garbage dumps, incinerators, burning of uncollected refuse, vehicular emissions
Smells	Municipal garbage dumps, porcine abattoirs, tobacco processing, Fumes from foundries, sewage works

Source: City of Harare, Health Department, annual reports 2002–2006.

MATERIALS AND METHODS

The Local Authorities are mandated under the Zimbabwe Atmospheric Pollution and Prevention Act (Chapter 20:03) and the Public Health Act (Chapter 15:17) which are both administered by the Ministry of Health and Child Care, to monitor air pollution levels and control emissions. The Air Pollution Control Unit (APCU), under the Environmental Health Division (City of Harare) monitored and controlled air pollution in Harare.

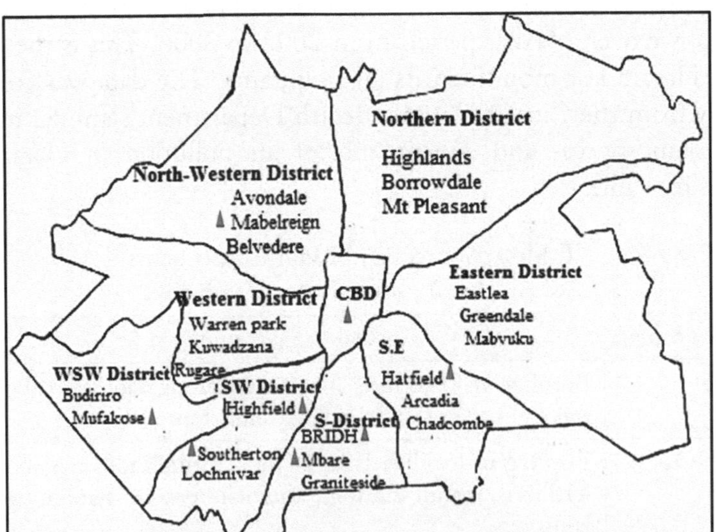

Figure 1: Map of the City of Harare showing the Monitoring Sites

Routine inspections for the monitoring of dust, smoke, fumes and odours were carried out in specified and non-specified areas. Routine monitoring of the ambient levels of suspended particulate matter, Sulphur dioxide and oxides of nitrogen were carried out from 2002 to 2006 from eight carefully selected strategic sites (Figure 1), each site being representative of a specific locality or type of area, where the air sampled would be most truly representative of that area as shown in Table 2.

Table 2: Strategic Sites for the Monitoring of Air Pollution in Harare, Zimbabwe

	Monitoring Site	*Reason for Selection*
1.	Town House	Central Business District with the most traffic
2.	Beatrice Road Infectious Diseases Hospital (BRIDH)	Major road demarcating industries and high density residential
3.	Southerton	Heavy Industries
4.	Mufakose	High density suburb windward of the heavy manufacturing industries
5.	Highfield	High density suburb windward side of the heavy industry
6.	Mabelreign	Low density suburb with no nearby sources of pollution
7.	Hatfield	Low density suburb with proximity to a fertilizer manufacturing company
8.	Mbare	Major bus terminus

Sulphur Dioxide Monitoring

The Bubbler method as illustrated in Figure 2, was the principle method used for the determination of SO_2. This consisted of drawing a measured volume of ambient air through a dilute solution of Hydrogen peroxide (H_2O_2) whose pH is known (e.g. 4.50). As the air stream bubbled through the solution, SO_2 in the air was absorbed and oxidized to Sulphuric acid (H_2SO_4).

$$SO_2 + H_2O_2 = H_2SO_4$$

The increase in the acidity of the solution reduced the pH of the solution. The sample was then titrated in the laboratory against a sodium borate solution using a pH meter to determine the end point which was pH 4.50. The SO_2 present in the sample was then calculated. The quantity of sodium borate used was the basis of the subsequent calculations that also took into account the volume of air sampled. The results were given as the weight of SO_2 in micrograms per cubic meter of air (μg/m^3).

The advantage of the Bubbler method was that it combined both the determination of SO_2 and the measurement of suspended particulate matter.

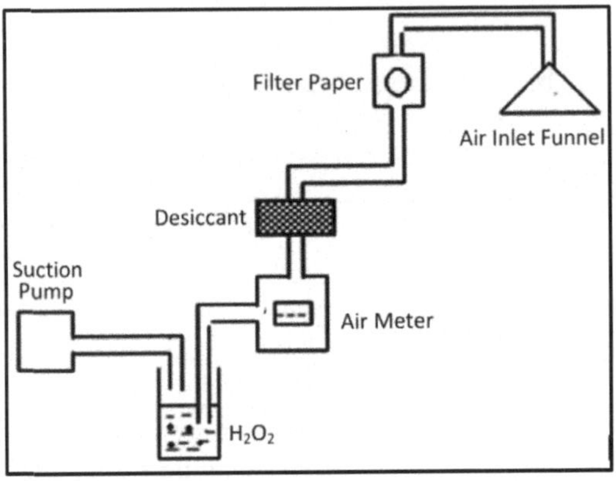

Figure 2: Illustration of the Bubbler Method for the Determination of SO_2 and SPM

Determination of Suspended Particulate Matter

The amount of SPM present in the air was measured by drawing a measured volume of air at a consistent rate through a white filter paper (Whatman No. 42), whose light transmission had previously been measured by a densitometer. The suspended air particles were filtered out onto the filter paper over a measured time period. The result was the soiling of the filter paper by the SPM. This resulted in the darkening of the filter paper. The degree of darkening was then measured using the densitometer. Calculations were done taking into account the weight of the filter paper before and after exposure, the volume of air sampled and the reference table from the densitometer. The results were given as the weight of the particles in micrograms per cubic meter of air (μg/m^3).

Determination of Nitrogen Dioxide

The method was based on the passive collection of nitrogen dioxide in a sampler in which the filter paper (Whatman No. 42), was impregnated with triethanolamine solution (TEA), to absorb the nitrogen dioxide over a period of time. Saltzman solution was used to wash the sample and absorbance was measured using a spectrophotometer. The time-weighted average of nitrogen dioxide was calculated from the mass of the nitric ion found in the sample and the results were given in micrograms per cubic meter ($\mu g/m^3$).

Smoke Observations and Complaints

Smoke plumes billowing from chimneystacks were a cause for concern. This smoke is linked to a variety of health problems. Dark smoke represents partially burned particles of fuel, the result of incomplete combustion. It can be dangerous because small particles are absorbed into the lungs. Besides the routine monitoring of the specific pollutants, the Air Pollution Control Unit also monitored smoke emissions and dealt with all public health complaints regarding air pollution.

Smoke was measured in terms of the degree of opacity using the Ringlemann chart system as indicated in Figure 3 and Table 3.

Table 3: The Ringlemann Scale

Shade	*Degree of Opacity (%)*	*Transmittance of Light*
0	0	100
1	20	80
2	40	60
3	60	40
4	80	20
5	100	0

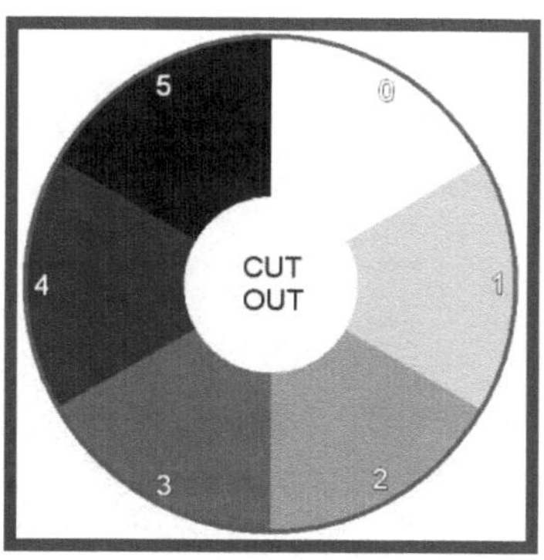

Figure 3: Ringlemann Chart

Plume opacity was the degree to which transmission of light was reduced as viewed through the cut out center of the chart. The maximum permissible density of smoke emitted from any chimney, except when soot blowing must be less than Ringlemann 3.[2]

Contravention of the above limits constituted an offence and was enforced by prosecution. Offenders were served with notices and prosecuted in accordance with the Zimbabwe Atmospheric Pollution Prevention Act Chapter 20.03 of 2001[2] and the Zimbabwe Public Health Act Chapter 15:09.[3]

The main reason for dark smoke emissions were given as poor coal quality, equipment breakdown, fire mismanagement, low chimney stacks, burning refuse and burning of tyres in home industries. Occasional complaints of dark smoke emissions came from municipal dump site fires which gave rise to thick black smoke and fumes.

RESULTS AND DISCUSSION

Sulphur Dioxide

The concentration of SO_2 was highest in Southerton and lowest at Town Hose in the 5-year period under review as shown in Figure 4. The trends for SO_2 levels remained similar in all areas over the 5-year period although the levels for Southerton showed a marked increase.

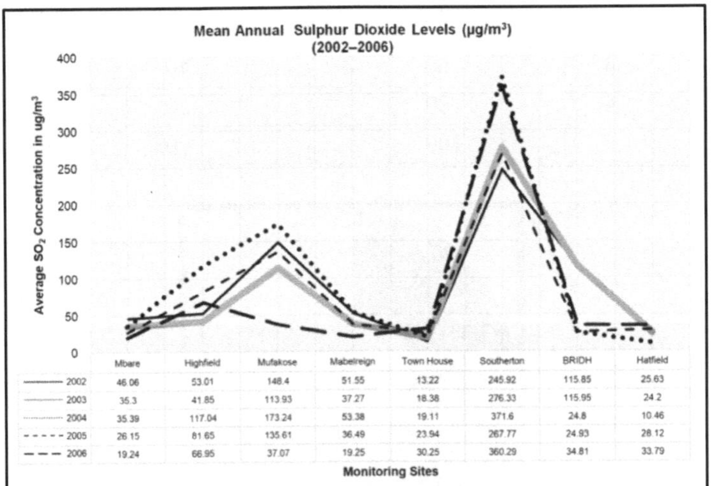

	Mbare	Highfield	Mufakose	Mabelreign	Town House	Southerton	BRIDH	Hatfield
2002	46.06	53.01	148.4	51.55	13.22	245.92	115.85	25.63
2003	35.3	41.85	113.93	37.27	18.38	276.33	115.95	24.2
2004	35.39	117.04	173.24	53.38	19.11	371.6	24.8	10.46
2005	26.15	81.65	135.61	36.49	23.94	267.77	24.93	28.12
2006	19.24	66.95	37.07	19.25	30.25	360.29	34.81	33.79

Figure 4: Mean Annual SO_2 Levels ($\mu g/m^3$) in Harare (2002–2006)

Source: Annual reports.

Southerton, Mufakose and Highfield had SO_2 levels higher than the WHO guidelines of 50 ug/m^3 per annum.[8] The Southerton site had the highest levels of SO_2 because of the heavy manufacturing industry which uses coal for firing boilers. Mufakose high density suburb had the second highest levels as it is on the windward side of the heavy industry. This shows that pollutants can be blown by the wind from the source of pollution to other areas. The same reason stands for the increased SO_2 levels in Highfield residential area.

Mbare, Hatfield, Town House and Mabelreign had SO_2 levels below the WHO guidelines.[9] However, the graph shows that the levels of SO_2 levels declined markedly in residential areas during the year 2006. In Southerton, and Hatfield the continued increase could be attributed to the use of subserviced machinery due to the harsh economic environment hence the increase in pollution.

Nitrogen Dioxide

The results in Table 4 show that the Nitrogen dioxide levels were generally below the WHO guidelines.[9] However, BRIDH and Town House in the year 2002, recorded levels slightly above the WHO guidelines with 40.62 μg/m^3 and 43.51 μg/m^3, respectively. This can be attributed to the high volumes of traffic that pass though these areas. The BRIDH site is along a major road leading to the main border post at Beitbridge and Town House is situated at the heart of the City with all traffic congestions and a main high density commuter omnibus rank one street away. Hatfield and Mbare sites air monitoring equipment broke down for three years and the NO_2 levels for these sites could not be measured during this period.

Table 4: Mean Annual NO_2 Levels (μg/m^3) in Harare (2002–2006)

Monitoring Site	*2002*	*2003*	*2004*	*2005*	*2006*
Mbare	0	0	0	21.06	20.25
Highfield	32.02	28.95	25.34	20.34	17.63
Mufakose	24.3	23	15.91	15.6	16.33
Mabelreign	24.72	23.41	18.34	16.04	15.54
Town House	43.51	37.55	29.98	28.01	29.6
Southerton	32.34	27.08	24.85	20.61	20.62
BRIDH	40.62	38.06	29.83	23.5	24.38
Hatfield	0	0	0	14.36	13.38
WHO Standard	40	40	40	40	40

Figure 5 gives a clear picture of the decline of the NO_2 levels. This may be explained by the decrease in the vehicular population due to the economic meltdown that also affected the APCU of the City of Harare. The unit failed to maintain the equipment due to the same reason; hence there is no further data beyond 2006.

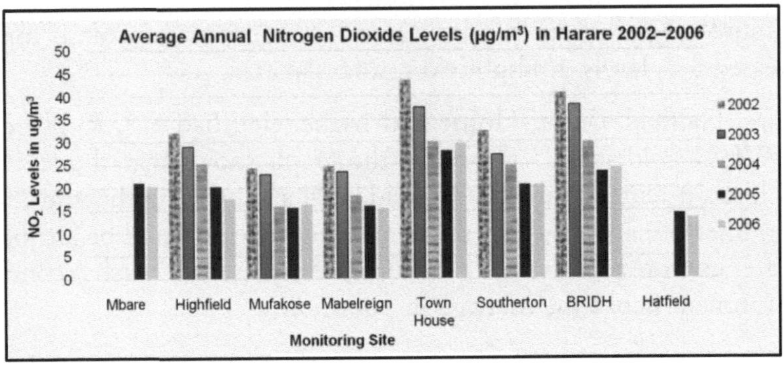

Figure 5: Average Annual NO_2 Levels ($\mu g/m^3$) in Harare (2002–2006)

Source: Annual reports.

Suspended Particulate Matter

The annual levels for the concentrations of SPM are given in Table 5. The concentrations of SPM in ambient air were highest in 2002. BRIDH had the highest SPM concentration of 89.38. There was general decrease in the level of SPM emissions over the following years with concentrations levels below the WHO guidelines.[9]

Table 5: Suspended Particulate Matter Annual Average Concentrations ($\mu g/m^3$) in Harare (2002–2006)

Monitoring Site	*2002*	*2003*	*2004*	*2005*	*2006*
Mbare	61.72	26.93	62.26	43.6	42.18
Highfield	63.33	34.22	50.28	47.9	36.03
Mufakose	35.46	21.98	30.37	40.94	35.1
Mabelreign	24.36	14.65	23.14	24.33	9.79
Town House	68.02	21.88	39.79	40.38	35.82
Southerton	35.02	24.39	33.05	50.4	41.97
BRIDH	89.38	44.02	39.81	69.39	42.48
Hatfield	26.46	12.77	18.86	20.17	15.02
WHO Standard	50	50	50	50	50

The highest Suspended Particulate Matter (SPM) was recorded at Beatrice Road Infectious Diseases Hospital sited between Mbare high-density suburb and the heavy industry and the lowest levels of SPM was recorded in the low density areas with no industrial activity as shown in Figure 6. Highfield and Mufakose high-density residential

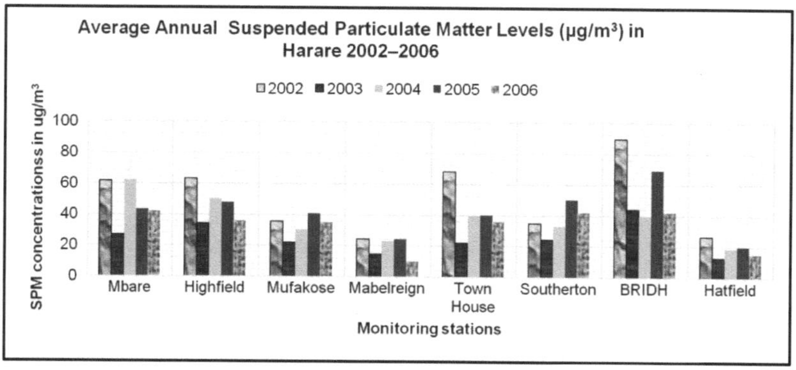

Figure 6: Average Annual SPM Levels (μg/m^3) in Harare 2002–2006

areas recorded higher levels of SPM than the low-density areas as they are on the windward side of the industry. The higher levels could also be attributed to the fact that residents in high-density areas sweep their yards more often than in the low-density areas and the yards in low-density areas are mostly covered with lawn and the roads in the low-density areas are tarred.

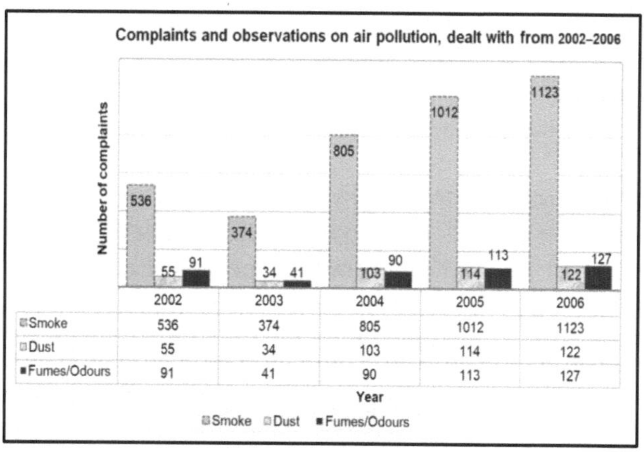

Figure 7: Complaints and Observations on Air Pollution Dealt with from 2002–2006

Source: Annual reports.

Air Pollution Complaints and General Observations

Smoke complaints and observations topped the list as shown in Figure 7. The other complaints on dust and odours, though they may look insignificant on Figure 7, confirm the existence of other forms of air pollution in Zimbabwe. The graph shows a marked increase in the number of complaints attended to especially for dark smoke.

Challenges

This report was developed with retrospective data from more than ten years. This data does not give the true image of the current air pollution trends in Zimbabwe.

Management Challenges

1. There was failure to allocate adequate funds and resources required for the maintenance of the air pollution monitoring equipment due to the lack of integrated management and bureaucracy within the organization.
2. Lack of coordination between the ministries and related organizations to reduce air pollution.
3. Lack of public awareness and participation to reduce air pollution.
4. Economic challenges.

These limitations notwithstanding, these findings however confirm the existence of air pollution challenges that need to be further researched and addressed.

Interventions

Monitoring Activities

Air pollution monitoring should be revived as a matter of urgency and should be improved and expanded from the old methods to the new technological methods making provisions for the necessary manpower and equipment in line with the global methods.

Supervision of air pollution by motor vehicles must be instituted in conjunction with the Ministry of Transport and special patrols set with the Police for the prevention of visible smoke emissions. Most vehicles in Zimbabwe are imported second hand vehicles. These vehicles have become unroadworthy because of poor maintenance due to the deteriorating economy and high cost of maintenance.

Cleaner Methods

Zimbabwe needs to implement cleaner methods of transportation to reduce emissions from vehicles which are a driving factor in air pollution. Promote the use of conventional buses as opposed to the small commuter omnibuses. Enhance the use of other forms of energy such as solar energy.

Research

Research in the field of air pollution control and monitoring is necessary in view of the paucity of data due to the closure of the air pollution control unit.

Resource Allocation

Taking into account that the mechanisms to control and manage pollution in the country are very limited and close to non-existence. Thus, Zimbabwe must invest more in its human and material resources to reduce air pollution, and to promote scientific research and the use of new technologies.

Integration and Coordination

The relevant ministries should coordinate and integrate the air pollution activities.

Education and Awareness

Looking after the environment is a collective responsibility and everyone should take responsibility. Thus, creating public awareness and education are a mandatory issue to be fulfilled.

CONCLUSION

The results, though retrospective, confirm the existence of some significant air pollution in Zimbabwe that needs to be addressed. Of the pollutants measured, SO_2 recorded the highest total levels of emissions. There appeared to be a decrease in the levels of NO_2 and SPM. Given the current scenario when the vehicles have increased, this picture might not be true for today. Zimbabwe has to realize how critical it is to know its enemy, air pollution. Continuous monitoring is necessary to clarify the air pollution challenges in Zimbabwe. Once the air quality situation of a city is defined, control and abatement strategies can be introduced using sound data.

The City's monitoring system did not measure all the pollutants exhaustively as some parameters such as lead, carbon dioxide, methane and other gases were not measured. Monitoring of air pollutants should provide the necessary baseline information as to which pollutants are of major concern and to identify their principal sources. Figures are necessary to track progress in air quality improvements and help to evaluate the effectiveness of policies aiming at reducing air pollution.

REFERENCES

[1] City of Harare Annual Reports (2002–2006).

[2] Government of Zimbabwe (1996). Atmospheric Pollution Prevention Act (CAP 20:03), Government Printer, Harare.

[3] Government of Zimbabwe (1996). Public Health Act (CAP 15:09), Government Printer, Harare.

[4] https://www.conserve-energy-future.com>top-most-polluted-countires-world

[5] https://www.who.int/gho/phe/outdoor_air_pollution/burden_text/en

[6] https://www.zimbabwemail.com>business>carimports-plunge-80pc

[7] Mujuru, M., Mcrindle, R.I., Gurira, R.C., Zvinowanda, C.M. and Maree, J. (2012). Air Quality Monitoring in Metropolitan Harare, Zimbabwe, *J. Environment Analytic Toxicol* 2:131.

[8] United Nations Environmental Project, African Regional Implementation review for the 14th Session of the Commission on Sustainable Development (CSD-14), Report on Air Pollution.

[9] World Health Organisation (2005) Air quality guidelines. Global update.

ABBREVIATIONS

ACPU	–	Air Pollution Control Unit
BRIDH	–	Beatrice Road Infectious Diseases Hospital
CO_2	–	Carbon Dioxide
H_2O_2	–	Hydrogen Peroxide
H_2SO_4	–	Sulphuric Acid
MW	–	Megawatts
NO_2	–	Nitrogen Dioxide
$PM_{2.5}$	–	Particulate Matter with a diameter of less than 2.5 micrometers
SO_2	–	Sulphur Dioxide
SPM	–	Suspended Particulate Matter
TEA	–	Triethanolamine
UNEP	–	United Nations Environmental Program
WHO	–	World Health Organisation

Long Term Monitoring of Air Pollution from Quarries with Health Effects—Case of Palestine

A. Rasem Hasan[1,2*], Amjad I.A. Hussein[3,4], Asmaa Al-Asmar[3], Abdelhaleem Khader[2], Hanan A. Jafar[1,2] and Tawfiq Saleh[1]

[1]Water and Environmental Studies Institute,
An-Najah National University, P.O. Box 7, Nablus, Palestine

[2]Civil Engineering Department, An-Najah National University,
P.O. Box 7, Nablus, Palestine

[3]Analysis, Poison Control and Calibration Center (APCC),
An-Najah National University, P.O. Box: 7, Nablus, Palestine

[4]Faculty of Medicine and Health Sciences,
An-Najah National University, P.O. Box 7, Nablus, Palestine

*E-mail: mallah@najah.edu

ABSTRACT: Stone industry is a major environmental pollutant in absence of proper protection. Palestine has more than 300 quarries as stone mines, and 1000 stone cutting facilities, with a total annual yield of 100 million tons of raw stone and 25 million square meters of high-quality façade stone. Also, stone industry contributes by 4.5% to the Gross National Product (GNP) and ranked top among all exported materials from Palestine. Unfortunately, due to the current political situation in the country and week law enforcement, stone industry cause serious pollution on the surrounding environment including air, water, soil, plant and human health. Particulate matter 10 (PM_{10}) was measured in the north Eastern Jerusalem and around quarries on a monthly basis during 2014–2019, by using Low Volume Sampler LVS3, Sven Leckel, Germany, other PM components were monitored by handheld TSI DustTrack DRX 8354. Excluding 2014 results; where the sampling was mainly within the quarries, the obtained PM_{10} results were between 15.5–177.7 µg/m^3 (24-hour average). While the first monitoring trials result in 2014 inside the quarries were as high as 397.1 µg/m^3. Air Quality Index ranges within 14.3–112.2, based on the U.S. Environmental Protection Agency (EPA) method. Health implications from exposure to PM were monitored by respirometry *measurements, being measured in 2019 using CONTEC SP10 respirometer. Subjects comprising of quarries' workers, truck drivers, and several residents in the surrounding area were tested for forced vital capacity (FVC), forced expiratory volume-one second (FEV1), and FEV1/FVC ratio. The results showed that 50% of subjects have severe to moderate lung abnormality (36–80% of the reference values for the same gender, age, cigarette and drug consumption), amongst all quarries' workers and drivers were in the most severe situation.*

Keywords: Air, Palestine, PM, Pollution, Quarries, Respirometry.

INTRODUCTION

Background

Air pollution is the introduction of foreign matter (that includes particulates, biological molecules, or other harmful materials) into atmosphere, possibly causing adverse effects to humans and the built and natural environment. Air pollution is of public health concern at three scales: micro (indoor air quality), meso (ambient air quality), and macro (global effects) (Davis and Masten, 2004).

Air pollution is considered as one of the main pollution problems around the world. According to the World Health Organization (WHO), this problem is a serious threat to the public health, and wellbeing (WHO, 2012). Many countries enforce strict regulations, among other solutions, with an objective to mitigate air pollution and alleviate its negative impacts as much as possible.

Air pollutants may get originated from either natural or anthropogenic sources. The pollutants originated from anthropogenic sources can be in three physical forms: gas, liquid, and solid. Solid air pollutants (also called particulate matter, PM), is believed to have great influence on human health risks (Khader and Martin 2019).

PM is well known as a common proxy indicator for air pollution with recorded direct negative effects on people more than any other pollutants (Mkpuma, 2015). PM consists of a complex mixture of solid and liquid particles of organic and inorganic substances suspended in air, mainly sulphate, nitrates, ammonia, sodium chloride, black carbon, mineral dust and water (Mkpuma, 2015). Particles with a diameter of 10 microns or less, (PM_{10}) can deeply penetrate and lodge inside the lungs, the even more health-damaging particles are those with a diameter of 2.5 microns or less, ($PM_{2.5}$).

Studies in the United States, Brazil, and Germany have related higher levels of particulates to increased risk of respiratory, cardiovascular, and cancer-related deaths, as well as pneumonia, lung function loss, hospital admissions, and asthma. Some investigations have pointed toward particle sizes smaller than 2.5 μm as a major contributor to elevated death rates in polluted cities (Davis and Masten, 2004; Watterson, 2012).

The WHO air quality guidelines (2005) recommended the annual mean of PM_{10} of 20 $\mu g/m^3$ and the 24-hour mean is 50 $\mu g/m^3$. The Palestinian Standard Institution (PSI), and in its ambient air quality standard PS801–2010 determined the PM_{10} maximum of 120 $\mu g/m^3$ on 24-hour mean and 70 $\mu g/m^3$ on annual mean. However, EPA still retains a 24-hours standard of 150 $\mu g/m^3$. The guidelines showed flexibility in the figures to indicate the global situation, which can balance the Palestinian situation as development country with special focus on the PM sources like quarries. From health perspective, reducing the PM_{10} pollution from 70 to 20 $\mu g/m^3$ can cut air pollution-related deaths by around 15%.

In Palestine, there are three possible anthropogenic sources of particulate air pollutants: quarries and stone cutting industry, transportation, and industrial activities, in addition to seasonal dust storms (Khader and Hasan, 2015).

Quarries in Palestine

Palestine has more than 300 quarries as stone mines, and 1000 stone-cutting facilities with a total annual yield of approximately 100 million tons of raw stone and 25 million square meters of processed stone (Sayara *et al.*, 2016). Quarrying is a form of land use and part of the local heritage where non-metallic rocks and aggregates are extracted from land (Ukpong, 2012). The final outputs of such industry are reduced-size stones, that can be used for several construction purposes (Nartey *et al.*, 2012). Final products of quarries range from façade stones to crushed stones used in concrete mix. Palestine represents 4% of the world's production of stone and marble (Sayara *et al.*, 2016). Regrettably, these industries cause significant pollution in the surrounding environment affecting air, water, soil, plant and human health. The WHO reported that around 92% of the world population suffers from poor air quality and so many industrial countries invest to reduce air pollution for their citizens (WHO, 2012).

Transportation

Despite the strategic importance of the transportation sector in modern societies, it is considered a major source of air pollution and a colossal energy consumer (Al-Sahili and Abu-Eisheh, 2002). In Palestine, transportation sector consumes more than 60% of total consumed energy, where the average age of privately owned automobiles is 16.5 years (Al-Sahili and Abu-Eisheh, 2006). These numbers indicate that there are many old cars that consume more fuel and emits more pollutants.

Industrial Activities

Due to the fact that the West-Bank is still under Israeli occupation and control, there is proof of negative environmental impacts in that area as Israeli authorities have relocated some of its polluting factories from Israel to the West-Bank. For instance, Geshuri Industries, a pesticides and fertilizers factory originally located in Israeli city of Kfar Saba, was closed for pollution violations and relocated to an area adjacent to Tulkarm city (Isaac, 2000).

Seasonal Dust Storms

Dust storms are among the most severe environmental problems in the Middle East. In Palestine, dust storms occur mainly in spring season (Khader and Martin 2019). Air pollution and increase of respiratory diseases are among the environmental impacts of dust storms. Other impacts include reduction of visibility, increased traffic accidents, reduction of soil fertility, and limited solar radiation (Kutiel and Furman, 2003).

QUARRIES' AIR QUALITY IN THE PALESTINIAN CONTEXT

Palestine is a fragmented country, consisting of 2 main separated regions; the West-Bank and Gaza strip. The country is inhabited by 4.8 million (PCBS, 2017), while all Palestinians in the world are estimated at 13 million, most live in diaspora. The Palestinian law of environment (law no. 7 for the year 1999) set the general rules, procedures, and sanctions for the protection of environment. The law and its regulations and bylaws have introduced the requirements for the new industries as to pass an Environmental Impact Assessment (EIA). For the existing industries before the issuance of the law, an Environmental Audit (EA) is to be conducted on frequent basis or as a response to a claim. Both the EIA and EA procedures require an assessment of air quality in and surrounding the facility. Most of the quarries in Palestine started operation before 1999. Country-wide and generally, quarries are not well monitored for their pollution effect on the surrounding environment, particularly air pollution, mainly due to two main reasons. First, capacity for monitoring air pollution is lacking in concerned governmental agencies, private consulting firms and laboratories. An-Najah National University in Nablus (in the north of the country) operates the lone air quality laboratory in Palestine that provides services for both academic research and community needs. Second reason is the current political framework, and its administration in the country that depends on geopolitical classification of lands. Based on Oslo II accord between the Palestinians and the Israeli in 1995, West-Bank lands are divided into 3 categories (Figure 1):

- Area A with full civil and security control by the Palestinian Authority (PA) – composed of 18% of West-Bank lands.
- Area B with Palestinian civil control and joint Israeli-Palestinian security control – composed of 22% of West-Bank lands.
- Area C with full Israeli civil and security control – composed of 60% of West-Bank lands.

In addition, the City of Hebron is divided into two categories (H1 and H20 to denote Palestinian or Israeli control.

This context (and the degree to which the numerous stakeholders approve of the division of power and responsibility) complicates project prioritization, permissions and project implementation.

Most of the quarries in the West-Bank are located in Area C, and so the Palestinian control over these quarries is limited. The Israeli authorities monitor the quality of such quarries only when pollution is impacting nearby Israeli citizens; mostly who are living in Israeli settlements inside the West-Bank.

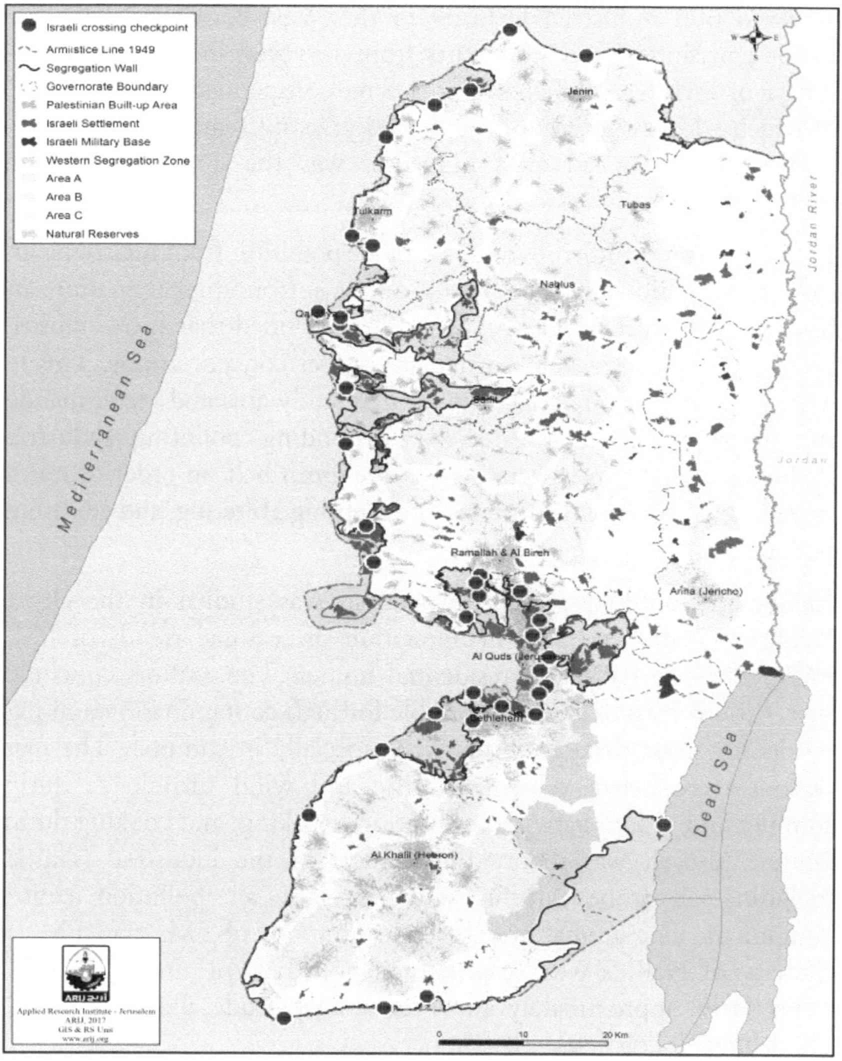

Figure 1: Geopolitical map of the West Bank, Palestine

(*Source:* ARIJ, 2017)

Environmental Impacts of Quarries

Besides the lack of air quality data at the national level in Palestine, there are limited number of published researches on air pollution and its impacts in the country (Jodeh *et al.,* 2017). Most of the activities that contribute to air pollution in Palestine are outdoors (Taneja *et al.,* 2008). The Applied Research Institute Jerusalem (ARIJ, 2015) has surveyed 600 industrial facilities in the West-Bank and quantified major emissions of air pollutants from surveyed industries as 3,749 ton/yr of PM_{10} and 6,341 ton/yr as Total Suspended Particles (TSP), of which 317 tonnes/yr and 18 tonnes/yr as NO_x and SO_x. Major air pollutants among surveyed industries was the quarries and stone cutting facilities.

In 2016, Sayara studied the impact of air pollution from quarrying and stone cutting industries in Palestine on the surrounding agriculture and biodiversity (Sayara *et al.*, 2016). They concluded that such industries produce elevated levels of particulate matter concentrations. This led to negative effects on agriculture in their study area and recommended the development of green belt surrounding polluting industries. Pollutant-tolerant trees were used in the green belt, in order to restrict spreading of quarrying dust via intercepting, filtering and absorbing pollutants.

Indoor and outdoor air quality analysis was studied in the city of Nablus in Palestine, with manifestation of seasonal trends of PM_{10}, $PM_{5.0}$, $PM_{2.5}$ and $PM_{1.0}$ in residential homes. The authors concluded that, various reasons were responsible for high contamination and high results for those particulate matters, especially in summer. The main reasons were heavy vehicular traffic and wind turbulence during summer and indoor activities of heating, smoking, and cooking during winter. Also, it was reported that most of the industrial facilities, including all stone cutting activities, lack air pollution control equipment. The annual average concentrations of PM_{10} and $PM_{2.5}$ in the city of Nablus were approximately 160 $\mu g/m^3$ and 111 $\mu g/m^3$, respectively, approximately an order of magnitude above the WHO Guidelines (Jodeh *et al.*, 2017).

Khader and Martin (2019) have reported values of outdoor PM_{10} and $PM_{2.5}$ in the same city in the range of 48.5 $\mu g/m^3$ and 38 $\mu g/m^3$, respectively. Similar results were reported in a country-wide context by Abdeen *et al.* (2014), where the difference in sampling locations and time of sampling affected such variation in the reported values.

Air Quality Index (AQI)

The AQI is an index for communicating and reporting daily air quality data to the public (EPA, 2014). The AQI focuses on health effects one may experience within a few hours or days after breathing unhealthy air (Figure 2). The AQI is calculated for four major air pollutants, namely, ground level ozone, particle pollution, carbon monoxide, and sulphur dioxide. For each of these pollutants, EPA has established national air quality standards to protect public health (EPA, 2014). Mkpuma (2015) used the below (equation 1) piecewise linear function of the pollutant concentration to calculate the AQI:

$$AQI = \frac{I_{\text{high}} - I_{\text{low}}}{C_{\text{high}} - C_{\text{low}}}(C - C_{\text{low}}) + I_{\text{low}} \qquad \ldots (1)$$

Where: AQI = air quality index, C = pollutant concentration, C_{low} = concentration breakpoint that is ≤ C, C_{high} = concentration breakpoint that is ≥ C, I_{low} = index breakpoint corresponding to C_{low}, I_{high} = index breakpoint corresponding to C_{high}. All breakpoint parameters listed in the EPA website.

Air Quality Index (AQI) Values	Levels of Health Concern	Colors
When the AQI is in this range:	*...air quality conditions are:*	*...as symbolized by this color:*
0 to 50	Good	Green
51 to 100	Moderate	Yellow
101 to 150	Unhealthy for Sensitive Groups	Orange
151 to 200	Unhealthy	Red
201 to 300	Very Unhealthy	Purple
301 to 500	Hazardous	Maroon

Figure 2: AQI Values and their Impacts (EPA, 2014; Mkpuma, 2015)

Air Pollution and Health

Around 8% of all deaths globally are due to diseases caused by air pollution, and mostly attributed to particulate matters (Cohen *et al.*, 2017; Tonne, 2017). WHO (2011) has quantified health effects due to exposure to air pollution, and listed all potential acute, chronic, and reproductive outcomes, including mortality, cardiovascular and respiratory diseases. EPA (2011) reported the proof of cost-effectiveness of mitigation measures regarding air pollution.

Epidemiological studies have been used to account for health effects of air pollution (Pascal *et al.,* 2013). This also plays a major role in identifying the hazards of air pollution as well as conducting quantitative risk assessments (Fann *et al.*, 2011). Nonetheless, epidemiological studies are costly, time consuming and depend on the availability of accurate data.

Since the respiratory system is the first to be affected by air pollution, an easy and quick method to evaluate effects of air pollution on the lung function is the spirometry test, invented by the English surgeon John Hutchinson in the 1840's. The test depends on the amount and speed of air that can be inhaled and exhaled by a person. A spirometry test is conducted by having a person inhale to total lung capacity and then exhaling as hard and as completely as possible. Forced expiratory volume (FEV) and forced vital capacity (FVC) are parameters to measure lung function during the spirometry test. FEV1 (exhaled amount of air during the 1st second of the test) is the most important measurement of lung function and used to diagnose asthma and pulmonary diseases. The ration of FEV1 to FVC represents the percentage of a person's vital capacity that they are able to expire in the first second of forced expiration to the full, with a normal value of approximately 70%.

Objectives of this Manuscript

This paper presents the results of the long-term monitoring of air pollution in the north-eastern Jerusalem's quarries as a result of collaboration between local authorities representing the community and the industries. Recommendations for preserving the surrounding environment and maintaining people's health are concluded. Among all communities in Palestine surrounded by quarries and stone cutting industries, the north eastern Jerusalem gave the bright image of co-operation between local communities and industries. This mutual cooperation endeavours to alleviate the effects of such industries on environment and people, through monitoring and implementation of proper solutions.

MATERIALS AND METHODS

Sampling Location

The north eastern Jerusalem in West-Bank of Palestine has several quarries surrounding the local communities (Figure 3). The selected study site is called Location 1 for confidentiality reasons. Location 1 is located downwind of most quarries in the region and the dominant wind in this area is western. This site was chosen to quantify air pollution and its effect on the population, as expected to be the most affected in north eastern Jerusalem region. On average the site is 540 meters above sea level, and it is 40 meters higher than its immediate surroundings.

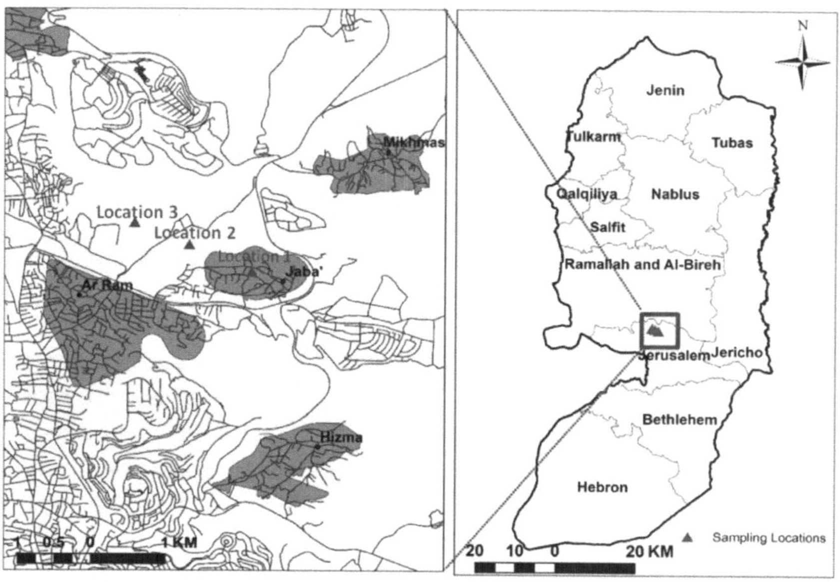

Figure 3: Sampling Locations in North East Jerusalem, Palestine

Air Analysis

Samples were taken for PM_{10} once a month during summer from May to October starting from 2015 till 2019, by using mini-vol, Low Volume Sampler LVS3, Sven Leckel, Germany. Besides the main sampling site (i.e. Location 1), a background sampling point was chosen randomly around 3 km upwind from the quarries' sites. All PM_{10} were calculated based on IO-2.3 (1999) and were corrected to standard conditions of 25°C and 760 mm Hg. Other PM components were measured by handheld TSI Dust Track DRX 8354. AQI was measured by using equation 1.

Spirometry Tests

To account for health effects, epidemiological procedures were found to be inaccurate due to lack of proper recording of health cases. Affected people were administered in health centres in surrounding cities, and mostly in Israeli health centres in Jerusalem. Instead, the lung functions due to prolonged exposure to air pollution was assessed based on the spirometry test by using CONTEC SP10 respirometer. This manuscript reports the results obtained in 2019 tests. A sample of 25 individuals were tested, 5 at a quarry in Location 2, 8 at a quarry in Location 3, and 12 at the Local Council of Location 1. Their ages ranged from 11 to 70 years (average age = 38.16, Median age = 34), 8 of them were smokers, and one of them suffered from asthma. Participants' lung efficiency was tested using a portable CONTEC SP10 spirometer. The tested parameters were forced vital capacity (FVC) and forced expiratory volume-one second (FEV1). The tests were repeated 3 times for everyone.

RESULTS AND DISCUSSION

It was found since 2014 that PM_{10} levels Palestine and inside quarries were surpassing the PSI limits around industrial areas and mostly during summer times. Based on this, several measures were put into practice to lessen air pollution, including wet operation with water recycling, stopping operation during hours when wind directs towards the residential areas. A decrease in PM_{10} level in the surroundings was recorded since 2016. Recommendations are made to control PM_{10} inside the quarries. It is also worth mentioning that around 90% of the PM_{10} are PM_5, and 75% are $PM_{2.5}$. So, health implications are anticipated.

Particulate Matter

Annual average PM_{10} measured in Location 1 during May to October 2015–2019, with maximum and minimum values are given in Figure 4. In general, the average concentration of PM_{10} recorded for all years is above the WHO guidelines' limit (50 μg/m^3 - 24-hour mean). Looking at the entire sampling period (2015–2019), 74% of the sampling tests recorded a concentration that is above the WHO limit (24-hour mean is 50 μg/m^3).

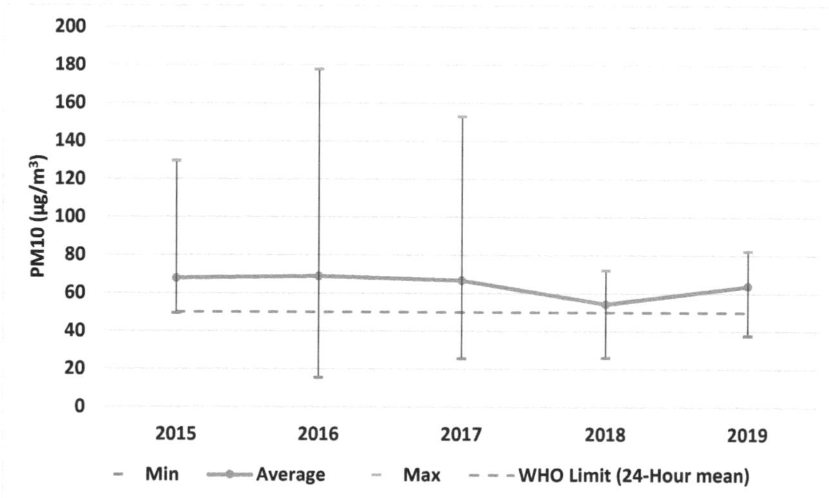

Figure 4: PM_{10} Average Concentration in Location 1, 2015–2019

It is worth mentioning that all quarries lack any equipment for air pollution minimization. The only measures taken are stopping of work at certain hours depending on prevailing weather conditions, mostly the wind speed and direction. The highest values of PM_{10} were recorded during July and August months (Figure 5), the driest and

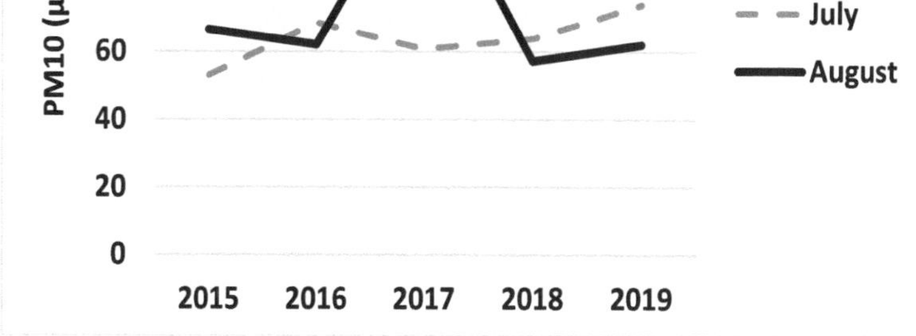

Figure 5: PM_{10} Concentration during July and August, 2015–2019

hottest months in Palestine, with low wind speeds (< 3 m/s most of the time). The highest value recorded over the period was 180 µg/m^3 during June 2016, while the highest in May was 120 µg/m^3, providing that certain times out of July–August might also show higher values.

For all PM_{10} values, 70.5% of the sampling tests show AQI between 51 to 100 (Figure 6), which means that the level of health concern is moderate. But, at certain times in the study area, AQI maximum values were almost close to 100 (Figure 7). Though, $PM_{2.5}$ was in the range of 65–75% of PM_{10}, and in conformity with Jodeh *et al.* (2017), and Kahder and Martin (2019). Although AQI is moderate most of the study time, such levels of PM_{10} warrant caution, since long-time exposure of children and adults to polluted air should be avoided.

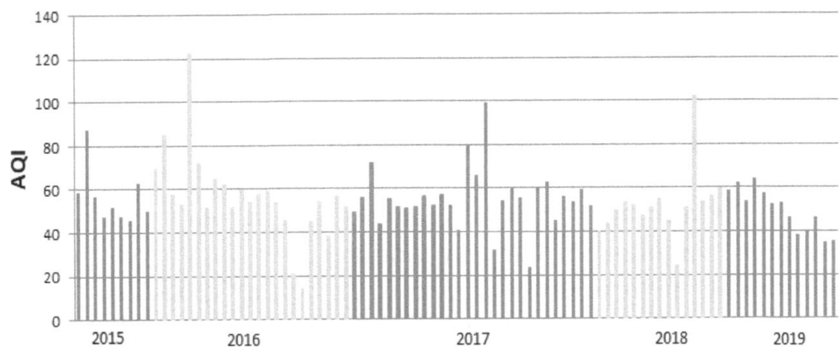

Figure 6: AQI in Location 1, 2015–2019

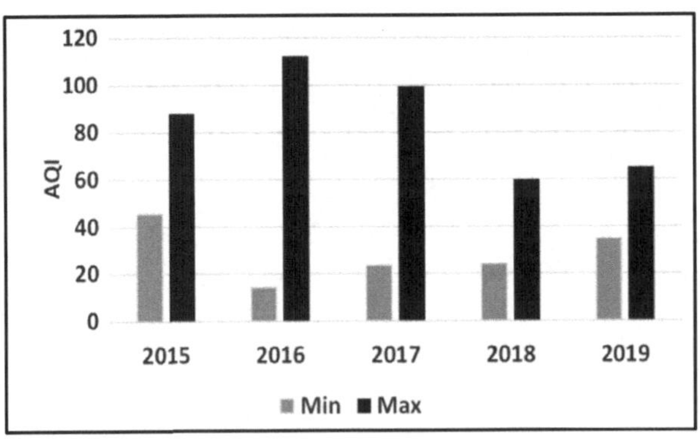

Figure 7: Minimum and Maximum AQI in Location 1, 2015–2019

Collaboration between industry owners and local village councils has led to reduced AQI levels in the recent years (Figures 6 and 7). This has been achieved by avoiding operation during mid-days and practicing of dust-watering, and continued cleaning of quarries sites from dust.

Spirometry

Spirometry results for the targeted sample (25 individuals) can be summarized in the following points:

- 16 out of the 25 individuals showed normal lung functions (FEV1/FVC ratio > 70%)
- 9 out of the 25 individuals showed severe to moderate lung abnormality
- 3 individuals showed mild abnormality (FEV1/FVC ratio 60–69%)
- 4 individuals showed moderate abnormality (FEV1/FVC ratio 50–59%)
- 2 individuals showed severe abnormality (FEV1/FVC ratio < 50%).

On an average, individuals in Location 1 Local Council showed healthier lung functions than the workers in quarries in Location 2 and Location 3, respectively.

The results show that 50% of sample have severe to moderate lung abnormality (36–80% of the reference values for the same gender, age, cigarette and drug consumption), with all quarry workers and drivers being in the most severe situation.

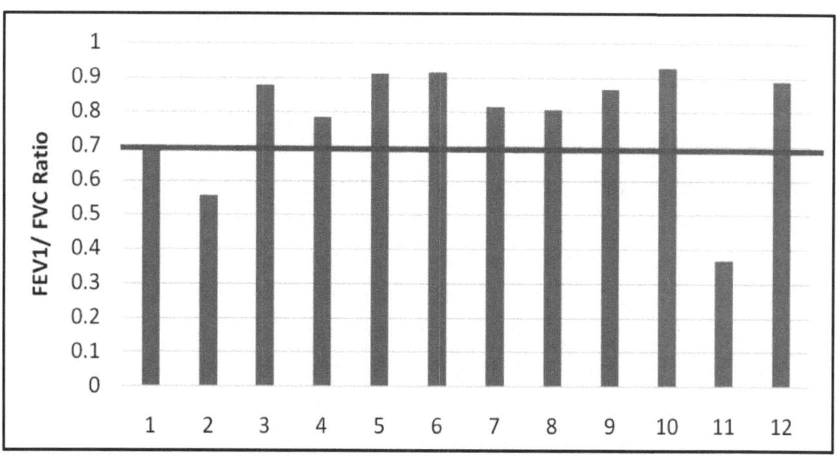

Figure 8: FEV1/FVC for Quarries Workers at Location 1

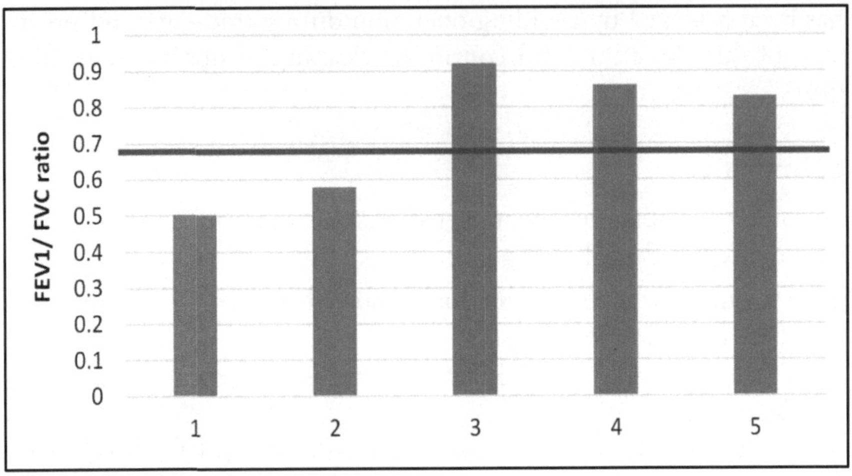

Figure 9: FEV1/FVC for Quarries Workers at Location 2

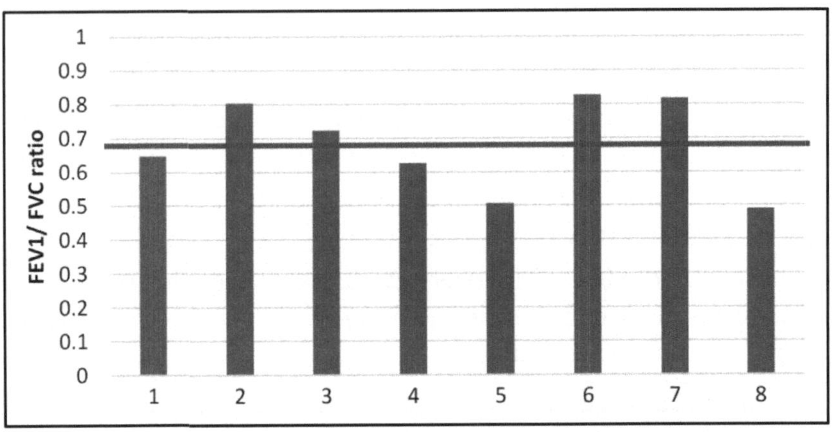

Figure 10: FEV1/FVC for Citizens of Location 3

CONCLUSIONS AND RECOMMENDATIONS

In absence of law enforcement, such as the case in area C of the West-Bank, collaboration between local industries and representatives of local population is needed to control pollution so as to reduce the health impact. In addition, collaborations also lead to continuous monitoring of polluting sites and residential places. In 2014, levels of PM_{10} recorded during summer months in Location 1 averaged at 397.1 μg/m^3. Later, and due to mutual agreement between the local councils and industrial facilities, certain measures were put in place to avoid polluting ambient air in North-Jerusalem, and this led to reduction in PM_{10} levels to a range of 15.5–177.7 $\mu g/m^3$ (24-hour average). Though AQI was moderate in the most affected village of North-Jerusalem, the spirometry tests revealed that health conditions of workers inside the quarries and drivers as well (though with less exposure) are most adversely affected, regarding their lung functions. This warrant industries to invest in more stringent control of air pollution generated inside quarries and to use closed system with air bags for crushing machines, electrostatic filters for screening machines and to synchronize work for mining with weather conditions.

REFERENCES

[1] Abdeen, Z., Qasrawi, R., Heo, J., Wu, B., Shpund, J., Vanger, A., Sharf, G., Moise, T., Brenner, S. and Nassar, K. (2014). Spatial and temporal variation in fine particulate matter mass and chemical composition: The middle east consortium for aerosol research study. *Sci. World J.*, DOI: 10.1155/2014/878704.

[2] Al-Sahili, K. and Abu-Eisheh, S. (2002). Traffic Systems and Environment in Palestine. Seminar: Decrease the Effects of Exhaust Gases, 21–26/12–2002. Ramallah, Palestine.

[3] Al-Sahili, K. and Abu-Eisheh, S. (2006). Report of: Decrease the Effects of Exhaust Gases in Palestine Project—Effect of Traffic Congestion and Public Transportation on Air Quality,15/6/2006, An-Najah National University, Nablus, Palestine.

[4] ARIJ (2017). Geopolitical Map of Palestine. Applied Research Institute (ARIJ). https://www.arij.org.

[5] ARIJ (2015). Status of the Environment in the State of Palestine. Applied Research Institute (ARIJ), Jerusalem. https://www.arij.org/files/arijadmin/2016/SOER_2015_final.pdf

[6] ARIJ (2008). Applied Research Institute—Jerusalem. Environmental profile of the West-Bank, Volume 5, Nablus district.www.arij.org

[7] Cohen, A.J., Brauer, M. and Burnett, R. (2017). Estimates and 25-year trends of the global burden of disease attributable to ambient air pollution: an analysis of data from the Global Burden of Diseases Study 2015. Lancet; 389: 1907–18.

[8] Davis, M. and Masten, S. (2004). Principles of Environmental Engineering and Science. McGraw-Hill, New York, USA.

[9] EPA (2014). A Guide to Air Quality and Your Health. US Environmental Protection Agency. https://www3.epa.gov/airnow/aqi_brochure_02_14.pdf

[10] EPA (2011). The benefits and costs of the Clean Air Act from 1990 to 2020. Final report rev A. https://www.epa.gov/sites/production/files/2015-07/documents/fullreport_rev_a.pdf

[11] Fann, N., Bell, M.L., Walker, K. and Hubbell, B. (2011). Improving the Linkages between Air Pollution Epidemiology and Quantitative Risk Assessment. *Environ Health Perspect.* doi: 10.1289/ehp.1103780.

[12] IO-2.3 (1999). Compendium Method, Sampling of Ambient Air for PM_{10} Concentration using The Rupprecht And Patashnick (R&P) Low Volume Partisol Sampler, Centre for Environmental Research Information Office of Research and Development U.S. Environmental Protection Agency Cincinnati, OH 45268.

[13] Isaac, J. (2000). The environmental impact of the Israeli occupation. *Applied Research Institute Jerusalem (ARIJ)*. www.arij.org.

[14] Jodeh, S., Hasan, A.R., Amarah, J., Judeh, F., Salghi, R. and Lgaz H. (2017). Indoor and Outdoor Air Quality Analysis for the City of Nablus in Palestine: Seasonal Trends of PM_{10}, $PM_{5.0}$, $PM_{2.5}$ & $PM_{1.0}$ of Residential Homes. Air Quality, Atmosphere and Health. https://doi.org/10.1007/s11869-017-0533-5.

[15] Khader, A. and Martin R. (2019). Use of Low-Cost Ambient Particulate Sensors in Nablus, Palestine with Application to the Assessment of Regional Dust Storms. Atmosphere, 10: 539.

[16] Khader, A.H. and Hasan, A.R. (2015). Air Quality Characterization in the city of Nablus-Palestine, The Sixth Jordanian International Civil Engineering Conference, Amman, Jordan, March 10–12, 2015.

[17] Kutiel, H. and Furman, H. (2003). Dust Storms in the Middle East: sources of origin and their temporal characteristics. *Indoor and Built Environment*, 12(6): 419–426.

[18] Mkpuma, R.O. (2015). Estimation of air quality status due to quarrying activities and its impacts on the environment and health of the people. *International Journal of Scientific & Engineering Research*, 6(9): 1828–1836.

[19] Nartey, V.K., Nanor, J.N. and Klake, R.K. (2012). Effects of Quarry Activities on Some Selected Communities in the Lower ManyaKrobo District of the Eastern Region of Ghana. *Atmospheric and Climate Sciences,* 2: 362–372.

[20] Pascal, M., Pascal, L., Bidondo, M.L, Cochet, A., Sarter, H., Stempfelet, M. and Wagner, V. (2013). A Review of the Epidemiological Methods Used to Investigate the Health Impacts of Air Pollution around Major Industrial Areas. *J Environ Public Health.* 1–17, doi:10.1155/2013/737926.

[21] PCBS (2017). Census of Palestinian people in the West-Bank and Gaza. Palestinian Central Bureau of Statistics. http://www.pcbs.gov.ps/census 2017.

[22] Sayara, T., Hamdan, Y. and Basheer-Salimia, R. (2016). Impact of air pollution from quarrying and stone cutting industries on agriculture and plant biodiversity. Resources and Environment, 6: 122–126.

[23] Taneja, A., Saini, R. and Masih, A. (2008). Indoor air quality of houses located in the urban environment of Agra, India. *Ann N Y Acad Sci.,* 1140, 228–245.

[24] Tonne, C. (2017). A call for epidemiology where the air pollution is. www.thelancet.com/planetary-health, Vol. 1.

[25] Ukpong, E.C. (2012). Environmental Impact of Aggregate Mining by Crush Rock Industries in Akamkpa Local Government Area of Cross River State, Nigerian. *Journal of Technology,* 31: 116–127.

[26] Watterson, T.L. (2012). Effects of Cache Valley Particulate Matter on Human Lung Cells. All Graduate Theses and Dissertations. Paper 1341. Utah State University, USA.

[27] WHO (2012). Ambient air pollution: A global assessment of exposure and burden of disease. https://apps.who.int/iris/bitstream/handle/10665/250141/9789241511353-eng.pdf;jsessionid=FEC4D95478E27B99A2C9F3CB4DDA882F?sequence=1 (accessed 2 Dec. 2019).

[28] WHO (2000). Quantification of the Health Effects of Exposure to Air Pollution, Report of a WHO Working Group Bilthoven, Netherlands 20–22 November 2000. http://www.euro.who.int/__data/assets/pdf_file/0011/112160/E74256.pdf

SECTION III

Examples of Action Plans from Developing World

Actions to Mitigate Health Effect Caused by Air Pollution in Cuba

Marisol Romeu Hernández

Biotechnology Unit, Delegation of the Ministry of Science Technology and Environment, Havana City, Cuba

E-mail: mdanyhdez@gmail.com

ABSTRACT: Epidemiological data support the association between air pollution and morbidity and mortality, mainly from respiratory and cardiovascular diseases, particularly in the elderly, infants, pregnant women and people with comorbidities. Most air pollutants produce harmful effects and this problem affects developed and developing countries alike. The industrial and urban growth of the last decade has led to an increase in air pollution and thus a greater risk to health. Urban development, changes in the Earth's surface and climate change are phenomena, resulting from the global population explosion, that are harming not only people but also animals and plants in the ecosystem. Between 1971 and 1980, Cuba experienced an accelerated process of industrial development and reduction of rural-urban inequalities, which led to the establishment of environmental legislation. Scientific entities in the country have carried out different projects to study the effects of exposure to environmental contaminants on the respiratory health of children and respiratory and cardiovascular diseases of adults. The Ministry of Science, Technology and the Environment and research entities, organize programs and projects that favour basic and applied research, with a multidisciplinary and cross-sector approach to these issues. The mitigation of damage to nature and the preservation of natural resources and the human species is an urgent challenge for humanity, which requires ethical behaviour based on humanistic, moral and ecological values, which must be integrated into environmental education with special emphasis on the development of environmental and human health.

Keywords: Air Pollution, Contamination, Environment, Health, Respiratory and Cardiovascular Diseases, Environmental Strategy.

INTRODUCTION

"*An important biological species is at risk of disappearing because of the rapid and progressive liquidation of its natural living conditions: man. We are now aware of this problem when it is almost too late to prevent. Hunger disappears and not the man.*" It was with this categorical phrase that Fidel Castro began his speech at the United Nations Conference on the Environment and Development, on June 12, 1992, in Rio de Janeiro, Brazil.[1] Almost 30 years have passed and the situation predicted by the Cuban leader has evidently worsened. A significant number of investigations have shed light on what is really contained in the air around us and how it affects our bodies. Without air there can be no life, but breathing polluted air condemns us to a life of illness and early death. According to the World Health Organization (WHO), air pollution causes about 7 million premature deaths each year. Overall, this problem is responsible for more deaths than other risk factors, such as malnutrition, alcohol use and physical inactivity.[2]

Clearly, polluted air is creating a global health emergency and Cuba is not exempt from it. In 1959, with the revolutionary triumph, began the attention to the country's environment and natural resources, based on government policies that promoted research and conservation actions.[3] In accordance with the socialist nature of the Cuban Revolution, modifications were made to the Constitution of the Republic, especially in Article 27, after the Rio Summit, in order to express the concept of the integration of the environment with sustainable economic and social development.[4] With the creation of the Ministry of Science, Technology and the Environment (Citma), environmental policy and management were promoted at the national level. However, despite the positive results achieved to date, difficulties persist in solving the main environmental problems, including air pollution caused, among other things, by the urban and industrial growth in recent decades.[3] The objective of this paper is to show the specific objectives and priority lines of action to improve environmental quality expressed in the National Environmental Strategy in accordance with Cuba's National Plan for Economic and Social Development, until 2030.[5]

MATERIALS AND METHODS

Historical-Logical Method

Documentary Analysis: Research reports and results reported by the Science, Technology and Innovation Entities, about those themes. Development of interface actions for the introduction of research results reported by research entities and their assimilation. Visits to Research and Service Centers located in the province and exchange with the national environmental authorities.

Also, we performed an electronic search on PubMed using the following terms: "ambient air pollution", "air pollutants" "climate change", "meteorological factors", "weather", "atmosphere", "outdoor", "particulate matter", "mortality", "human health", "health effects", "infectious disease", "cardiovascular disease", "ischemic heart disease", "cancer", and "respiratory disease". The references of all retrieved original articles and reviews were assessed for additional relevant items.

We also reviewed recent reports on the relationship between climate and human health from non-biomedical journals. Such as regulatory documents by environmental and health produced in Cuba.

RESULTS AND DISCUSSION

Air Pollution and Health: General Considerations

The environment is defined, according to Law No. 81 of the Environment of Cuba as the system of abiotic, biotic and socio-economic elements with which man interacts, while adapting to it, transforming it and using it to satisfy his needs.[6] Pollution could be defined as any undesirable modification of the environment, caused by the introduction of physical, chemical or biological agents (contaminants) in quantities greater than natural, which is harmful to human health, damages natural resources or alters the ecological balance.[6–9]

Air pollution or contamination of the air is therefore one of the main ways in which part of the environment can be degraded or affected. Yassi A. describes it as the emission of hazardous substances into the air at a rate that exceeds the capacity of natural processes in the atmosphere to transform them.

Health goes beyond a simple focus on the lack of disease in humans. The World Organization has defined it as "a state of complete physical, mental and social well-being and not merely the absence of disease or infirmity". It has an adverse relationship with pollution.[10]

In more than 100 countries, the right to a healthy environment enjoys constitutional status, which is the strongest form of legal protection available. At least 155 states have legislated, through treaties, constitutions and laws, to respect, protect and fulfil the right to a healthy environment. The right to clean air is also implicit in the Universal Declaration of Human Rights and the International Covenant on Economic, Social and Cultural Rights, and is fully enshrined in the Sustainable Development Goals—the global plan for achieving peace and prosperity.[11]

Air pollution is a problem that affects developed and developing countries alike. The urban and industrial growth in recent decades has led to an increase in air pollution and therefore a greater risk to health. Most air pollutants produce harmful effects depending on their concentration, so it is necessary to identify their existence and control the levels of each of these potentially dangerous elements. The main sources of pollution from human origin are transport, energy production and industrial activities, as well as inorganic gaseous compounds, which are extremely dangerous to the health of the population. Air pollution by different chemicals has a strong global connotation today, given the harmful effects on ecosystems and in particular on human health.[7, 8]

Urban development, the modification of the Earth's surface and climate change are phenomena resulting from a global population explosion and are altering the composition of the air, harming not only to people but also the animals and plants in the ecosystem.

Background in Cuba

The Constitution of the Republic of Cuba states: "The State protects the environment and the natural resources of the country. It recognizes its close link to sustainable economic and social development in order to make human life more rational and ensure the survival, well-being and security of present and future generations. It is the responsibility of the competent bodies to implement this policy. It is the duty of citizens to contribute to the protection of water, the atmosphere, and the conservation of soil, flora, fauna and all the rich potential of nature".[4]

From 1971 to 1980, Cuba experienced an accelerated process of industrial development and reduction of inequalities between the countryside and the city. This led to the establishment of environmental legislation, as part of the institutionalization process. The Law 33, first law endorsed about the environmental protection, represented an early and important normative expression of the principles of Cuban environmental policy that laid the foundations for the development of the national legal system in this area.[12] In accordance with the socialist nature of the Cuban Revolution, after the Rio Summit, modifications were made to the Constitution, especially to Article 27, in order to express the concept of the integration of the environment with sustainable economic and social development.

In March 1985, after the Vienna Convention for the Protection of the Ozone Layer, Cuba adopted, under the auspices of the United Nations the measures to protect health and the environment from the effects of stratospheric ozone depletion and was established the National Office to control the use of CFCs in companies and organizations.

The creation in 1994 of the Ministry of Science, Technology and Environment gave a boost to environmental policy and management at the national level. This meant at the time a strengthening of the Cuban institutional framework, in circumstances where the international trend was towards deregulation of environmental policies, as result of the prevailing neoliberalism. This transcendental institutional change imposed, in turn, the need to revise the country's strategic and regulatory frameworks in environmental matters.[3] For those reasons, the Law 33 was reverse and was approved in 1997, the Law 81, established the principles of environmental policy and the basic rules to regulate the environmental management of the State. Beside of the actions of citizens and society in general, in order to protect the environment and contribute to achieving the objectives of sustainable development of the country.

Conservation and rational use of the environment is our duty; the systematic fight against the causes that originate its deterioration. Furthermore, the corresponding rehabilitation actions; the constant increase of the citizens' knowledge about the interrelations of human beings, nature and society; the reduction and elimination of environmentally unsustainable production and consumption modalities.

Besides, the promotion of demographic policies that are adequate to the territorial conditions.[6]

Based on those requirements, the National Environmental Strategy (NES) was developed, the first cycle of which was approved in 1997 and subsequently revised in 2007 and 2011. The NES has been a key tool in the national environmental effort and in promoting sustainable development in Cuba, bringing favourable results that exceeded projected expectations in several areas.[3]

In Cuba there is an environmental policy articulated to the process of general development, oriented to the protection of the environment and has as a center of attention the man, based on the protection of the ecosystems, according to levels of priorities determined by science, supported by technology, and sustained in principles of responsibility. It corresponds to the Ministry of Science, Technology and Environment, in its condition of Organism of the Central Administration of the State rector of the environmental policy, to propose it and to control its implementation. The rest of Organism of the Central Administration of the State have the obligation to incorporate, in their development policies, elements of environmental protection in correspondence with the established public policies.

Other actors in environmental policy and management are companies, foreign investment modalities, cooperatives, small farmers, self-employed workers and other forms of non-state management. The Local Bodies of People's Power must guarantee the application of environmental policy and establish the adaptations and priorities corresponding to the characteristics of each territory.

Although the country has important achievements to show in the environmental context, it is not exempt from environmental problems. There is still a long way to go to achieve a healthy environment, since human and financial resources, adequate inter-institutional coordination and agreement, and technical-organizational measures are required. All this, without forgetting that man, according to his level of knowledge and environmental awareness, is the key factor for the solution of these problems.[3]

The strategic line of environmental health that Cuba has developed over the years has contributed to increasing the population's standard of living and reducing the morbidity and mortality linked to certain diseases, thus ratifying the close link between the environment and human health. It is therefore important to promote research in the broad field of environmental health, with the aim of establishing a set of rights, duties and obligations that will help to preserve human health and the environment.[13]

Several studies were carried out on air pollution and its relationship to respiratory diseases, such as Bronchial Asthma, Acute Respiratory Diseases (ARI) and Lung Cancer, among others.[9, 14–17] The research carried out by researchers from the National Institute of Hygiene, Epidemiology and Microbiology, belonging to the ministry of Public Health, focused on the study of acute and chronic respiratory diseases in children and adults and their environmental determinants in different cities of the country.[7, 8]

Researchers from the National Institute of Meteorology, belonging to the Ministry of Science, Technology and Innovation, have also conducted studies on pollution levels that could be associated with bronchial asthma in Havana municipalities, and have carried out studies on thermodynamic aspects of meteorology and its influence on the dispersion of atmospheric pollutants. The results of these have shown that air pollution has grown alarmingly in recent decades, as a result of the discharge into the air of increasingly abundant quantities of fumes, toxic gases and chemicals.[18, 19]

Growing urbanization, vehicle congestion and the high cost of control measures have made urban air pollution a crucial problem in Cuba. The responsibility to protect and promote health is not only of health personnel, it is of all people who influence the environment, in one way or another. The concentration of the population, industry and automotive transport in urban areas has produced an ecological imbalance with serious effects on environmental hygiene. It is necessary to work in the protection of the immediate environment to work and home, to the regional, national and global level; therefore, each individual has the duty to care for and maintain a clean and safe environment, which guarantees the survival of present and future generations.

Cuba's National Environmental Strategy 2016–2020

The NES 2016–2020 defines seven main environmental problems in Cuba, based on considering the greatest impact and which take place on a broader national scale, taking into account, among other factors, the effects on the health and quality of life of the population.

Many factors contribute to this environmental problem. It includes technological obsolescence and the high level of deterioration of the existing industrial park, insufficient coverage for the treatment of liquid waste, solid waste and gaseous emissions. Other factors are technological indiscipline and failure to comply with repair and maintenance cycles, the low level of introduction of preventive practices and strategies aimed at reducing the generation of waste and emissions at the source.

Cuba has been blocked by United State of America for more than 60 years. It limits material and financial resources for the execution of actions aimed at solving those problems and the low level of execution and effectiveness of investments aimed at preventing, reducing and controlling pollution. In general, the capacities for the systematic monitoring and evaluation of the contamination are insufficient, which affects in a remarkable way the knowledge about the tendencies of this problem in the national scope.

In the particular case of the occurrence of noise situations, the fundamental causes are the social and administrative indiscipline, in addition to the problems derived from technological obsolescence and the lack of search for alternatives to reduce them, with a low level of demand in the application of existing regulations by the competent authorities.

In recent years, efforts have been increased to reverse the national situation by implementing short-, medium- and long-term programs to solve the problems of a group of contaminating sources. The general strategic objectives of the NES are harmonizing with the proposed general objectives of the strategic axis "Natural Resources and Environment of the National Plan of Economic and Social Development until 2030: Proposal of Vision of the Nation, Axes and Strategic Economic Sectors":[20]

1. To guarantee a rational use of natural resources and the conservation of ecosystems, as a basis for sustainable development.
2. To reduce pollution as a way to improve environmental quality.
3. To effectively implement actions to address climate change, prioritizing adaptation measures.

4. Improve and develop the instruments of environmental policy and management as a support for decision making at different levels.

3.4 Improvement of Environmental Quality

Objectives	*Actions Axis*
Prevent, reduce and control the pollution caused by the inadequate disposal of liquid and solid waste and the emission of atmospheric pollutants.	• Gradual solution of the polluting sources included in the prioritized programs, with special emphasis on the ecosystems of environmental interest (main cities, bays, mountains, basins). • Introduction of preventive practices and strategies aimed at reducing the generation of waste and emissions at source. • Updating of the inventory and estimation of emissions from sources that generate atmospheric pollution. • Guarantee the inclusion in the economy's plans, based on the availability of resources and the established priorities, of the necessary actions to face pollution and ensure the effectiveness of solutions. • Polluting sources included in the prioritized programs, with special emphasis on the ecosystems of environmental interest (main cities, bays, mountains, basins). • Introduction of clean, modern and efficient technologies in the projection and execution of new investments carried out at the national level.
Prevent, reduce and control pollution caused by the inadequate management of chemical products and hazardous wastes, ensuring the integrated management of these components throughout their life cycle.	• Implementation of the Chemical Safety Decree-Law. • Preparation, establishment and execution of action plans for the adequate management of waste with high environmental impact. • Execution of the schedule for the elimination of ozone-depleting substances. • Awareness and dissemination of risks associated with other hazardous products and wastes that are managing on a daily basis in the domestic sector.

Increase the capacity to adapt to and mitigate climate change in the economic and service sectors, as well as in Cuban society in general.	• Incorporation of the adaptation dimension into programs plans and projects. • Identification and establishment of action plans corresponding to the control of contaminating sources, particularly those that compromise human health and increase the risk of disasters. • Evaluation and incorporation of the mitigation of greenhouse gas emissions into investment programs, plans and projects. • Evaluation and introduction of the use of modern and efficient technologies, based on sustainability • Inclusion in the annual plans of the actions, financing and resources required for the implementation of the Directives for the Confrontation of Climate Change 2016–2020. • Execute research and introduce its results aimed at supporting the policy of confronting climate change.
Reduce the vulnerability of ecosystems and priority sectors to the effects of climate change.	• Introduce in the economy plans the results of science and technological innovation for the reduction of vulnerabilities. • Identification of the current or future impacts that climate change may have on the country's biological diversity, in order to be able to design effective adaptation strategies. • Strengthening of monitoring, surveillance and early warning systems to assess systematically the state and quality of the coastal zone, water, forest, and human, animal and plant health.
Increase awareness of the impacts of climate change at all levels of Cuban society, as well as participation in actions aimed at addressing climate change.	• Introduce the perception of the impacts of climate change as a dimension of national, sectorial and local plans and programs; with emphasis on environmental education and dissemination processes. • Systematization of results of science, technology and innovation and good practices. • Strengthening of coordinated work and knowledge management related to the perception of the impacts of climate change.

CONCLUSIONS

1. The concentration of the population, industry and automotive transport in urban areas has produced an ecological imbalance with serious effects on environmental hygiene. Cuba, despite being an underdeveloped country, economic growth and Industrialization has led to consistent measures for environmental protection and sustainable development.
2. The responsibility to protect, heal and promote health does not only belong to the health personnel. It belongs to all the people who influence the environment in one way or another. In Cuba, programs are developed to reduce atmospheric pollutants that affect human health, basis of a National Environmental Strategy in accordance with Cuba's National Plan for Economic and Social Development, until 2030.
3. The main objectives in the National Environmental Strategy is aimed to prevent, reduce and control the pollution caused by the inadequate disposal of liquid and solid waste and the emission of atmospheric pollutants, the inadequate management of chemical products and hazardous wastes. Furthermore, to increase the capacity to adapt to and mitigate climate change and increase awareness of their impacts at all levels of Cuban society.

REFERENCES

[1] Castro Ruz, F. Speech given in Rio de Janeiro at the United Nations Conference on Environment and Development, 12 June 1992 [Internet], 1992.

[2] WHO. Report of the panel on urbanization of the WHO commission on health and environment. Geneva; WHO (unpublished document: WHO/EHE/92.5).

[3] National environmental strategy of Cuba 2016/2020. Official Document.

[4] Article 27 of the Constitution of the Republic of Cuba on the protection of the environment July 11, 1997. Havana: Official Gazette of the Republic of Cuba; 1997.

[5] Cuba's National Plan for Economic and Social Development, until 2030. Official Document.

[6] Law 81 on the Environment in Cuba: Environment Available at URL: http://www.medioambiente.cu/legislacion/leyes/L-81.htm

[7] Molina, E. and Meneses, E. Epidemiological evaluation of the impact of air pollutants. Methodological proposal. Rev Cubana Hig Epidemiol. 2003;

41(2–3). Available at URL: http://scielo.sld.cu/scielo.php?script=sci_issuetoc&pid=0253-175120030002&lng=es&nrm=iso

[8] Romero Placeres M, Más Bermejo P, Lacasaña Navarro M, Téllez Rojo Solís MM, Aguilar Valdés J, Romieu I. Atmospheric pollution, bronchial asthma and acute respiratory infections in minors in Havana Sal Público Mex. 2004;46:222–3.

[9] Romieu I, Meneses F, Ruiz S, Sienra JJ, Huerta J, White MC, Etzel RA. Effect of air pollution on the respiratory health of asthmatic children living in Mexico City. Am J Respir Crit Care Med. 1996; 154:300–7

[10] Yassi, A., Kjellstrom, T. and de Kok, T., Guidotti. Salud Ambiental Básica (Spanish version made at INHEM). Mexico DF. UNEP, 2002.

[11] United Nations. Programme for Development. Research on Human Development in Cuba, 1996.

[12] Law 33. "On Environmental Protection and the Rational Use of Natural Resources", 1981, Official Document.

[13] Law 41 on Public Health (Chapter 3): Hygiene and Epidemiology. Section five. Environmental Health Control. Available at URL: http://www.medioambiente.cu/legislacion/leyes/L-41.htm

[14] More P. Environmental Health, Human Development and Quality of Life. Chapter 20. Internet monograph. National Institute of Ecology of Mexico, 2005. Available at URL: http://www.ine.gob.mx/ueajei/publicaciones/libros/363/cap20.html#top

[15] National Bureau of Standards. Environmental Management Standards. Air Quality. Hygienic and sanitary requirements. NC 39. 1999.

[16] Pan American Health Organization. La salud y el medio ambiente en el desarrollo sostenible (scientific publication No. 572). Washington DC: PAHO; 2000, p. 49–90.

[17] Cuesta, O., Wallo, A., Collazo, A., Arriba, A., González, M., Labrador, R., Sánchez, P. and Ananias, G. Methods for the presage of bronchial asthma and sharp breathing infections and their contribution to the system of epidemic surveillance in the riverside this of the bay of Havana, 2001.

[18] Cuesta Santos, O., Arnaldo Collazo Aranda, Yosdany González and cols. Inventory of atmospheric emissions of the main fixed sources of Havana, 2012.

[19] Cuesta, O.; Sánchez, P.; López, C.; Pérez, R.; Marín, M.; Biart, R. and y Villarroel, J. (2009). State of the Atmospheric Contamination in Cuba. In GEO CUBA. *Evaluation of the Cuban environment.* ISBN 978-959-300-002-4, La Habana, 2009.

[20] National Plan of Economic and Social Development until 2030. Official Document.

Role of Universities in Air Pollution Control: An Action Plan Taken by Ferdowsi University of Mashhad, Iran

Mohammad Kafi[1] and Seyed Yaser Banihashemi[2]

[1] Ferdowsi University of Mashhad, Iran

[2]Green Office's Steering Committee, Ferdowsi University of Mashhad, Iran

E-mail: [1]m.kafi@um.ac.ir; [2]y.banihashemi@um.ac.ir

ABSTRACT: Air pollution is one of the major urban environmental problems in many parts of the world including the NAM countries. In each big city in the world, there may be one or more higher education institute that are training future managers and leaders of their society, and at the same time they are a part of their society that play a role in air pollution or remediation of air pollution in that city. Mashhad is the second largest city of Iran with more than 3 million people and Ferdowsi University of Mashhad is the largest University in this city with seventy years of successful academic performance, 300 ha campus in the heart of the city and 30,000 students and staffs. We started to be a green University about 20 years ago, hoping to reduce the impact of environmental problems on the campus and community. Our understanding was to be an educational institution to train future generations to meet their own needs and save natural resources, energy, water, materials and not to add CO_2 to the ambient except sequiesterit as carbihydrate inside plant and tree biomass. To materialize this aim we established an office in the name of Green Office using a team of sustainability professionals. First of all, we have established seven different committees including, greenery and landscape, green culture, education and research, green business, green infrastructure and equipments, waste management, safety and crisis management and traffic management. Finally, our Green Office team manage environmental initiatives to help us reduce our negative impacts on air pollution by conserving energy and water, making finite resources last longer, and reducing our contribution to air *pollution and climate change and at the same time prepare our citizens to reduce air pollution. The primary objective of this paper is to highlight how universities, in general, can combat rising air pollution by studying and analyzing the initiatives taken by the Ferdowsi University of Mashhad in this regard.*

Keywords: Air Pollution, Carbon Sequestration, Higher Education, Green Office, Iran.

INTRODUCTION

Air pollution is one of the current concerns. Human health is facing a serious risk due to air pollution, as a consequence of urbanization (Ketabi Yazdi *et al.*, 2016). According to the World Health Organization (WHO), 1.5 million premature deaths occur per year, an average of one out of 10 deaths worldwide, caused by air pollution annually. Also, this report stated that air pollution is the 5th highest cause of premature deaths in the world after blood pressure, smoking, poor diet and obesity (WHO, 2016).

WHO recognizes motor vehicles, power plants, building heating systems, agriculture/waste incineration and industry as the major sources of ambient air pollution and suggests policies for investments in cleaner transport, energy-efficient buildings, renewable energy power plants and better waste management strategies (WHO, 2016).

In Iran, polluted air is the 5th highest cause of death which is responsible for about 13,000 lives lost every year and swallows up to 7% of the country's Gross Domestic Product (GDP). Mashhad, the 2nd largest city in Iran, faces the same challenge that needs to be addressed by experts (Yearbook statistical Khorasan Razavi Province, 2015).

Situation of Iran and Mashhad Regarding Air Pollution

The process of industrialization and increment of vehicles imposed a great impact on air quality of big cities in Iran. The Air Quality Index AQI values of some of the big cities of Iran such as Tehran, Isfahan, Arak, Mashhad and Tabriz were higher than the permissible values of the Iranian Environmental Agency for more than 80 days in a year (Miri *et al.*, 2016). Also, the particles smaller than 10 microns is the main reason of pollution in these cities (Miri *et al.*, 2016). The AQI value was above the standard level at 25 days of the year in Urmia city (Khorsandi *et al.*, 2011) and particulate matters were responsible pollutants of this city. Studies carried out in some major cities around the world, such as Beijing in China; Delhi, India and some Malaysian cities in 1997, showed that for more than 80 percent of the days, in a year, air was polluted and particulate matters less than 10 microns were dominant there (Ahmad *et al.*, 2006; Mohan and Kandya, 2007; Kumar and Goyal, 2011; Mu *et al.*, 2014). These cities are similar to most metropolis of Iran in terms of industrialization, transportation of vehicles and population and most of the air pollution were caused by industries and urban traffic there (Miri *et al.*, 2016).

The air pollution problems, which have the fifth rank among the mortality factors, devours between 2.4 and 7 percent of countries' GDP annually to offset its adverse effects. In Iran, 2.4% of GDP consumes for management of *different aspects of air pollutions* (World Bank Group, 2010).

Mashhad, being located in the Northeast of Iran, is the second largest city in this country and the second most major destination for Muslim pilgrims. An increasing number of inhabitants apart from an overload of tourists has made this city overcrowded and has also caused traffic-related problems, making routine urban travels annoying for residents and visitors (Atarodi *et al.*, 2018).

Mashhad is facing air pollution due to motor vehicle emissions and industries. As it is a religious city, the vehicle traffic of pilgrims (Over 20 million pilgrims and tourists per year) alongside industries, poses a threat to the air quality. It has a population of 3,001,184 (2016 census) with the 351 Square Kilometer area and also more than 400,000 cars pass on the road of this city every day (Statistical yearbook of Khorasan Razavi province, 2016). These numbers of vehicles as well as industries are regarded as the primary sources of air pollution in Mashhad. Based on AQI, Mashhad air quality was classified in groups of good, moderate, unhealthy and dangerous for the sensitive people, unhealthy, very unhealthy people, respectively.

In a cross-sectional study, the instantaneous concentrations of air pollutants including O_3, CO, SO_2, NO_2, $PM_{2.5}$ and PM_{10} were measured at the three stations of Mashhad during 2015–2016 (Katabi *et al.*, 2016). The results of the AQI indicated that the air quality in 46 days of the year was exceeded the standard limit. The air quality category included 12.5% of unhealthy days for sensitive groups and 74.5% of healthy days. According to this research, $PM_{2.5}$ was determined as the main responsible pollutant in non-standard conditions of Mashhad air. These researchers proposed that optimizing public transportation, considering environmental standards at the urban transportation management and launch intelligent traffic control are the most appropriate strategies to control Mashhad air pollution.

The Importance of Universities in Air Pollution Issue

Universities impose a key influence on society in a different manner: firstly, they train and educate human power that they will manage and administrate the society in the future. Secondly, they are a part of the cities particularly big and polluted cities for example 7% of Mashhad population are University students and thirdly, academicians participating in governance at the international, national, regional and local levels. Therefore, universities have actual and potential roles in fulfilling educational, research, governance, and economic development functions, as well as facilitate and mediate functions in development and life style of their surrounding society. In a study Sedlacek (2013) outlined seven hypotheses from the literature and used to develop an analytical framework for considering the three main functions of Universities: education, research and governance. The framework distinguishes the individual and the societal educational functions, knowledge creation and knowledge transfer within the research function, and finally the internal and external governance functions. This framework has been applied to a particular case study—the University of Graz, which hosts a Regional Centre of Expertise on Education for Sustainable Development and concluded that particular bridging organizations such as this regional centre are only a few of the factors that were found to be facilitative in the case under study.

In spite of increasing number of universities becoming engaged with sustainable development and using green campus initiatives to reduce air pollution and other environmental issues, most higher education institutions continue to be managed traditionally and maybe lagging behind some top companies in helping societies become more environment friendly. Lozano *et al.* (2013) analyses the texts of eleven declarations, charters, and partnerships developed for higher education institutes, and proposed that for Universities to become sustainability leaders and change drivers, they must ensure that the needs of present and future generations be better understood and built upon. Therefore, university leaders and staff must be empowered to suggest and implement new paradigms, and ensure that sustainable development is the 'Golden Thread' throughout the entire University system.

Given the growing global interest in universities' role towards promoting sustainability and environmental conservation, an increasing number of universities are committing themselves to be green and clean Universities. But many university stakeholders and academicians are unaware of environmental principles. Nejati and Nejati (2013) investigated the perceptions of major stakeholders within the university-setting on the role of the university in contributing to sustainability and they reported a four-dimensional structure for the key factors of a sustainable university from the perspective of students, including 1) community outreach, 2) sustainability commitment and monitoring, 3) waste and energy, and 4) land use and planning.

Nowadays relationship between Universities and society and their interaction is most necessary but there is a lack of methodology and indices for evaluation of university-society collaboration. The Swedish Governmental Agency for Innovation Systems (VINNOVA) has been commissioned by the Swedish government to present an evaluation model for university–society collaboration, Their results suggest that there is broad knowledge on the complexity of university–society collaboration, and of the difficulties associated with evaluation and they suggest to focus on the construction of relevant indicators, while there is a widespread lack of discussion and agreement on the objectives and goals of university–society collaboration, as well as discussions on how to define the concept (Bölling, *et al.*, 2016)

Ferdowsi University of Mashhad (FUM) as a Green University

Ferdowsi University of Mashhad, which is named after Abulghasem-e-Ferdowsi, the great Iranian poet who composed the Shahnameh, is the largest and most prestigious university in the city of Mashhad and the eastern part of Iran, with 70 years of successful academic performance, 300 ha campus area in the heart of the city and about 30,000 students and staffs. We started to be a Green University about 20 years ago, with an idea that we should be the leader of our community in tackling the environmental challenges. To materialize this aim, a Green Office was established by a group of our professionals to work on sustainability issues including air pollution. The office is directed by a steering committee chaired by the university president. First of all, the Green University indicators were outlined based on the instructions proposed by the Iranian Ministry of Science, Research and Technology, and then, seven specialized committees were established to address the relevant indicators. The governance structure of the Green Office is demonstrated in Figure 1.

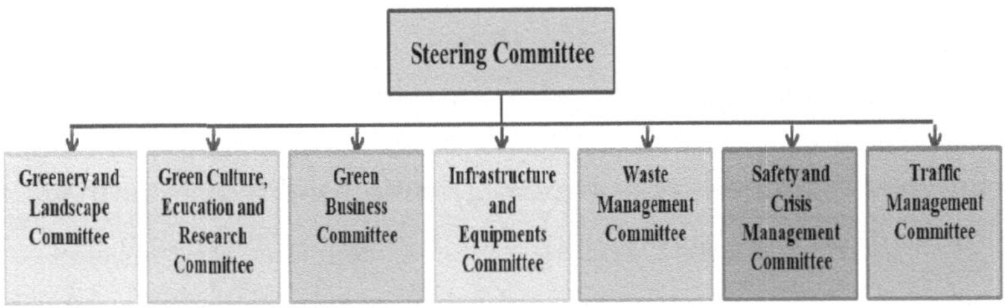

Figure 1: Green Office's Governance Structure

Each committee is composed of professionals and executive members and holds responsibility for improving a number of assigned indicators. While some indicators such as waste separation and recycling are very specific to one or two committees, others such as carbon footprint are general and need to be dealt with by more than 4 or 5 committees.

Some of the main selected actions implemented in recent years in each committee are listed as follow:

Greenery and Landscape Committee

1. Identifying and locating on-campus vegetation and animal habitat to preserve endemic species and conserve biodiversity;
2. Developing and conserving low-input and endemic plants throughout the campus;
3. Replacing grass and other high-water consumption vegetation with sand mulch to save water resources;
4. Organizing workshops on planting and plant conservation.

Green Culture, Education and Research Committee

1. Organizing workshops on sustainable and green life-style for students, staffs, and school children.
2. Developing an application named "EcoPaya" to use gamification techniques in promoting and encouraging green behaviours on campus.
3. Designing and introducing a representative character named "Dr. Green" for spreading the green culture throughout the campus.
4. Organizing occasions/campaigns on collecting waste materials such as plastics or cigarette butts on and off-campus.
5. Collaborating with other universities for exchanging green experiences and findings.
6. Encouraging graduate students to carry out research on the Green University problems and devise practical solutions.

Green Business Committee

1. Identifying green businesses and start-ups and introducing them to each other to promote the green ecosystem.
2. Communicating Green Office's identified and prioritized problems with green businesses to find novel solutions.
3. Supporting green businesses to commercialize their products or services.
4. Introducing green businesses to the market and facilitating their branding process.

Infrastructure and Equipments Committee

1. Installing metering equipment to monitor energy consumption in different nodes of water, electricity and gas networks.
2. Developing a Web-GIS[1] Smart Infrastructure Management System to facilitate integrated management of the university's infrastructure.
3. Replacing high consumption energy appliances and devices with low-power devices.
4. Developing photovoltaic power plants on-campus.

[1]Geospatial Information System.

Waste Management Committee

1. Trying to offer well-balanced, healthy and sustainable catering to reduce waste.
2. Collecting leaves from all around the campus and producing leaf soil to use it again as fertilizer.
3. Providing all colleges and administrative buildings with waste separation bins.
4. Identification of special wastes in all labs and workshops and the way they are preserved and disposed.

Safety and Crisis Management Committee

1. Creating safety photo ID for all buildings, labs and work spaces to know the safety gaps and develop an action plan to fill the gaps.
2. Collaborating with Center for Crisis Management in Mashhad city to educate new students and get them ready to rescue themselves and others on campus at the time of disasters or other types of incidents.

Traffic Management Committee

1. Providing infrastructure for sustainable modes of transport such as bicycles and electric buses on-campus.
2. Providing facilities to make green travel easier and more attractive.
3. Installing GPS on university buses to monitor their mobility and their arrivals to bus stops.

These are the actions that apart from the size and responsibility of any organization, which could be applied for preventing air pollution or absorb more pollutants from the air.

Prospective of Ferdowsi University of Mashhad

1. We are living in a country with relatively low oil prices; therefore some other motivator should persuade people to use public transport. Our goal is to encourage our students and staff to use public transport from home to University and use biking or walking inside the campus.
2. For supporting a plan, all beneficiaries should be involved. Since in a University majority of the society is student, thus, the

University authority will try to share them in decision making in green lifestyle.

3. University should be a role model for the society, therefore we will try to have more impact to the society through our staff and students to extend the green lifestyle.
4. Central and state Universities might be involved in many committees and meetings for making decision and approve a plan to apply in the society. We will try to highlight our presentation and affect more than the past for saving the environment.
5. Humankind disturbed the nature to prepare a better life for themselves but now is the time to popularise the "one health" concept, and universities could voluntarily respect this concept.

CONCLUSIONS

Polluted air is the 5th highest cause of death in Iran and Mashhad the 2nd largest city of Iran where Ferdowsi University being located, faces the same challenge. Since Universities train and educate human resources that they will manage and administrate the society in the future as well as they are participating in governance of society in different levels, their role in establishment of a value and behavior is unique. We believe that interaction between Universities and society is most necessary but there is a lack of methodology and indices for evaluation of university-society collaboration. The Green University indicators were outlined based on the instructions proposed by the Iranian Ministry of Science, Research and Technology at FUM, and then, seven specialized committees were established to address the relevant indicators including greenery and landscape, green culture, education and research, green business, infrastructure and equipment, safety and crisis, water and traffic management. University authority hope that by implementing these strategies not only the production of the pollutant will be reduced inside the University campus but also our graduates will help to extend these values to the society and have a better environmentally friendly life style in the rest of their life.

REFERENCES

[1] Ahmad, A., Hashim, M., Hashim, M.N., Ayof, M.N. and Budi, A.S. (2006). The use of remote sensing and GIS to estimate Air Quality Index (AQI) Over Peninsular Malaysia. *Malaysia: Proc. Map Malaysia,* 2006. P. 5.

[2] Atarodi, Z., Karimyan, K., Gupta, V.K., Abbasi, M. and Moradi, M. (2018). Evaluation of indoor air quality and its symptoms in office building—A case study of Mashhad, *Iran. Data in Brief*, (2018), 20 74–79. DOI: 10.1016/j.dib.2018.07.051.

[3] Ketabi, D., Esmaili, R., Alidadi, H., Peirovi, R. and Joulaai, F. (2016). Evaluation of Mashhad City Air Quality based on Air Quality Index (AQI). *Iranian Journal of Research in Environmental Health*, 2: 228–236.

[4] Khorsandi, H., Amini Tapok, F., Cargar, H. and Mousavi Moughanjogi, S. (2013). Study of Urmia city air quality according to the air quality index (AQI). *Urmia Medicinal Journal*, 23:767–75.

[5] Kumar, A. and Goyal, P. (2011). Forecasting of daily air quality index in Delhi. *Sci Total Environ*, 409:5517–5523.

[6] Bölling, M. and Eriksson, Y. (2016). Collaboration with society: The future role of universities? Identifying challenges for evaluation, Research Evaluation, 25(2), 209–218, https://doi.org/10.1093/reseval/rvv043

[7] Miri, M., Mohammadi, A., Nemati, S., Abdolahnejad, A. and Nikoonahad, A. (2016). The trend of air quality index (AQI) variations in Mashhad metropolis using GIS. *Journal of health research in community*, 2, 12–20.

[8] Mohan, M. and Kandya, A. (2007). An analysis of the annual and seasonal trends of air quality index of Delhi. *Environ Monit Assess*, 131:267–77.

[9] Nejati, M. and Nejati, M. (2013). Assessment of sustainable university factors from the perspective of university students. *Journal of Cleaner Production*, 48: 101–107. https://doi.org/10.1016/j.jclepro.2012.09.006 Get rights and content.

[10] Mu, H, Otani, S., Okamoto, M., Yokoyama, Y., Tokushima, Y., Onishi, K., *et al.* (2014). Assessment of effects of air pollution on daily outpatient visits using the air quality index. *Yonago Acta Med.*, 57:133–136.

[11] Lozanoa, R., Lukmanb, R., Lozanoc, F.J., Huisinghd, D. and Lambrechtse, W. (2013). Declarations for sustainability in higher education: becoming better leaders, through addressing the university system. Journal of Cleaner Production. 48:10–19. https://doi.org/10.1016/j.jclepro.2011.10.006

[12] Sedlacek, S. (2013). The role of universities in fostering sustainable development at the regional level. *Journal of Cleaner Production*, 48: 74–84. https://doi.org/10.1016/j.jclepro.2013.01.029

[13] World Bank Group (2016). The Cost of Air Pollution Strengthening the Economic Case for Action, https://creativecommons.org

[14] World Health Organization (WHO) (2016). Air Quality and Health, Retrieved from. http://www.who.int/mediacentre/news/releases/2014/air-pollution/en

[15] Yearbook statistical Khorasan Razavi Province (2015). Statistical Center of Iran. Available at: URL:http://nnt.sci.org.ir/sites/Apps/yearbook/Lists/year_book.

A Study on the Status of Air Pollution and its Impacts in Nepal

Bimala Devi Devkota Paudel[*] and Pawan Kumar Neupane
Nepal Academy of Science and Technology (NAST),
Khumaltar, Lalitpur, Nepal
*E-mail: diku98@gmail.com

ABSTRACT: Air pollution, both indoor and outdoor, is a serious problem in developing countries including Nepal. This study revised the available publication on air pollution and synopsized the information to analyze the air pollution status, its impacts and government efforts in controlling/minimizing the air pollution in Nepal. The published literature has revealed that air quality is continuously degrading causing multiple impacts including human health. Based on real time data of some representative urban areas, Air Quality Index (AQI) is in continuously degrading state mainly because of increasing number of vehicles, industrial activities, and least adoption of environmental safeguard measures. In rural areas, firewood smoke, burning agriculture wastes, uses of solid fuel, pesticides are becoming major sources of air pollution. Studies on effects of air pollution on ecosystem health, trans-boundary pollutants and regular monitoring of indoor air pollution are still inadequate and country's endeavors in controlling air pollution are yet to be invigorated. Currently, real time monitoring stations are installed in limited locations and efforts are ongoing to expanding both rural and urban areas. Scientific studies along with the introduction of advance technologies, stringent implementation of existing laws and rules and capacity building of existing institutional arrangements are imperative for controlling and monitoring air pollution in Nepal.

Keywords: Air Pollution, Air Quality Index (AQI), Degradation, Human Health, Real Time Data, Nepal.

INTRODUCTION

Ambient Air Pollution in Nepal

Air pollution is becoming a serious problem in developing countries including Nepal, (area: 147,181 square kilometers and population:

26.5 million people in 2011) with Human Development Index 0.579 ranking 147 out of 189 countries and territories (UNDP 2019). The country is facing increasing burden of air pollution as seen over most of the subcontinent for the last decades (MoEST 2005). The energy consumption pattern of the country showed an increase by a factor of 1.6 (from 2001 to 2016), while the emissions of various species increased by a factor of 1.2–2.4 (Sadavarte *et al.,* 2019) ranking the country at the last position, among the 180 countries in air quality determined on the basis of three indicators viz. household solid fuels, $PM_{2.5}$ exposure and exceed in Environmental Performance Index (EPI) report, 2018 (Wendling *et al.,* 2018). This indicates the most degraded state of both indoor and outdoor air quality in Nepal. Also, the regional and global air pollution has significant effects in Nepal as the typical regional atmospheric transport process supports in trans-boundary movement of air pollutants (Carrico *et al.,* 2003).

Indoor air pollution is the major problem in the rural household, as in other parts of developing world where around 50% of people still depends on crude biomass fuels in the form of wood, dung, charcoal and crop residues for domestic energy (Bruce *et al.,* 2000). While in urban areas growing number of vehicles, brick industries; unscientific construction activities etc. causes the outdoor pollution. A total of 8.9 Teragram (Tg) of CO_2, 110 Gigagram (Gg) of CH_4, 2.1 Gg of N_2O, 64 Gg of NOx, 1714 Gg of CO, 407 Gg of NMVOCs, 195 Gg of $PM_{2.5}$, 23 Gg of BC, 83 Gg of OC and 24 Gg of SO_2 emissions were estimated in 2011 from the five energy-use sectors viz. Residential, Industry, Transport, Commercial and Agriculture (Sadavarte *et al.,* 2019). SO_2 and NO_X emission intensity of Nepal is lower putting the country in 87th position in the EPI report.

The incomplete combustion of such crude biomass emits a significant amount of the fine particulate matter $PM_{2.5}$ and ozone precursor emissions, which have been linked to degrading air quality, adverse health impacts, climate change and effects on the cryosphere (Fiore *et al.,* 2015; Shakya *et al.,* 2016, Sadavarte *et al.,* 2019). The average concentration of $PM_{2.5}$ is 54.2 μg/m^3 ranking 8th out of 73 countries (of its capital Kathmandu is 54.4 μg/m^3 ranking 7th polluted city) (IQAir AirVisual 2018) and annual median concentration including both urban and rural area is 64 μg/m^3 (33–123 μg/m^3) while including urban area only it is 74 μg/m^3(39 –140 μg/m^3)) (WHO 2016). Since, the Indo-Gangetic Plain is one of the most polluted regions in the world (Saikawa *et al.,* 2019); Nepal is also vulnerable to the risks posed by transboundary air pollutants. However, the detail study of impacts of such pollutants is yet to be conducted in Nepal. This study intends to revise the available publications on air pollution and synopsize the information to analyze the air pollution status, its impacts and government efforts in controlling/minimizing the air pollution in Nepal.

STUDY METHOD

The published literatures were reviewed to find the air pollution status in both rural and urban areas of Nepal. The impacts on human health, government interventions on policies and technological programs for minimizing/controlling air pollution were analyzed. The real time air pollution data measured by the stations located in major urban hubs of Nepal viz. Ratnapark-Kathmandu, US Embassy-Kathmandu, Bhaisipati-Kathmandu, Pulchwok-Lalitpur, Bhaktapur, Dhulikhel, Bharatpur, Lumbini, Pokhara, Butwal, Birgunj and Itahari were retrieved (www. pollution.gov.np) and real time data of Air Quality Index (AQI) for these hubs on a single day on 16 January 2020 were taken for the representative purpose (www.aqicn.org).

Air Pollution in Urban and Rural Areas

There is an increasing trend of urban population in Nepal. Kathmandu Valley, the capital city, has population of 2.5 million with annual population growth rate of 4.2% and population density 2799.8/km^2 (CBS 2011). Most of the urban centers of Nepal embed topographic features like deep valleys, narrow river basin and plain areas which are generally associated with adverse meteorological conditions allowing merely a poor air pollution carrying capacity and may generate severe air pollution level even where the local emissions are low (Regmi, 2013).

Industrial activities (though in small scale), vehicular emission, improper management of solid waste, deficient safeguard measures are the some of the causes for air quality degradation in the cities of Nepal. Till, 2016/17, 2783428 vehicles (444259 in 2016/17 only) have been registered in Nepal (CBS 2019). With this rapid increase in vehicles registration along with the road conditions and quality of fuel are reasonable for the air pollution caused by vehicular emission in the Kathmandu valley and other cities of Nepal. In addition, hospital and other sectors waste burning, coal burning and open burning of stuff like tires and tubes, plastics, papers are the complementary sources of air pollutants in these areas.

Kathmandu, the bowl-shaped valley covers three districts (Kathmandu, the capital city, Lalitpur and Bhaktapur district) stands very high in particulate ambient air pollution with levels five times more than the acceptable guidelines (Shrestha 2018). Construction activities (buildings, roads, excavations etc.) along with increasing number of vehicles are responsible for the degrading air quality of Kathmandu. Brick kilns (40%), motor vehicles (37%) and biomass/garbage burning (22%) have been also identified as the major sources of elemental carbon (black carbon) in the Kathmandu (Kim *et al.,* 2015).

Similarly, Lumbini, in southern Nepal, is a United Nations Educational, Scientific and Cultural Organization (UNESCO) world heritage site and a touristic destination site also has the degraded state of air quality as in other urban cities of Nepal. A study conducted by Rupakheti *et al.,* 2017 had observed that the 24 h average $PM_{2.5}$ and PM_{10} concentrations exceeded the World Health Organization (WHO) guideline very frequently (94% and 85% of the sampled period, respectively), inferring substantial health risks for the residents and visitors in the region. Brick kilns and cement factories in the Lumbini-Bhairahawa corridor produce pollutants which may be the cause of air pollution in the Lumbini premises. These two instances of urban areas air pollution in Nepal represent the overall scenario of urban air pollution in Nepal. The one-day real time data on AQI at the stations as illustrated in Table 1 and Figure 1 depicts that the air quality of the particular day remain unhealthy in all urban areas of Nepal which reveals the fact that air quality in the urban areas are in steadily degraded state.

Most of the districts in Nepal had a high burden of indoor air pollution and its markers (Ghimire *et al.,* 2019). Majority of people living in the rural areas of Nepal are victimized from indoor air pollution as they depend on biomass fuel for cooking and heating (CBS 2019). Such crude fuel produces pollutants like carbon monoxide, sulfur oxide, particles, and volatile organic, which are responsible to produce different disease and illness in human beings (Harrison *et al.,* 2002). However, the regular monitoring of air

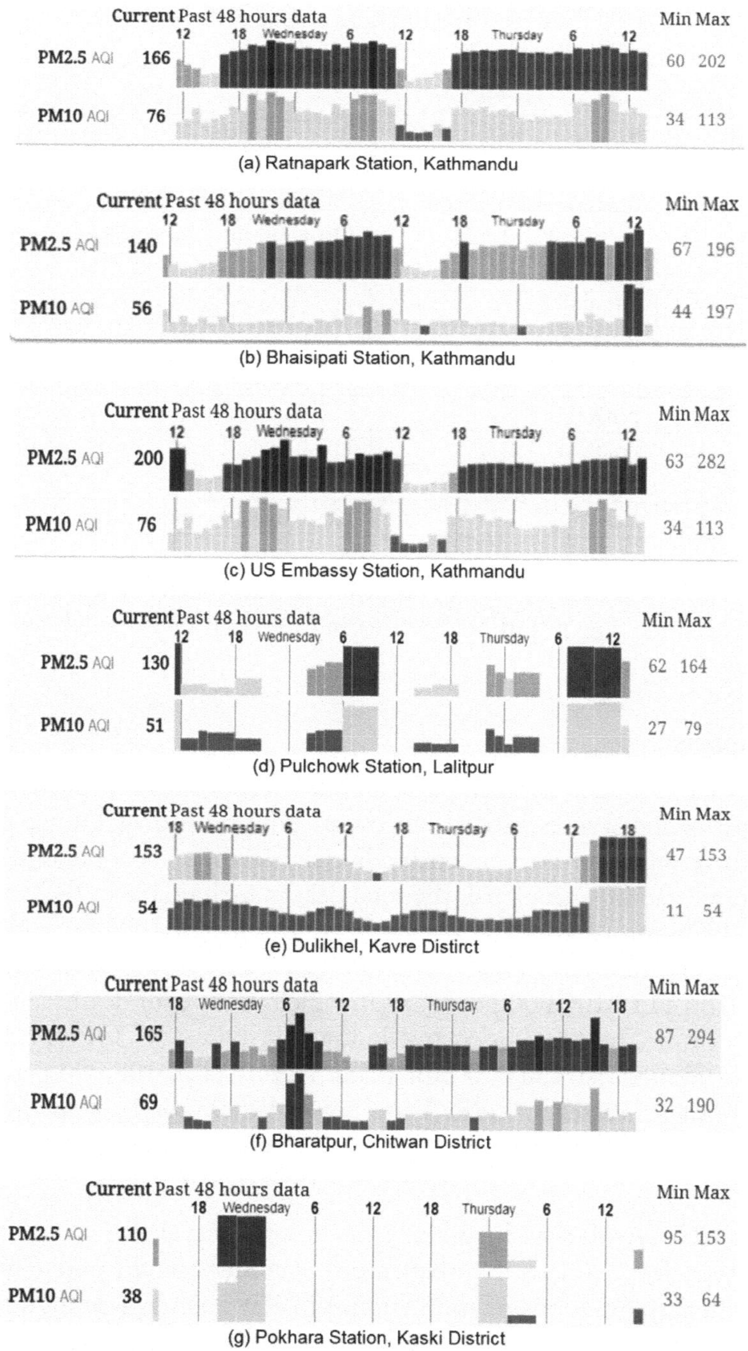

Figure 1 (a-g): AQI at Urban Hubs of Nepal Adopted from www.aqicn.org on 16 January 2020

Table 1: AQI Ranking of Urban Areas of Nepal Observed on 16 January 2020

S.N.	*Urban Areas*	*US AQI*
1.	Ratnapark, Kathmandu	166-Unhealthy
2.	US Embassy-Kathmandu	200-Unhealthy
3.	Bhaisipati-Kathmandu	135-Unhealthy for sensitive groups
4.	Pulchwok-Lalitpur	130-Unhealthy for sensitive group
5.	Bhaktapur	157-Unhealthy
6.	Dhulikhel	115-Unhealthy for sensitive groups
7.	Bharatpur	139-Unhealthy for sensitive groups
8.	Lumbini	122-Unhealthy for sensitive groups
9.	Butwal	112-Unhealthy for sensitive groups
10.	Pokhara	110-Unhealthy for sensitive groups
11.	Birgunj	160-Unhealthy
12.	Itahari	135-Unhealthy for sensitive groups

pollution whether indoor or outdoor in rural areas is still lacking in Nepal since the pollution monitoring are mainly focused on urban areas.

Impacts

Air pollution has visible impacts on human as well as ecosystem health. Most of the studies carried out on air pollution focus human health with very few on impacts on biodiversity, climate, economic cost and natural resource. In Nepal, the total deaths attributable to the Ambient Air Pollution (AAP) in 2012 was 9,943 deaths per 100,000 capita and Years of Life Lost (YLL) attributable to AAP in the same year was 296130 YLLs 100,000 per capita in which number of deaths and YLLs only from Chronic Obstructive Pulmonary Disease (COPD) accounts 1,770 deaths per 100,000 capita and 37,949 YLLs 100,000 per capita and other diseases includes Acute Lower Respiratory Disease (ALRI), Lung cancer, Stroke and Ischeamic Heart Disease (IHD) (WHO 2016).

Research on diseases in urban areas reveals prevalence of respiratory diseases like COPD in urban areas like Kathmandu and Chitwan. Pulmonary functions of traffic police in Kathmandu was found to be significantly weakened (Shrestha *et al.,* 2015) indicates that the continuous exposure in the urban areas of Nepal would pose serious respiratory problems. Similarly, brick kiln workers in Kathmandu are severely affected by the airborne particulate matter (Sanjel *et al.,* 2017). Studies on rural communities have shown that households using solid biomass fuel as a primary source of fuel are attributed to respiratory diseases and like Acute Respiratory Infection (ARI), Acute Lower Respiratory Infection (ALRI), pneumonia, asthma etc. and deficit in lung functions (Dhimal *et al.,* 2010; Dhimal *et al.,* 2016). Likewise, people using traditional mud stoves and unrefined biomass reported health problems like tearing of eyes, difficulty in breathing and productive cough (Chhabi *et al.,* 2015). Though a very few studies have been carried out focusing on indoor pollution and health impacts in rural communities, most of the studies has elucidated that the direct impact of indoor pollution are on children, young adults and women health with more exposure on traditional practice of biomass burning (Pandey 1984; Reid *et al.,* 1986; Pandey *et al.,* 1989; Pokhrel *et al.,* 2005, Shrestha and Shrestha 2005; Joshi *et al.,* 2009; Dhimal *et al.,* 2010; Kurmi *et al.,* 2013; Dhimal *et al.,* 2016; Chhabi *et al.,* 2015).

Based on the bitty evidence, especially from the newspapers it can be explained that along with human health impacts, air pollution in Nepal may also have affected the ecosystem health. It has been reported that there is growing negative effects of air pollution on biodiversity (Grub *et al.*, 1999).

Policy Intervention and Pollution Monitoring

Following the constitution of Nepal 2015 which has stated that every citizen shall have the right to live in a clean and healthy environment, government of Nepal is amending existing acts and regulation on environment sectors as well. The main law Environmental Act 2019 has been authenticated and draft of Environmental Protection Regulation 2020 has been prepared. The important acts, policies and strategies that are linked with air pollution control are National Climate Change Policy 2019, National Environmental Policy 2019, National Low Carbon Economic Development Strategy, National Pollution Control Strategy and Action Plan, Transport Policy 2001, Transport Management Act 2049 (Nepalese calendar year), National Indoor Air Quality Standards 2009, National Ambient Air Quality Standards 2012 (Updated from 2003), Nepal Vehicle Mass Emission Standards (NVMES) 2012, Environment Friendly Vehicle and Transport Policy 2014, Vehicles and Transport Management Rules

2054 (Nepalese calendar year). Nepal Government has also formed various committees to deal with the air pollution problem like Task Force on Air Pollution Control in Kathmandu Valley, 2073 (Nepalese calendar year), High Level Committee on Probing and Solving the Issues on 20 year Old Vehicles, 2058 (Nepalese calendar year), Committee on Implementation of the Order of Supreme Court on Phase out of 20 Year Old Vehicles, 2058/59 (Nepalese calendar year), Committee on Review of Vehicle Emission Standard and Monitoring Mechanism 2060 (Nepalese calendar year), Technical Committee on the Relocation of Brick Industries from Kathmandu Valley 2060 (Nepalese calendar year). Air quality Management Action Plan for Kathmandu Valley has been drafted and submitted for approval. These legislative frameworks and policies formulated by government of Nepal targets sustainability, energy conservation, low carbon development and pollution control.

Government has created institutional arrangement for the implementation acts, rules and regulation, policies, strategies and action plan that intends to control the environmental pollution in Nepal. Department of Environment has been established under the Ministry of Forest and Environment to monitor and regulate the environmental standards of the country. Governments has established 19 air pollution monitoring stations (Figure 2) mainly in the urban areas that provide

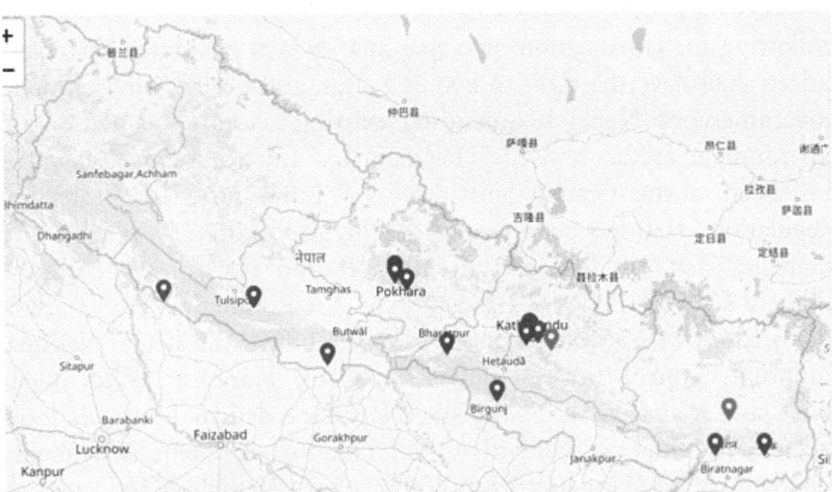

Figure 2: Air Pollution Monitoring Stations in Nepal

Source: www.pollution.gov.np, Jan. 20, 2020

data on air quality index, $PM_{2.5}$, PM_{10} and Total Suspended Particulates (TSP) through the website www.pollution.gov.np and also the database are linked with worldwide real time database website like www.aqicn.org and www.waqi.org.

EFFORTS, EXISTING GAPS AND WAY FORWARD

The existing knowledge on the air pollution reveals that Nepal has to take serious control measures immediately. However, the country's endeavors in controlling air pollution still are not sufficient.

Monitoring of transboundary air pollutants and its impacts is lacking. Snow sampling from the Himalayas can help to find out the traces of air pollutant sintering into the country. Furthermore, it is also imperative to expand sufficient and reliable monitoring stations to cover the rural areas as well. The political leaderships shall understand the value of data as well as scientific publication and apply in formulation and implementation of policies. Based on the database system and scientific evidence, policy makers shall take quick decisions and implement essential mechanisms in controlling the pollution, thus minimizing the risk of both human and ecosystem health. The existing institutional arrangement has to be invigorated by involving local governments in maintaining air quality. Scientific analysis of air pollution data as well as research from academia needs to be groom up to support evidence-based policy making.

Along with the effective traffic management, clean fuel, clean vehicles and advanced technologies to control vehicular exhaust pollution needs to be introduced immediately. Punishment and reward system as envisaged by the existing laws shall be implemented properly. Road vegetation is an effective way to mitigate road generated particulates. Green belts with high Air Pollution Tolerance Index (APTI) can be the effective way to mitigate road side generated particulates (Dhamala *et al.,* 2018). Provision of tax subsidy to the industries with environmentally friendly technologies can help in controlling air pollution from industrial sectors. Awareness along with alternatives and technologies in rural communities can reduce indoor air pollution in rural communities. Construction activities shall be conducted in proper way to reduce suspension and re-suspension of dust particles problems.

Prime minister of Nepal vision on *Mask free Kathmandu City* has drawn the national attention and showed the government's concern for controlling the air pollution. However, scientific studies along with the introduction of advance technologies for controlling and monitoring air pollution are imperative to achieve such ambitious vision. Public involvement is also equally important in controlling air pollution by making environmentally friendly behavior and creating pressure to the government for the effective implementation of policies and meaningful coordination between related institutions.

ACKNOWLEDGEMENTS

We are indebted to Dr. Sunil Babu Shrestha, Vice-Chancellor and Dr. Mahesh Kumar Adhikari, Secretary of Nepal Academy of Science and Technology (NAST) for a nomination to participate in the workshop on *Air Pollution and Public Health: Challenges and Interventions* organized by NAM S&T Center in New Delhi. We are also thankful to Dr. Tista Prasai Joshi, Scientific Officer, NAST for her support in collecting literatures and related popular articles.

REFERENCES

[1] Bruce, N., Perez-Padilla, R. and Albalak, R. (2000). Indoor air pollution in developing countries: a major environmental and public health challenge. *Bulletin of the World Health Organization,* 78 (9):1078–1092.

[2] Carrico, C.M., Bergin, M.H., Shrestha, A.B., Dibb, J.E., Gomes, L. and Harris, J.M. (2003). The importance of carbon and mineral dust to seasonal aerosol properties in the Nepal Himalaya. *Atmospheric Environment,* 37 (20): 2811–2824.

[3] CBS (2019). Environment Statistics of Nepal 2019. Central Bureau of Statistics, Kathmandu.

[4] CBS (2011). National Population and Housing Census. Central Bureau of Statistics, Kathmandu.

[5] Dhamala, M., Kunwar, K. and Byanju, R. (2018). Air pollution tolerance index: An approach towards the effective green belt around Kathmandu metropolitan city, Nepal. *Nepal Journal of Environmental Science,* 4: 23–29.

[6] Dhimal, M., Karki, K.B., Aryal, K.K., Dhakal, P., Joshi, H.D., Pande, A.R., Puri, S., Gyawali, P., Mahotra, N.B., Sharma, A.K. and Kurmi, O. (2016). Indoor Air Pollution and its effects on Human Health in Ilam District of Eastern Nepal. Kathmandu: Government of Nepal, *Nepal Health Research Council (NHRC).*

[7] Dhimal, M., Dhakal, P., Shresth, N., Baral, K. and Maskey, M. (2010). Environmental burden of acute respiratory infection and pneumonia due to indoor smoke in Dhading. *Journal of Nepal Health Research Council,* 16: 1–4.

[8] Fiore, A.M., Naik, V. and Leibensperger, E.M. (2015). Air quality and climate connections. *J. Air Waste Manage,* 65: 645–685.

[9] Ghimire, S., Mishra, S.R., Sharma, A., Siweya, A., Shrestha, N. and Adhikari, B. (2019). Geographic and socio economic variation in markers of indoor air pollution in Nepal: evidence from nationally-representative data. *BMC Public Health,* 19(1): 195.

[10] Grub, A., Bungener, P., Contat, F., Nussbaum, S., Endtner, V. and Fuhrer, J. (1999). Air pollution and floristic biodiversity. *Pollution Atmospherique,* 42–48.

[11] Gurung, A. and Bell, M.L. (2013). The state of scientific evidence on air pollution and human health in Nepal. *Environmental Research,* 124: 54–64.

[12] Harrison, R., Thornton, C., Lawrence, R., Mark, D., Kinnersley, R. and Ayres, J. (2002). Personal exposure monitoring of particulate matter, Nitrogen dioxide and Carbon monoxide including susceptible groups. *Occup Environ Med.,* 59 (10): 67.

[13] IQAir AirVisual (2018). 2018 World air quality report: Region and city PM2.5 Ranking. IQAir Group.

[14] Joshi, H., Pandeya, R. and Dhakal, B. (2009). Health impact of indoor pollution. *J. Nepal Health Res Counc.,* 7 (15): 69–75.

[15] Kim, B., Park, J.-S., Kim, S.-W., Jeon, H.J., Cho, C.Y., Kim, J.-H., Hong, S.K., Rupakheti, M., Panday, A., Park, R., Rokjin, Hong, J.Y. and Yoon, S.-C. (2015). Source apportionment of PM10 mass and particulate carbon in the Kathmandu Valley, Nepal. *Atmospheric Environment,* 123: 190–199.

[16] Kurmi, O.P., Devereux, G.S., Cairns, W., Smith, S., Semple, S., Steiner, M.F.C., Simkhada, P., Lam, K.B.H. and Ayres, J.G.. (2013). Reduced lung function due to biomass smoke exposure in young adults in rural Nepal. *Eur Respir J.,* 41: 25–30.

[17] MoEST (2005). Ambient air quality of Kathmandu Valley, Then, Ministry of Environment, Science and Technology, Singh Durbar, Kathmandu.

[18] Pandey M.R., Neupane, R.P., Gautam, A. and Shrestha, I.B. (1989). Domestic smoke pollution and acute respiratory infections in a rural community of the hill region of Nepal. *Environment International,* 15(1–6): 337–340.

[19] Pandey, M.R. (1984). Domestic smoke pollution and chronic bronchitis in a rural community of the hill region of Nepal. Thorax, 39: 337–339.

[20] Pokhrel, A.K., Smith, K.R., Khalakdina, A., Deuja, A. and Bates, M.N. (2005). Case-control study of indoor cooking smoke exposure and cataract in Nepal. *Int J Epidemiol.,* 34 (3): 702–708.

[21] Regmi, R.R. (2013). Pollution: air. In: Environment and Natural Resources (eds.) P.K Jha, F.P. Neupane, M.L. Shrestha and I.P. Khanal. *Publ. Nepal Academy of Science and Technology, Khumaltar, Lalitpur,* pp. 398–403.

[22] Reid, H., Smith, K.R. and Sherchand, B.J. (1986). Indoor smoke exposures from traditional and improved woodstoves: comparisons among rural Nepali women. *Mountain Res Dev.,* 6: 293–304.

[23] Rupakheti, D., Adhikary, B., Praveen, P.S., Rupakheti, M., Kang, S., Mahata, K.S., Naja, M., Zhang, Q., Panday, A.K. and Lawrence, M.G. (2017). Pre-monsoon air quality over Lumbini, a world heritage site along the Himalayan foothills, *Atmos. Chem. Phys.,* 17: 11041–11063.

[24] Sadavarte, P., Rupakheti, M., Bhave, P., Shakya, K. and Lawrence, M. (2019). Nepal emission inventory - Part I: Technologies and combustion sources (NEEMI-Tech) for 2001–2016. *Atmospheric Chemistry and Physics,* 19 (20): 12953–12973.

[25] Saikawa, E., Panday, A., Shichang, K., Gautam, R., Zusman, E., Cong, Z., Somanathan, E. and Adhikary, B. (2019). Air pollution in the Hindu Kush Himalaya, Chapter 10 in P. Wester, A. Mishra, A. Mukherji, A.B. Shrestha (eds) (2019) The HinduKush Himalaya Assessment—Mountains, Climate Change, Sustainability and People. *Springer Nature Switzerland AG, Cham.*

[26] Sanjel, S., Khanal, S.N., Thygerson, S.M., Carter, W.S., Johnston, J.D. and Joshi, S.K. (2017). Respiratory symptoms and illnesses related to the concentration of airborne particulate matter among brick kiln workers in Kathmandu valley, Nepal. *Annals of Occupational and Environmental Medicine,* 29: 9.

[27] Shakya, K.M., Rupakheti, M., Aryal, K. and Peltier, R. (2016). Respiratory effects of high levels of particulate exposure in a cohort of traffic police in Nepal. *J. Occup. Environ. Med.,* 58: 218–225.

[28] Shrestha, I.L. and Shrestha, S.L. (2005). Indoor air pollution from biomass fuels and respiratory health of the exposed population in Nepalese household. *Int J Occup Environ Health,* 11:150–160.

[29] Shrestha, S.L. (2018). Assessment of Ambient Particulate Air Pollution and its Attribution to Environmental Burden of Disease in Kathmandu Valley, Nepal: A Review, 4 (4).

[30] Shrestha H.S., Nepal, O., Khanal, K. and Kapoor, B.K. (2015). A cross sectional study of lung functions in traffic police personnel at work in Kathmandu Valley, Nepal. *Annals of Clinical Chemistry and Laboratory Medicine 1* (1): 42–48.

[31] UNDP (2019). Human Development Report 2019: Beyond income, Beyond averages, Beyond today: Inequalities in Human Development in the 21st Century, UN, New York.

[32] Wendling, Z.A., Emerson, J.W., Esty, D.C., Levy, M.A. and de Sherbinin, A. (2018). The Environmental Performance Index, 2018 Yale Center for Environmental Law & Policy.

[33] WHO (2016). Ambient air pollution: A global assessment of exposure and Burden of Diesease. *Word Health Organization.*

New Delhi RESOLUTION

New Delhi Resolution on “Air Pollution and Public Health: Challenges and Interventions”

NEW DELHI RESOLUTION
on
Air Pollution and Public Health: Challenges and Interventions

We the participants of the three-day International Workshop on '**Air Pollution and Public Health: Challenges and Interventions**', held in New Delhi, India during 5–7 February, 2020.

Representing the governments, academic and research institutions and S&T agencies from Bhutan, Cambodia, Cuba, Egypt, India, Indonesia, Iran, Malaysia, Myanmar, Nepal, Palestine, Sri Lanka, Zambia and Zimbabwe.

Expressing gratitude to the Centre for Science and Technology of the Non-Aligned and Other Developing Countries (NAM S&T Centre) and Sri Venkateswara College (University of Delhi), India, for organizing the workshop and the industry sponsors for their support.

Recognizing that air pollution is a multifaceted global environmental issue because the air pollutants from stationary and mobile sources are frequently transferred over great distances and across national borders, with serious implications on 'One Health', and a lot needs to be done to find workable and sustainable solutions to control the same.

Being deeply concerned about the existing and potential damage to the natural and man-made environment and the increasing evidence of effects on human health resulting directly or in combination with the other factors, from emissions of major air pollutants from fossil fuel combustion and their conversion products.

Being Conscious that some of the effects of air pollution on the environment are cumulative over time, and may be difficult or impossible to reverse.

Further acknowledging that a number of NAM and other developing countries have committed themselves to implement reduction of annual carbon and sulphur emissions.

Unanimously resolve and recommend the following:

- Appropriate policies should be adopted and programmes initiated to control air pollution resulting from emissions of oxides of carbon, sulphur & nitrogen; hydrocarbons and particulate matter, from stationary and mobile sources in order to achieve acceptable levels of ambient air quality.
- Efforts should be made to have continuous monitoring systems of primary and secondary pollutants and their sources to provide precise information about the level of air pollution in a region-specific manner, including hotspots, to get a clear idea about the intensity and magnitude of the problem.
- Member countries should share the information about emissions of different categories of air pollutants from various sources.
- In order to precisely establish the cause-effect relationship, efforts should be made to continuously monitor and record seasonal fluctuations in air pollution levels and correlate with increased incidents of associated health issues.
- Governments, industries as well as research and healthcare institutions should encourage trans-disciplinary, holistic research to study the effects of air pollution on human, animal and plant health.
- The understanding of concept of 'One Health' should be promoted especially in the context of air pollution.
- Special emphasis should be given to conduct in-depth analysis of the sources, types, causes and effects of indoor and ambient air pollution.
- Capacity building programs, workshops and training sessions should be conducted to sensitize people about the causes and harmful effects of indoor and ambient air pollution.
- Governments should encourage educational institutions to adopt pedagogical tools in curriculum in environment and climate change to increase awareness and sensitize students on harmful effects of air pollution and their role in minimizing the effects.

- Local health workers and medical practitioners should be encouraged to be involved in epidemiological programs, to create database of various health issues related to air pollution.
- Governments should encourage and fund start-ups and researchers engaged in designing environment friendly solutions in sectors such as transportation, household cooking, construction etc. to minimize emission of air pollutants.
- Appropriate emission control strategies should be adopted on national and international basis, which may include emission standards for various categories of polluters, within a legal framework, and other effective and efficient means of reducing the pollution levels.
- Internationally coordinated research and development programmes should be supported and undertaken aimed at better understanding of the effects of air pollution on man and the environment, and improving technologies for fossil fuel combustion and control of pollutant emissions.
- Governments in NAM and other developing countries should encourage the use of biofuels and other sources of renewable energy. Focus should also be directed towards waste management and biotreatment of polluted environments.
- All major industrial and infrastructure projects should undergo a human health impact assessment as a part of the approval process.
- Industries should be mandated to adopt 'Cleaner Technologies' in their production processes to reduce air pollution at source. Industries should be encouraged with incentives to implement 'Process Optimization' and 'Process Integration' for mitigation of air pollution.

It was also agreed that the above set of recommendations should be integrated as a part of the already existing air pollution control policies of various NAM and other developing countries.

Thus, unanimously adopted on this day, the 7th February 2020 in New Delhi, India.

Author Index